Periodic Table of the Elements with the Gmelin System Numbers

1 H 2																1 H 2	2 He 1
3 Li 20	4 Be 26											5 B 13	6 C 14	7 N 4	8 O 3	9 F 5	10 Ne 1
11 Na 21	12 Mg 27											13 Al 35	14 Si 15	15 P 16	16 S 9	17 Cl 6	18 Ar 1
19 * K 22	20 Ca 28	21 Sc 39	22 Ti 41	23 V 48	24 Cr 52	25 Mn 56	26 Fe 59	27 Co 58	28 Ni 57	29 Cu 60	30 Zn 32	31 Ga 36	32 Ge 45	33 As 17	34 Se 10	35 Br 7	36 Kr 1
37 Rb 24	38 Sr 29	39 Y 39	40 Zr 42	41 Nb 49	42 Mo 53	43 Tc 69	44 Ru 63	45 Rh 64	46 Pd 65	47 Ag 61	48 Cd 33	49 In 37	50 Sn 46	51 Sb 18	52 Te 11	53 I 8	54 Xe 1
55 Cs 25	56 Ba 30	57** La 39	72 Hf 43	73 Ta 50	74 W 54	75 Re 70	76 Os 66	77 Ir 67	78 Pt 68	79 Au 62	80 Hg 34	81 Tl 38	82 Pb 47	83 Bi 19	84 Po 12	85 At	86 Rn 1
87 Fr	88 Ra 31	89*** Ac 40	104 71	105 71													* NH_4 23

**Lanthanides 39	58 Ce	59 Pr	60 Nd	61 Pm	62 Sm	63 Eu	64 Gd	65 Tb	66 Dy	67 Ho	68 Er	69 Tm	70 Yb	71 Lu
***Actinides	90 Th 44	91 Pa 51	92 U 55	93 Np 71	94 Pu 71	95 Am 71	96 Cm 71	97 Bk 71	98 Cf 71	99 Es 71	100 Fm 71	101 Md 71	102 No 71	103 Lr 71

A Key to the Gmelin System is given on the Inside Back Cover

Gmelin Handbook of Inorganic Chemistry

8th Edition

Gmelin Handbook of Inorganic Chemistry

8th Edition

Gmelin Handbuch der Anorganischen Chemie

Achte, völlig neu bearbeitete Auflage

Prepared and issued by	Gmelin-Institut für Anorganische Chemie der Max-Planck-Gesellschaft zur Förderung der Wissenschaften Director: Ekkehard Fluck

Founded by	Leopold Gmelin
8th Edition	8th Edition begun under the auspices of the Deutsche Chemische Gesellschaft by R. J. Meyer
Continued by	E.H.E. Pietsch and A. Kotowski, and by Margot Becke-Goehring

Springer-Verlag Berlin Heidelberg GmbH 1985

Gmelin-Institut für Anorganische Chemie
der Max-Planck-Gesellschaft zur Förderung der Wissenschaften

Gmelin Handbook of Inorganic Chemistry

8th Edition

At
Astatine

With 135 illustrations

AUTHORS

Klara Berei, Central Research Institute for Physics Hungarian Academy of Sciences, Budapest, Hungary

Siegfried H. Eberle, Institut für Radiochemie, Universität Karlsruhe, and Kernforschungszentrum Karlsruhe

H.W. Kirby, Mound Facility, Monsanto Research Corporation, Miamisburg, Ohio, USA

Helmut Münzel, Technische Hochschule Darmstadt

Kurt Rössler, Kernforschungsanlage Jülich

Arnulf Seidel, Institut für Genetik und für Toxikologie von Spaltstoffen, Kernforschungszentrum Karlsruhe

László Vasáros, Central Research Institute for Physics Hungarian Academy of Sciences, Budapest, Hungary

CHIEF EDITORS

Hans Karl Kugler, Gmelin-Institut, Frankfurt am Main

Cornelius Keller, Schule für Kerntechnik, Kernforschungszentrum Karlsruhe

System Number 8 a

Springer-Verlag Berlin Heidelberg GmbH 1985

LITERATURE CLOSING DATE: JUNE 1984
IN SOME CASES MORE RECENT DATA HAVE BEEN CONSIDERED

Library of Congress Catalog Card Number: Agr 25-1383

ISBN 978-3-662-05870-1 ISBN 978-3-662-05868-8 (eBook)
DOI 10 1007/978-3-662-05868-8

Originally published by Springer-Verlag Berlin · Heidelberg · New York · Tokyo in 1985
Softcover reprint of the hardcover 8th edition 1985

Preface

Astatine is — besides radon and francium — the only natural radioelement which only has short-lived isotopes, thereby excluding experiments with "weighable" amounts of the element. This implies that all available data on physics and chemistry of this element are based on experiments on the tracer scale with 10^{-12} to 10^{-16} g — and this will also not change in future because no longer-lived isotopes as yet known are to be expected.

Due to the fact that the only isotope of At occurring in the natural decay series, ^{219}At, results from the 0.005% α-branching of ^{223}Fr which itself is produced by the only 1.38% α-branching of ^{227}Ac — a member of the ^{235}U series — there is no chance to recover substantial amounts of ^{219}At from natural sources for scientific research of At. All studies, therefore, are being done with the isotopes ^{209}At to ^{211}At having half-lives in the few hours region and being obtained by irradiation of bismuth with α-particles via (α, xn) reactions or by proton irradiation of heavy elements via spallation reactions. The mostly used isotope is 7.22 h ^{211}At. The fast separation of the obtained At isotopes is no very difficult procedure and is either being done by wet adsorption-precipitation techniques or making use of its high volatility by destilling in air.

Astatine is a halogen but has marked metallic properties and, in some respects, seems to resemble polonium more than it does iodine, analogies between At and I are likely to be questionable at best. Any anomalous behavior of At may be attributed, at least in part, to reactions with unknown trace impurities.

Five oxidation states of At are known: At^-, At^0, At^+, AtO_3^- and AtO_4^-, even if the true chemical nature of them is not quite clear. This is especially valid for the monovalent astatine cation At^+, for which also a notation as an unspecified "hydrated" $At(\Theta)^+$ is being used. There are also some indications for an intermediate oxidation state between At^{I+} and At^{V+}, may be At^{III} in form of AtO^+ or AtO_2^-. Several interhalogen compounds are known, compounds like AtCl, AtBr or AtI are extracted by carbon tetrachloride in the presence of free halogens whereas anions AtX_2^- (Cl, Br, I) are formed in the aqueous phase.

A large number of organic astatine compounds were prepared using, e.g., halogen exchange or recoil astatination. There composition mostly was verified by radiogas chromatography and high-pressure liquid chromatography. A number of organic ^{211}At compounds is being used in nuclear medicine for special purposes, e.g., for anticancer drugs or other local irradiations of malignant diseases.

Presently we have a pretty good knowledge of the chemical and physical behavior of this rare heaviest halogen element. But on the other hand there are a lot of important open questions, like the true nature of At species in solution or its complex-forming behavior. This first Gmelin volume on this chemical highly interesting element can act as a guide for further research, both for the applied and scientific fields.

I want to express my thanks to all the authors who have presented contributions of a very high scientific quality. I also wish to thank the Gmelin Institute for the excellent cooperation, especially to Dr. Kugler and Prof. Dr. Fluck.

Karlsruhe
June 1985

Cornelius Keller

Table of Contents

Page

Astatine

Atomic Number 85

1 History

H.W. Kirby
Mound Facility
Monsanto Research Corporation
Miamisburg, Ohio, USA

1.1 Introduction

Astatine At, element 85, has no stable isotopes. The longest-lived isotope is ^{210}At, a β^+-emitter with a half-life of 8.3 h, produced by the bombardment of ^{209}Bi with high-energy α-particles. The longest-lived naturally occurring isotope is 0.9-min ^{219}At, an α-emitter resulting from the rare α-branching of 22-min ^{223}Fr. Chemically, At behaves as a halogen but the analogy between At and I, its next lower homologue, is not as close as one might expect from their relative positions in the Periodic Table. General reviews of astatine chemistry are given by [1 to 16].

In 1913, the discovery of atomic numbers [17] by Henry Gwyn Jeffreys Moseley (1887 to 1915) made possible, for the first time, "a veritable roll-call of the elements" [18]. Moseley found that, when the known elements, from Al to Au, were arranged in numerical order, according to their atomic weights, the integer corresponding to the rank of each element was directly proportional to the square root of the frequency of its K or L X-ray emission. Moseley's work left vacancies for elements of atomic numbers 43, 61 and 75, vacancies that were subsequently filled by the discoveries of technetium [19], promethium [20] and rhenium [21, 22], respectively. See also "Technetium" Suppl. Vol. 1, 1982, pp. 1/11, "Rhenium" 1941, pp. 1/10, and the history of promethium in "Seltenerdelemente" B1, 1976, pp. 2/12.

In 1916, Siegbahn and Friman [23] examined the X-ray spectra of the elements from Ta (number 73) to U (number 92). Their results left vacancies at numbers 85, 86, 87, 89 and 91. However, it was already evident from radiochemical considerations [24] that numbers 86, 89 and 91 belonged to Rn, Ac and UX_2, respectively. (UX_2, "brevium" [25, 26] is now known as ^{234m}Pa.) Numbers 85 and 87, homologues of I and Cs, respectively, remained unaccounted for. In a theoretical study, Wagner [27] noted that element 85 must fall between Po (84) and Rn (86) while element 87 would fall between Rn (86) and Ra (88). He concluded that both elements were probably radioactive.

1.2 The Search for a Stable or Long-Lived Isotope

The prehistory of astatine resembles and is closely associated with that of francium. Indeed, some of the same investigators who erroneously claimed the discovery of "ekacesium" (francium) also claimed to have discovered "ekaiodine" (astatine). (Eka, the Sanskrit word for "one", was first used by Dimitri Ivanovich Mendeleev (1834 to 1907), in his classic paper on the Periodic Table [28], to suggest the existence of an unknown heavy homologue of a known element to whose name it was attached as a prefix. This convention was widely adopted by later researchers. However, unlike "ekacesium", which Mendeleev foresaw in his 1872 paper, the first explicit allusion to "ekaiodine" occurs in the 1906 edition of his textbook, Principles of Chemistry [29].)

The earliest attempts to find ekaiodine experimentally appear to have been made by Loring and Druce [30 to 32], who reported seeing lines attributable to the L_α and L_β radiations of element 85 in the X-ray spectra of impurities isolated from commercial manganese sulfate. However, the lines were so faint that even the authors regarded them as uncertain evidence; cf. [33]. Unsuccessful attempts to detect the X-ray emissions of ekaiodine in natural materials were also reported by [34 to 36].

Of course, not all researchers acknowledged failure in their search for a stable or long-lived isotope of ekaiodine. For example, Rajendralal De, of the University of Dacca, reported the discovery of not one but two new elements in Travancore monazite. One of them, he said, was ThF, an α-emitter with a half-life of $>10^{10}$ years, analogous to RaF (^{210}Po), which he proposed to name "gourium". The other was stable ekaiodine, for which he suggested the name "dakin" [37].

Perhaps the most interesting of the erroneous claims was that of Fred Allison and his collaborators at the Alabama Polytechnic Institute. In 1930, Allison and Murphy [38] described a "magneto-optic" method of chemical analysis which, they claimed, could detect and quantify inorganic compounds in aqueous solutions at concentrations as low as 10^{-11} g/g of water. Even more remarkable was their claim that the method could determine the number of isotopes of the metallic element in the compound.

In 1930, Allison and co-workers announced the discovery of element 87, ekacaesium, for which they later proposed the name "virginium" (symbol: Va) [39, 40]. In a related study, the same group claimed to have found element 85, ekaiodine, in sea water, fluorite, apatite, monazite, kainite, KBr, HF and HBr [41]. They suggested the name "alabamine" (symbol: Am) for the new element and described the working up of 100 pounds of Brazilian monazite to yield a solution of peralabamic acid, $HAmO_4$. They reduced the $HAmO_4$ to HAm with SO_2 and converted it to $\sim 2.5 \times 10^{-6}$ g of Am as LiAm. They said that Am was easily oxidized in either acid or alkaline solution and prepared several of its oxygen compounds, of which peralabamate was the most stable [42]. Both Va and Am (renamed Vi and Ab) were briefly enshrined in a 1941 version of the Periodic Table [43], although only one other laboratory [44] ever succeeded in duplicating the Allison group's results. The controversial magneto-optic effect (now generally discredited) is described in detail by [45 to 49]; see also "Francium" 1983, pp. 2/4.

1.3 The Search for Short-Lived Ekaiodine in Natural Materials

It was apparent, from the radioactive displacement laws [50 to 52], that either the α-decay of the as yet undiscovered element 87 or the β^--decay of an isotope of element 84 (Po) would yield element 85 [36, 53]. As noted above, the probability that ekaiodine would be radioactive was recognized as early as 1920 [27], but no serious effort to detect it by its radioactivity was undertaken until the late 1930's.

Minder [54], in 1938, and Turner [55], in 1940, predicted that RaA (^{219}Po), the α-emitting daughter of ^{222}Rn, would also be somewhat β^--active. To test this hypothesis, Minder [56] followed the growth of β^--activity in ^{222}Rn and found that the initial increase in β^--activity was 1.5 times greater than that predicted by theory for the ingrowth of the known β^--emitters, RaB (^{214}Pb) + RaC (^{214}Bi). The discrepancy, he said, could only be accounted for by assuming β^--branching in RaA, equivalent to somewhat more than 15% of the β^--radiation of RaB.

On the basis of this evidence, Minder claimed the discovery of element 85 and proposed the name "helvetium" (symbol: Hv), in honor of Switzerland. Karlik and Bernert [57], repeated Minder's experiments and concluded that he had been misled by contamination of his Rn stream with RaB and RaC and by secondary electrons. Labhart and Medicus [58] were unable to detect β^--activity in RaA, and concluded that, if β^--branching occurs, the β^-/α ratio must be $<3.5\%$. On the other hand, Hulubei and Cauchois [59, 60] reported the observation of three X-ray lines attributable to an excited state of element 85 among the decay products of ^{222}Rn.

Turner [55] had also predicted that ThA (^{216}Po), the α-emitting daughter of thoron (^{220}Rn), would exhibit β^--branching. In 1942, Minder, in collaboration with Leigh-Smith [61], reported the isolation of another isotope of element 85, presumably the product of ThA β^--decay, by sublimation from the active deposit of thoron. They named it "anglo-helvetium" in honor of both their countries. However, an attempt by Karlik and Bernert [62] to reproduce this work gave negative results; see also [63, 105].

1.4 Discovery of Astatine by Nuclear Synthesis

The first unequivocal identification of element 85 was made in 1940 by the team of Dale Raymond Corson (born 1914), Kenneth Ross McKenzie (born 1912) and Emilio Segrè (born 1905). Working with the 60-inch cyclotron of the Crocker Radiation Laboratory at the University of California (Berkeley) [64 to 66], they bombarded ^{209}Bi with 32-MeV α-particles and obtained a product that was α-radioactive. Two groups of α-particles were observed, of which 60% had a range of 6.55 cm and 40% a range of 4.52 cm. The two groups were not genetically related. The product also showed X-rays which had the absorption characteristics of the K X-rays from Po. In chemical reactions, all these radiations separated together and all had a half-life of 7.5 h.

The identity of the unknown α-emitter as an isotope of ekaiodine, was established by elimination of all other possibilities: The bombarded Bi was dissolved in HNO_3 and adjusted to an acid concentration of 0.25 N. Pb and Tl were precipitated as chlorides with HCl; no activity was found in the precipitates. When Bi was precipitated with $SnCl_2$ in alkaline solution, the precipitate was found to be inactive. In fractional precipitation of Bi_2S_3 from 6 N HCl and then from solutions of progressively lower acidity, the activity concentrated in the first fraction. Fractional hydrolysis of Bi concentrated the activity in the last fraction.

The chemical properties of the unknown substance closely resembled those of Po. To establish that the unknown activity was not Po, a standard Po tracer was prepared by bombarding Bi with deuterons and dissolving the Bi in the same way as the Bi containing the unknown. Known amounts of the Po tracer were then mixed with the unknown and several experiments were performed to determine the degree to which the unknown would follow the chemistry of Po: When a piece of solid Bi was placed in the solution, the Po deposited on the Bi but the unknown did not. In the fractional sulfide and fractional hydrolytic precipitations described above, the ratio of unknown activity to Po activity was different in each fraction. KI was added to an HNO_3 solution containing both activities and the I^- was extracted with CCl_4. The I^- was reduced with SO_2 and precipitated with $AgNO_3$. The precipitate contained only the 7.5-h activity; the Po was left in solution. However, the extraction of the unknown activity was incomplete. KI was added to another sample and the I was distilled off. In this case, the unknown followed the I (though not quantitatively), but all the Po remained behind.

Thus, it seemed definite that the unknown α-emitter was not an isotope of Tl, Pb, Bi, Po or any other of the known elements up to U. Other experiments ruled out the possibility that it might be Te, Se, Hg, Cu or a fission product. The inevitable conclusion was that it must be element 85 (ekaiodine), the first artificially produced α-emitter.

Further evidence was adduced by a comparison of the metabolism of element 85 with that of iodine. Hamilton and Soley [67] demonstrated that both halogens concentrated in the thyroid tissues of guinea pigs and were excreted in a similar manner; see also [68, 69], cf. Chapter 8 in this volume.

With regard to the two groups of α-particles, analysis of data from coincidence counting, photographic emulsion tracks and critical absorption measurements led to the conclusion that the 6.55 cm α-particles came from AcC′ (^{211}Po). This suggested that a ^{209}Bi (α, 2n) reaction had produced ^{211}85, which decayed both by α-emission (4.52 cm range) to ^{207}Bi and by orbital electron capture to ^{211}Po.

In 1947, Corson, MacKenzie and Segrè [70] proposed the name astatine (symbol: At) for element 85, from the Greek, astatos (unstable). The name and symbol were recognized in 1949 by the Commission on Inorganic Nomenclature of the International Union of Pure and Applied Chemistry [71].

1.5 The Discovery of Naturally Occurring Element 85

In 1943, Karlik and Bernert [72], working rapidly with intense sources of RaA (^{218}Po), prepared from ~100 mCi of ^{222}Rn, observed a group of α-particles with a range of 5.53 cm, equivalent to an energy of 6.75 MeV, alongside the rapidly growing 7.7 MeV α-particles of RaC′ (^{214}Po). They attributed this group to an α-emitting isotope of element 85, produced by the β^--branching of RaA. The α-particle ratio, 85/RaC′, which was as high as four after the first 50 seconds of growth, declined rapidly with time as the growth of RaC′ from RaB + RaC became predominant. Karlik and Bernert estimated the branching ratio of RaA to be 3.3×10^{-4} and the half-life of ^{218}85, based on the Geiger-Nuttall relationship [73], to be ~2 s. The discovery of ^{218}85 by Karlik and Bernert was confirmed by Walen [74], who estimated that the RaA β^-/α ratio was $(2.00 \pm 0.05) \times 10^{-4}$% and the upper limit of the half-life was ~1.5 to 2 s. (The currently accepted value of the branching ratio is 0.0185% [75].) Walen also reported a weak β^--branch (< ~0.1%) in ^{218}85, yielding 1.3-s ^{218}Rn. (The currently accepted half-life of ^{218}Rn is 35 ms [75].) See also [76, 77].

In analogous experiments with the actinon (^{219}Rn) from a weak source of ^{227}Ac, Karlik and Bernert [62, 78] observed a previously unreported 8.4 MeV α-particle group, which they ascribed to ^{215}85 arising from a 5×10^{-4}% β^--branching in AcA (^{215}Po). Their conclusions were criticized on energetic grounds by Feather [79], but Avignon [80], using a much larger source of An, repeated the work with essentially the same results except that the α-energy was 8.04 MeV and the β^-/α ratio was $(2.3 \pm 0.2) \times 10^{-4}$%; see also [81].

The results obtained with ThA (^{216}Po) in equilibrium with thoron (^{220}Rn) were less convincing. Karlik and Bernert [62, 82] reported the observation of 7.72 MeV α-particles, which decayed with the half-life of Tn (55.6 s). The abundance, relative to ThA, was 1.35×10^{-4}. They postulated (by analogy to RaA and AcA) that this α-particle group resulted from the β^--branching of ThA to yield ^{216}85. However, doubts were expressed by [79, 83, 84], and by the authors themselves [62] because of a significant energy imbalance implied by this explanation. These questions have never been resolved and modern compilations list ^{216}At only as a member of the artificial ^{228}Pa collateral series [75, 81, 85]. (Note, however, that the currently accepted α-energy of ^{216}At is 7.800 MeV, only 80 keV higher than that estimated by Karlik and Bernert.)

In 1953, Hyde and Ghiorso [86] reported the isolation of 0.9-min ^{219}At, produced by a 0.004% α-branching in AcK (^{223}Fr). This discovery became possible when it was found that large amounts of ^{227}Ac could be produced in a nuclear reactor by the $^{226}Ra(n, \gamma)$ reaction [87]. The AcK daughter was isolated from a 20-mCi source of ^{227}Ac and purified by a method previously developed by Hyde [88]. The method consisted of the coprecipitation of AcK with free silicotungstic acid from a solution of Ac in saturated HCl, followed by separation of the AcK from the carrier on a Dowex-50 cation exchange column. Carrier-free and radioactively pure AcK could be prepared in 40 min or less. To isolate the At, the AcK was evaporated on a Pt filament, which was then heated by an electric current to a temperature just high enough to volatilize the At without also volatilizing Fr, Pb or Bi. During a 15-second milking period, the volatilized At was collected on a Pt disc mounted 1 to 2 mm above the heated filament. The collector disc was placed in a differential pulse height analyzer within one half minute after the collection period. A new α-group was observed, which decayed with a half-life of 0.9 min and an energy of 6.27 MeV. The AcA (^{215}Po) peak was observed to grow to a maximum in 4 to 6 min and then to decay. This was the behavior to be expected if the new At isotope decayed to a β^--emitting Bi isotope; i.e.,

$$^{219}\text{At} \xrightarrow[0.9\ \text{min}]{\alpha} {}^{215}\text{Bi} \xrightarrow[8\ \text{min}]{\beta^-} {}^{215}\text{Po} \xrightarrow[0.00183\ \text{s}]{\alpha}$$

However, efforts to obtain quantitative agreement between the observed growth of ^{215}Po and its theoretical growth based on this decay chain were unsuccessful, as were several attempts to isolate the postulated ^{215}Bi.

In 1952, Peppard and co-workers [89] isolated trace amounts of ^{225}Ac from U refinery wastes and from Th derived from Brazilian monazite. ^{225}Ac, the grandparent of 32-ms ^{217}At, is a member of the synthetic $4n+1$ radioactive series which was not previously thought to occur in nature because of the geologically short half-life of its longest-lived member, ^{237}Np. However, an α-spectrum of the isolated ^{225}Ac clearly showed the presence of ^{217}At, indicating that it, too, may be regarded as "naturally occurring." The natural origin of the $4n+1$ series is discussed in Chapter 2, p. 10.

1.6 Discovery of Astatine Isotopes

Since its discovery in 1940, isotopes of astatine have been identified with mass numbers ranging from 196 to 219. They are listed in Table 1/1 with the names of their discoverers, the date of their discovery and the methods by which they were first produced. The half-lives given are the currently accepted values [75]. In some cases, because of wartime secrecy, the actual date of discovery may not be the year in which the isotope was first reported in the open literature.

Table 1/1
Discovery of Astatine Isotopes.

mass number	half-life [75]	class [a)] [75]	mode of first preparation	discoverers and date of discovery	Ref.
196	0.3 s	C	$^{185}Re(^{20}Ne, 9n)$	Treytl, Valli; 1967	[90]
197	0.4 s	A	$^{185}Re(^{20}Ne, 8n)$	Treytl, Valli; 1967	[90]
198	4.9 s	A	$^{185}Re(^{20}Ne, 7n)$	Treytl, Valli; 1967	[90]
198m	1.5 s	A	$^{185}Re(^{20}Ne, 7n)$	Treytl, Valli; 1967	[90]
199	7.2 s	A	$^{185}Re(^{20}Ne, 6n)$	Treytl, Valli; 1967	[90]
200g	42 s	A	$^{209}Bi(\alpha, 13n)$	Barton, Ghiorso, Perlman; 1951	[91]
200m	4.3 s	C	$^{185}Re(^{20}Ne, 5n)$	Treytl, Valli; 1967	[90]

Table 1/1 [continued]

mass number	half-life [75]	class a) [75]	mode of first preparation	discoverers and date of discovery	Ref.
201	1.50 min	A	^{209}Bi (α, 12n)	Barton, Ghiorso, Perlman; 1951	[91]
202	3.0 min	A	$^{197}Au(^{12}C, 7n)$	Latimer, Gordon, Thomas; 1961	[92]
203	7.3 min	A	$^{197}Au(^{12}C, 6n)$	Miller, Hamilton, Purnam, Haymond, Rossi; 1950	[93]
204	9.3 min	A	^{209}Bi (α, 9n)	Barton, Ghiorso, Perlman; 1951	[91]
205	26.2 min	A	$^{197}Au(^{12}C, 4n)$	Miller, Hamilton, Purnam, Haymond, Rossi; 1950	[93]
206	31.4 min	A	^{209}Bi (α, 7n)	Barton, Ghiorso, Perlman; 1951	[91]
207	1.80 h	A	^{209}Bi (α, 6n)	Templeton, Ghiorso, Perlman; 1948	[103]
208	1.63 h	A	daughter ^{212}Fr	Hyde, Ghiorso, Seaborg; 1950	[95]
209	5.42 h	A	^{209}Bi (α, 4n)	Barton, Ghiorso, Perlman; 1951	[91]
210	8.3 h	A	^{209}Bi (α, 3n)	Kelly, Segrè; 1947	[96, cf. 94]
211	7.214 h	A	^{209}Bi (α, 2n)	Corson, MacKenzie, Segrè; 1940	[64]
212	0.315 s	A	^{209}Bi (α, n)	Weissbluth, Putnam, Segrè; 1948	[104]
212m	0.122 s	A	^{209}Bi (α, n)	Jones; 1963	[97]
213	0.11 µs	A	descendant ^{225}Pa	Keys; 1951	[98]
214	~2 µs	B	^{218}Fr α-decay	Meinke, Ghiorso, Seaborg; 1949	[99]
215	0.10 ms	A	natural ^{215}Po β^--decay	Karlik, Bernert; 1944	[78]
216	0.30 ms	A	natural ^{216}Po β^--decay	Karlik, Bernert; 1943	[82]
216m	unknown	F	descendant ^{224}Ac	Briand, Chevallier, Touati; 1971	[100]
217	32.3 ms	A	^{221}Fr α-decay	Hagemann, Katzin, Studier, Ghiorso, Seaborg; 1947	[101]
				English, Cranshaw, Demers, Harvey, Hincks, Jelley, May; 1947	[102]
218	1.5 to 2.0 s	B	natural ^{218}Po β^--decay	Karlik, Bernert; 1943	[72]
219	0.9 min	B	natural ^{223}Fr α-decay	Hyde, Ghiorso; 1953	[86]

a) A = element and mass number certain,
B = element certain; mass number probable,
C = element probable, mass number certain or probable,
F = insufficient evidence.

References:

[1] E. Anders (Ann. Rev. Nucl. Sci. **9** [1959] 203/20). — [2] E.H. Appelman (UCRL-9025 [1960] 1/113; N.S.A. **14** [1960] No. 11501). — [3] E.H. Appelman (NAS-NS-3012 [1960] 1/29; N.S.A. **14** [1960] No. 20201). — [4] E.H. Appelman (MTP [Med. Tech. Publ. Co.] Intern. Rev. Sci. Inorg. Chem. Ser. One **3** [1972] 181/98; C.A. **76** [1972] No. 120925). — [5] A.H.W. Aten (Advan. Inorg. Chem. Radiochem. **6** [1964] 207/23).

[6] A.J. Downs, C.J. Adams (in: J.C. Bailar, H.J. Emeléus, R. Nyholm, A.F. Trotman-Dickenson, Comprehensive Inorganic Chemistry, Vol. 2, Pergamon, Oxford 1973, pp. 1573/

94). — [7] R. Dreyer, F. Rösch, G.J. Beyer (Wiss. Fortschr. **32** [1982] 251/5; C.A. **97** [1982] No. 151718). — [8] C.S. Garner (in: A.C. Wahl, N.A. Bonner, Radioactivity Applied to Chemistry, Wiley, New York 1951, pp. 179/243). — [9] M. Haïssinsky (in: P. Pascal, Nouveau Traité de Chimie Minérale, Vol. 16, Masson, Paris 1960, pp. 659/66). — [10] E.K. Hyde (J. Phys. Chem. **58** [1954] 21/6).

[11] E.K. Hyde (J. Chem. Educ. **36** [1959] 15/21). — [12] V.A. Khalkin, E. Herrmann (Isotopenpraxis **11** [1975] 333/40). — [13] V.A. Khalkin, E. Herrmann, J.V. Norseev, I. Dreyer (Chemiker-Ztg. **101** [1977] 470/81). — [14] A.K. Lavrukhina, A.A. Pozdnyakov (Analytical Chemistry of Technetium, Promethium, Astatine and Francium, Ann Arbor Sci., Ann Arbor 1970, pp. 227/60). — [15] J. Sedlet (Treatise Anal. Chem. II **6** [1964] 487/501).

[16] V.D. Nefedov, Yu.V. Norseev, M.A. Toropova, V.A. Khalkin (Usp. Khim. **37** [1968] 193/215; Russ. Chem. Rev. **37** [1968] 87/98; JINR-P-2895 [1966] 1/36; C.A. **66** [1967] No. 70899, **68** [1968] No. 100679; N.S.A. **21** [1967] No. 6). — [17] H.G.J. Moseley (Phil. Mag. [6] **26** [1913] 1024/34, [6] **27** [1914] 703/14). — [18] F. Soddy (Ann. Rept. Progr. Chem. [Chem. Soc. London] **11** [1915] 266/92). — [19] C. Perrier, E. Segrè (Nature **140** [1937] 193/4; Atti Reale Accad. Lincei Classe Sci. Fis. Mat. Nat. [6] **25** [1937] 723/30; J. Chem. Phys. **5** [1937] 712/6; C.A. **1937** 6964, 7745, **1938** 3689). — [20] J.A. Marinsky, L.E. Glendenin, C.D. Coryell (J. Am. Chem. Soc. **69** [1947] 2781/5).

[21] W. Noddack, I. Tacke (Naturwissenschaften **13** [1925] 567/71). — [22] O. Berg, I. Tacke (Naturwissenschaften **13** [1925] 571/4). — [23] M. Siegbahn, E. Friman (Phil. Mag. [6] **32** [1916] 39/49). — [24] F. Soddy (Ann. Rept. Progr. Chem. [Chem. Soc. London] **10** [1914] 262/88; The Chemistry of the Radio-Elements, Pt. 2, Longmans, Green, London 1914). — [25] K. Fajans, O. Göhring (Physik. Z. **14** [1913] 877/84).

[26] O. Göhring (Diss. Karlsruhe 1914). — [27] E. Wagner (Z. Elektrochem. **26** [1920] 260/2). — [28] D. Mendelejeff [Mendeleev] (Liebigs Ann. Chem. Suppl.-Vol. **8** [1872] 133/229). — [29] D.I. Mendeleev (Osnovy Chimii, 8th Ed., Moscow 1906 from [13, Ref. No. 5]). — [30] J.G.F. Druce (Chem. News **131** [1925] 273/7).

[31] F.H. Loring, J.G.F. Druce (Chem. News **131** [1925] 305, 321). — [32] F.H. Loring (Chem. News **131** [1925] 338/41). — [33] W. Prandtl (Z. Angew. Chem. **39** [1926] 1049/51). — [34] J.N. Friend (Nature **117** [1926] 789/90). — [35] Herszfinkiel (Compt. Rend. **184** [1927] 968/70; C.A. **1927** 2097).

[36] G. Hevesy, R. Hobbie (Z. Anorg. Allgem. Chem. **208** [1932] 107/12). — [37] R. De (Twin Elements in Travancore Monazite, Bani Press, Dacca 1937, pp. 1/18; Indian J. Phys. **13** [1939] 407/8; C.A. **1937** 6103, **1940** 4660). — [38] F. Allison, E.J. Murphy (Phys. Rev. [2] **35** [1930] 124/5, [2] **36** [1930] 1097/8; J. Am. Chem. Soc. **52** [1930] 3796/806; J. Chem. Educ. **7** [1930] 2411/3; Ind. Eng. Chem. Anal. Ed. **4** [1932] 9/12). — [39] F. Allison, E.J. Murphy (Phys. Rev. [2] **35** [1930] 285). — [40] F. Allison, E.R. Bishop, A.L. Sommer, J.H. Christensen (J. Am. Chem. Soc. **54** [1932] 613/5).

[41] F. Allison, E.J. Murphy. E.R. Bishop, A.L. Sommer (Phys. Rev. [2] **37** [1931] 1178/80). — [42] F. Allison, E.R. Bishop, A.L. Sommer (J. Am. Chem. Soc. **54** [1932] 616/20). — [43] R.F. Trimble (J. Chem. Educ. **52** [1975] 585). — [44] J.L. McGhee, M. Lawrenz (J. Am. Chem. Soc. **54** [1932] 405/6). — [45] F. Allison (J. Chem. Educ. **10** [1933] 71/8).

[46] F.G. Slack (J. Franklin Inst. **218** [1934] 445/62; C.A. **1935** 1735). — [47] S.S. Cooper, T.R. Ball (J. Chem. Educ. **13** [1936] 210/5, 278/83, 326/8). — [48] G.C. Comstock (Phys. Rev. [2] **51** [1937] 776/7). — [49] P. Rémy-Gennеté (Bull. Soc. Chim. France [4] **53** [1933] 140/4; Rev. Gen. Sci. **44** [1934] 393/9; C.A. **1933** 4451). — [50] A.S. Russell (Chem. News **107** [1913] 49/52).

[51] K. Fajans (Physik. Z. **14** [1913] 131/42, 257). — [52] F. Soddy (Chem. News **107** [1913] 97/9). — [53] E.B. Andersen (Kgl. Danske Videnskab. Selskab Math. Fys. Medd. **16** No. 5 [1938] 1/22; C.A. **1939** 474). — [54] W. Minder (Helv. Phys. Acta **11** [1938] 497/506). — [55] L.A. Turner (Phys. Rev. [2] **57** [1940] 950/7, **58** [1940] 181/2).

[56] W. Minder (Helv. Phys. Acta **13** [1940] 144/52). — [57] B. Karlik, T. Bernert (Naturwissenschaften **30** [1942] 685/6; Sitz.-Ber. Akad. Wiss. Wien Math. Naturw. Kl. IIa **151** [1942] 255/65; C.A. **1944** 4190, **1949** 4128). — [58] H. Labhart, H. Medicus (Helv. Phys. Acta **16** [1943] 225/6, 392/406). — [59] H. Hulubei, Y. Cauchois (Compt. Rend. **209** [1939] 39/42, **210** [1940] 696/7; C.A. **1939** 7191, **1940** 7736). — [60] H. Hulubei (J. Chim. Phys. **44** [1947] 225/9).

[61] A. Leigh-Smith, W. Minder (Nature **150** [1942] 767/8). — [62] B. Karlik, T. Bernert (Anz. Akad. Wiss. Wien Math. Naturw. Kl. **81** No. 1 [1944] 2/3; Z. Physik **123** [1944] 51/72; C.A. **1949** 7331, **1948** 5768). — [63] A. Leigh-Smith, O. D'Agostino (Rend. Ist. Super. Sanita [Ital. Ed.] **10** [1947] 523/4; C.A. **1948** 3258). — [64] D.R. Corson, K.R. MacKenzie, E. Segrè (Phys. Rev. [2] **57** [1940] 459, 1087, [2] **58** [1940] 672/8). — [65] D.R. Corson, K.R. MacKenzie (Phys. Rev. [2] **57** [1940] 250).

[66] E. Segrè, K.R. MacKenzie, D.R. Corson (Phys. Rev. [2] **57** [1940] 1087). — [67] J.G. Hamilton, M.H. Soley (Proc. Natl. Acad. Sci. U.S. **26** [1940] 483/9). — [68] J.G. Hamilton, C.W. Asling, W.M. Garrison, K.G. Scott, D. Axelrod-Heller (Proc. Soc. Exp. Biol. Med. **73** [1950] 51/3; C.A. **1950** 4525). — [69] A.M. Friedman (Ther. Nucl. Med. Symp., Hartford 1977 [1978], pp. 139/44; C.A. **91** [1979] No. 188739). — [70] D.R. Corson, K.R. MacKenzie, E. Segrè (Nature **159** [1947] 24).

[71] Anonymous (Chem. Eng. News **27** [1949] 2996/9, 3093), E.J. Crane (Chem. Eng. News **27** [1949] 3779). — [72] B. Karlik, T. Bernert (Naturwissenschaften **31** [1943] 298/9; Sitz.-Ber. Akad. Wiss. Wien Math. Naturw. Kl. IIa **152** [1943] 103/10; C.A. **1944** 19, **1949** 1648). — [73] H. Geiger, J.M. Nuttall (Phil. Mag. [6] **22** [1911] 613/21, [6] **23** [1912] 439/45, [6] **24** [1912] 647/54). — [74] R. Walen (Compt. Rend. **227** [1948] 1090/2; J. Phys. Radium **10** [1949] 95/103; C.A. **1949** 3287, 4951). — [75] C.M. Lederer, V.S. Shirley (Table of Isotopes, 7th Ed., Wiley, New York 1978).

[76] B. Karlik, T. Bernert (Naturwissenschaften **33** [1946] 23). — [77] B. Karlik (Monatsh. Chem. **77** [1947] 348/51). — [78] B. Karlik, T. Bernert (Naturwissenschaften **32** [1944] 44). — [79] N. Feather (Nucleonics **5** No. 1 [1949] 22/41). — [80] P. Avignon (J. Phys. Radium [8] **11** [1950] 521/3).

[81] A. Ghiorso, W.W. Meinke, G.T. Seaborg (Phys. Rev. [2] **74** [1948] 695/6). — [82] B. Karlik, T. Bernert (Naturwissenschaften **31** [1943] 492). — [83] S. Flügge, A. Krebs (Naturwissenschaften **32** [1944] 71/2). —[84] I. Perlman, A. Ghiorso, G.T. Seaborg (Phys. Rev. [2] **77** [1950] 26/50). — [85] E.K. Hyde, I. Perlman, G.T. Seaborg (Nuclear Properties of the Heavy Elements, Vol. 2, Prentice-Hall, Englewood Cliffs 1964).

[86] E.K. Hyde, A. Ghiorso (Phys. Rev. [2] **90** [1953] 267/70). — [87] F. Hagemann (J. Am. Chem. Soc. **72** [1950] 768/71). — [88] E.K. Hyde (J. Am. Chem. Soc. **74** [1952] 4181/4). — [89] D.F. Peppard, G.W. Mason, P.R. Gray, J.F. Mech (J. Am. Chem. Soc. **74** [1952] 6081/4). — [90] W. Treytl, K. Valli (Nucl. Phys. A **97** [1967] 405/16).

[91] G.W. Barton, A. Ghiorso, I. Perlman (Phys. Rev. [2] **82** [1951] 13/9). — [92] R.M. Latimer, G.E. Gordon, T.D. Thomas (J. Inorg. Nucl. Chem. **17** [1961] 1/5). — [93] J.F. Miller, J.G. Hamilton, T.M. Purnam, H.R. Haymond, G.B. Rossi (Phys. Rev. [2] **80** [1950] 486). — [94] G.T. Seaborg, I. Perlman (Rev. Mod. Phys. **20** [1948] 585/667). — [95] E.K. Hyde, A. Ghiorso, G.T. Seaborg (Phys. Rev. [2] **77** [1950] 765/70).

[96] E.L. Kelly, E. Segrè (Phys. Rev. [2] **75** [1949] 999/1005). — [97] W.B. Jones (Phys. Rev. [2] **130** [1963] 2042/3). — [98] J.D. Keys (Diss. McGill 1951). — [99] W.W. Meinke, A. Ghiorso, G.T. Seaborg (Phys. Rev. [2] **75** [1949] 314/5). — [100] J.P. Briand, P. Chevallier, A. Touati (J. Phys. [Paris] **32** [1971] 101/5).

[101] F. Hagemann, L.I. Katzin, M.H. Studier, A. Ghiorso, G.T. Seaborg (Phys. Rev. [2] **72** [1947] 252). — [102] A.C. English, T.E. Cranshaw, P. Demers, J.A. Harvey, E.P. Hincks, J.V. Jelley, A.N. May (Phys. Rev. [2] **72** [1947] 253/4). — [103] D.H. Templeton, A. Ghiorso, I. Perlman (Unpublished data [1948] from [94, p. 633]). — [104] M. Weissbluth, T.M. Putnam, E. Segrè (Unpublished data [1948] from [94, p. 633]). — [105] C.W. Martin (Nature **151** [1943] 309).

2 Natural Occurrence

H.W. Kirby
Mound Facility
Monsanto Research Corporation
Miamisburg, Ohio, USA

Astatine is the rarest of all the elements found in nature. The extremely short half-lives of all At isotopes preclude the possibility of their independent existence. Isotopes of At occur naturally in both the 4n+2 (^{238}U) and the 4n+3 (^{235}U) decay chains. ^{217}At, a member of the artificial 4n+1 (^{237}Np) series, has also been detected in natural materials.

There are no At isotopes in the 4n (^{232}Th) decay series; however, because of the geochemical similarity of U and Th [1], all Th minerals are contaminated by significant amounts of ^{230}Th (ionium), a member of the ^{238}U decay chain. Therefore, Th minerals may be expected to contain minute quantities of ^{218}At, a descendant of ^{230}Th.

For the natural abundance and distribution of U, see [1 to 11].

For calculated amounts of astatine in model stellar atmospheres of Sirius and G, F, A, B and D stars see [16, 17].

2.1 Astatine-218 ($T_{1/2}$ ~2 s)

^{218}At, the most abundant of the naturally occurring At isotopes, is produced in nature by the 0.0185% β^--branching of ^{218}Po (**Fig.** 2-**1**).

The mass ratio of ^{218}At to ^{238}U, calculated from their respective half-lives and mass numbers, is 1.8×10^{-21}. It is estimated that the average crustal abundance of U is 2.7 ppm [11], of which 99.289 wt% is ^{238}U [12]. It follows, therefore, that the crustal abundance of ^{218}At is ~4.8×10^{-21} ppm. If one assumes that the crustal mass, to a depth of 35 km, is 2.5×10^{25} g (calculated from the data in [13]), the global inventory of ^{218}At is ~160 mg, cf. [14].

2.2 Astatine-215 ($T_{1/2}$=0.10 ms). Astatine-219 ($T_{1/2}$=0.9 min)

^{215}At, produced by the 2.3×10^{-4}% β^--branching of ^{215}Po (**Fig.** 2-**2**, p. 12), has a crustal abundance of 1.8×10^{-28} ppm and a global inventory of ~4.5 ng.

^{219}At, the longest-lived At isotope in nature, is produced by the rare (~0.005%) α-branching of ^{223}Fr, which is itself produced by the 1.38% α-branching of ^{227}Ac, a descendant of ^{235}U (Fig. 2-2, p. 12). Since ^{235}U represents only 0.711 wt% of natural U [12], the crustal abundance of ^{219}At is only 3×10^{-23} ppm and the global inventory is ~750 μg.

2.3 Astatine-217 ($T_{1/2}$=32.3 ms)

The 4n+1 radioactive series is not, in the usual sense, considered to be "naturally occurring", because its longest-lived member, ^{237}Np has a geologically short half-life of only 2.4×10^6 a. Thus, any ^{237}Np that might have been present when the earth was formed, 4.6×10^9 years ago, would have long since disappeared. However, minute traces of ^{237}Np and of ^{225}Ac, the grandparent of ^{217}At (**Fig.** 2-**3**, p. 13), have been detected in U refinery wastes and ^{225}Ac has been found in ^{232}Th isolated from Brazilian monazite [15]. It is believed that these nuclides are being formed continually in nature by the bombardment of natural Th and U with neutrons produced by one or more of the following processes; the (α, n) reaction on light elements, spontaneous fission of ^{238}U, induced fission of ^{235}U and cosmic rays.

Fig. 2-1

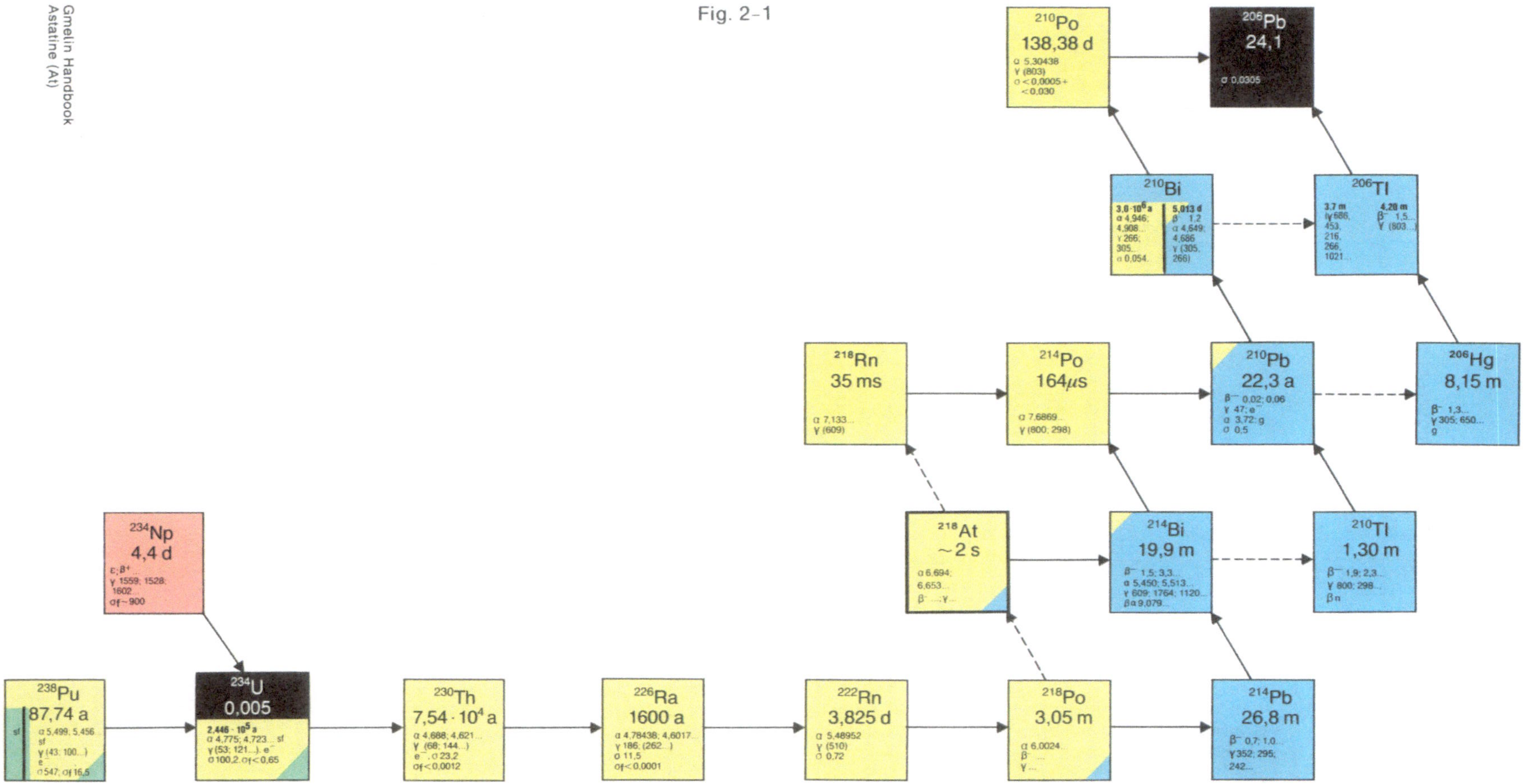

Genetic Relationships of ^{218}At (explanations on pp. 14 and 19).

Fig. 2-2

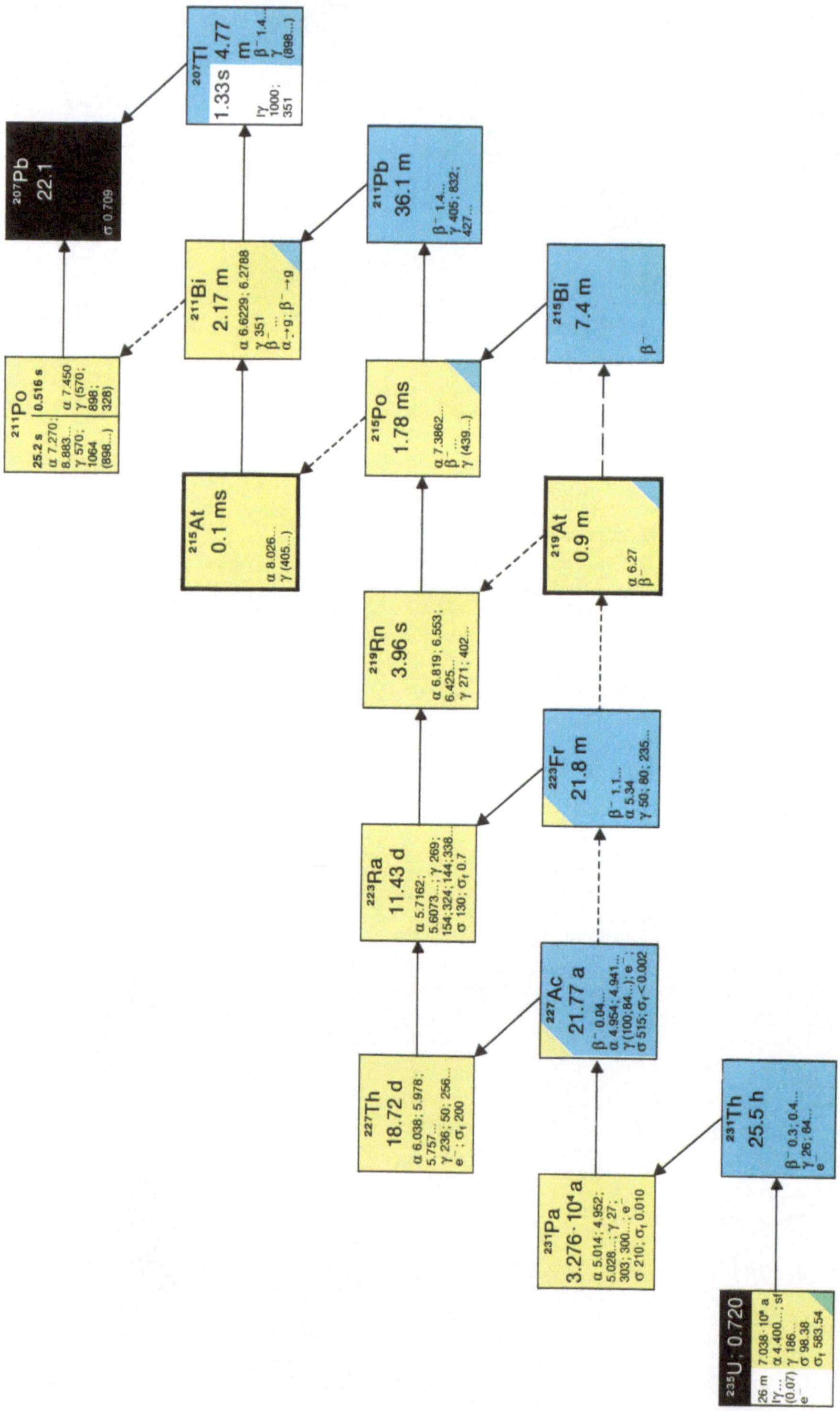

Genetic Relationships of ^{215}At and ^{219}At (explanations on pp. 14 and 19).

Fig. 2-3

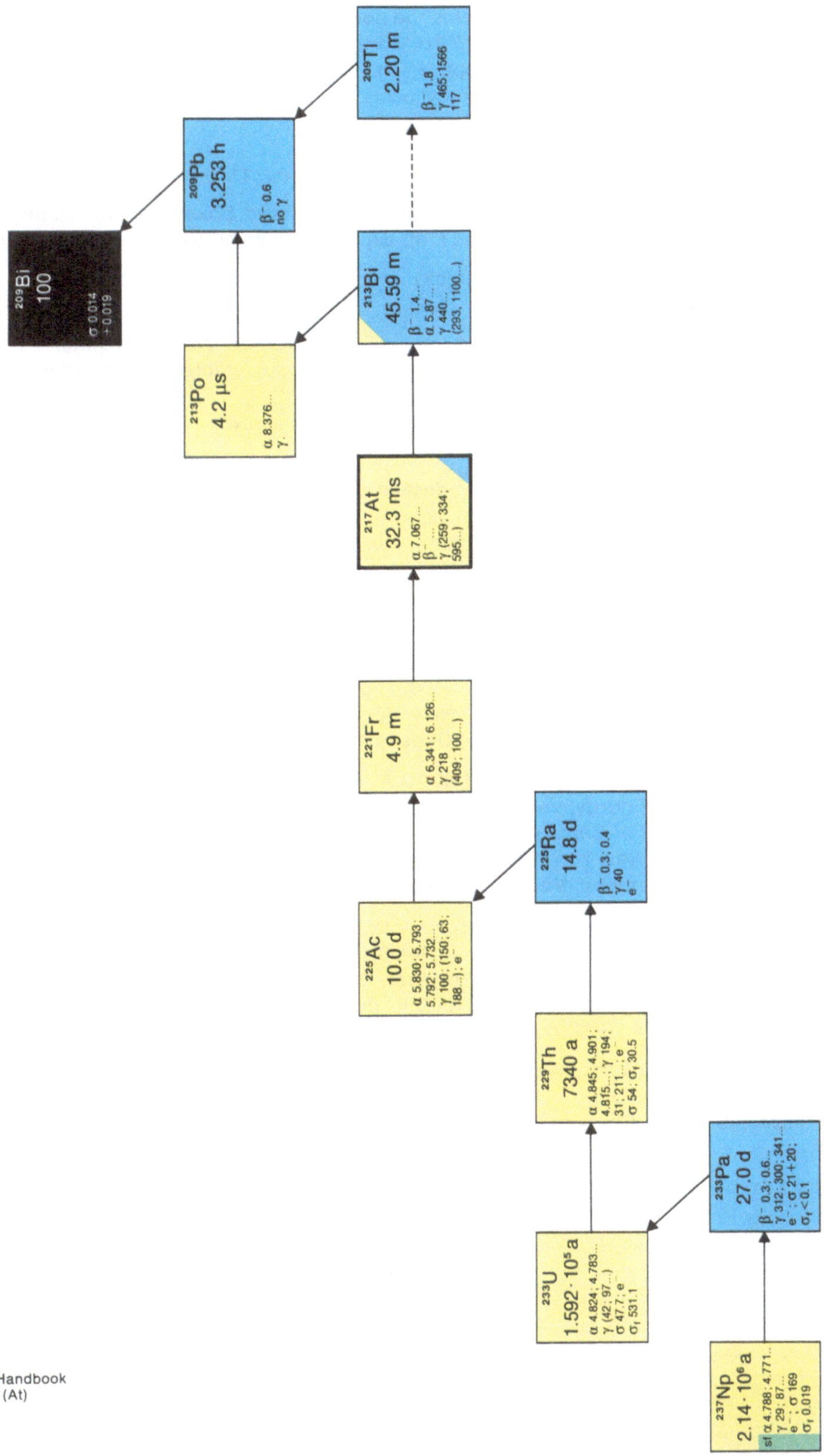

Genetic relationships of ^{217}At (explanations on pp. 14 and 19).

Peppard et al. [15] have estimated that the equilibrium mass ratios in nature are 4×10^{-11} for $^{229}Th/^{232}Th$ and 1.3×10^{-13} for $^{233}U/^{238}U$. To the extent that these estimates are reliable, they imply that the "synthetic" isotope, ^{217}At, has a natural abundance of 5×10^{-23} ppm – about the same as that of "naturally occurring" ^{219}At – and much higher than that of "natural" ^{215}At!

Explanations for Decay Series (Figg. 2-1, 2-2, and 2-3, pp. 11/3)

The abbreviations and the arrangement in the colored boxes are the same as those used in the Karlsruhe Chart of Nuclides [18], explained in some examples given on p. 19.

The colors designate the mode of decay:

yellow	α-decay	green	spontaneous fission
red	β^+-decay and electron capture (ε)	white	isomeric transition
blue	β^--decay	black	stable or primordial nuclide

Relative decay probabilities are designated by different arrows:

-------→ <5% of the decays
— — — —→ from 5% up to 95% of the decays
————→ >95% of the decays

direction of the arrows

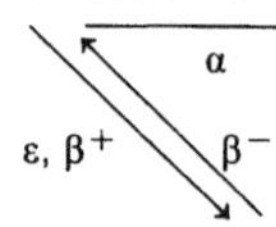

References:

[1] J.A.S. Adams, J.K. Osmond, J.J.W. Rogers (Phys. Chem. Earth **3** [1959] 298/348). – [2] J. Chervet (in: P. Pascal, Nouveau Traité de Chimie Minérale, Vol. 15, Pt. 1, Masson, Paris 1960/61, pp. 52/94). – [3] "Uran" 1936, pp. 1/37. – [4] J.J. Katz, E. Rabinowitch (The Chemistry of Uranium, McGraw-Hill, New York 1951, pp. 68/108). – [5] P.F. Kerr (Proc. Intern. Conf. Peaceful Uses At. Energy, Geneva 1955 [1956], Vol. 6, pp. 5/59).

[6] H.W. Kirby (MLM-2111 [1974] 1/82; C.A. **82** [1975] No. 142685). – [7] R.D. Nininger (Minerals for Atomic Energy, Van Nostrand, New York 1954). – [8] K. Rankama, T.G. Sahama (Geochemistry, Univ. of Chicago Press, Chicago 1950). – [9] K. Rankama (Progress in Isotope Geology, Wiley, New York 1963). – [10] J.J.W. Rogers, J.A.S. Adams (in: K.H. Wedepohl, Handbook of Geochemistry, Vol. 2, Pt. 5, Springer, Berlin 1972, Chapter 92-B/C. I/II, D/O).

[11] S.R. Taylor (Geochim. Cosmochim. Acta **28** [1964] 1273/85). – [12] B.R. Grundy, A.N. Hamer (J. Inorg. Nucl. Chem. **23** [1961] 148/50). – [13] A. Heydemann (in: K.H. Wedepohl, Handbook of Geochemistry, Vol. 1, Springer, Berlin 1969, pp. 376/412). – [14] I. Asimov (J. Chem. Educ. **30** [1953] 616/8). – [15] D.F. Peppard, G.W. Mason, P.R. Gray, J.F. Mech (J. Am. Chem. Soc. **74** [1952] 6081/4).

[16] J.W. Fowler (Astrophys. J. **188** [1974] 295/307). – [17] R.L. Kurucz (Astrophys. J. Suppl. Ser. **40** [1979] 1/31). – [18] W. Seelmann-Eggebert, G. Pfennig, H. Münzel, H. Klewe-Nebenius (Karlsruhe Chart of Nuclides, 5th Ed., Gesellschaft für Kernforschung, Karlsruhe 1981).

3 Nuclear Properties of Astatine Isotopes

H. Münzel
Technische Hochschule Darmstadt
Darmstadt, Federal Republic of Germany

The structure of nuclides as well as their decay properties are discussed in many books and monographs, see for instance [1 to 14]. Brief descriptions are also found in two volumes of this handbook [15, 16], which should be consulted for further explanations of special notations used for instance in decay schemes.

Wapstra and Bos [30] evaluated from all published experimental data for decay and reactions energies the ground state masses of nuclides. The values obtained in this extensive analysis for the At isotopes are given in the respective sections below. The masses are given in atomic mass units $u = (1.660566 \pm 0.000009) \times 10^{-27}$ kg $\triangleq 931501.6 \pm 2.6$ keV. The corresponding values for the mass excess Δm are listed in Table 3/1, p. 16.

There are several sets of mass excess values for the At isotopes available, which were calculated using different theoretical approaches [18 to 26]. The models are based either on collective or on microscopic descriptions of the nuclei. Short introductions to most of these models are given in the compilation edited by Maripuu [17], see also the very brief comments in [16]. As an example the numerical results of the calculation of Liran et al. [21] are also listed in Table 3/1, p. 16. Several of the available calculated data sets are compared with each other in **Fig.** 3-**1**, p. 17, where the differences to the values given by Liran et al. [21] are shown. The sets apparently agree with each other reasonably well in that region of the nucleon number A where experimental data for Δm are available. This is because all mass formulae contain an adjustable parameter, for instance 178, in the case of Liran et al. [21], for which numerical values are obtained by a fit to the experimental data. Which mass formula predicts the true trend outside this A range is still an open question.

Most of the mentioned mass formulae include the effect of nuclear deformation explicitly. Accordingly, values for the equilibrium deformation in the ground state, radial proton and/or neutron distribution as well as quadrupol moments were deduced from the mass calculation. The values agree in general quite well with experimental results [18, 26, 27].

Generally, the prominent mode of the decay of excited nuclei is the emission of γ rays with an energy of $E_\gamma = E_i^* - E_f^*$, i.e., the difference of the excitation energy of the initial and final state. However, conversion electrons may also be emitted. The energy of the conversion electrons is approximately $E_e = E_i^* - E_f^* - E_B$, where E_B denotes the binding energy of the emitted electron. Values for E_B are listed in Table 3/2, p. 18. The emission of electrons from the K shell are favored if this transition is energetically possible. The conversion coefficient, i.e., the ratio of the emission probability of conversion electrons and γ rays ($\alpha = e/\gamma$), depends on the transition energy $E_i^* - E_f^*$, the multiple order, and the proton number of the nuclide. Calculated conversion coefficients for the emission of electrons from the K shell (α_K) are shown in **Fig.** 3-**2**, p. 18. The agreement between different sets of calculated values [66 to 73] for α_K as well as with the experimental data is, in general, very good. Experimental values for α or α_K are given below, together with the other decay data for the Fr isotopes. The holes left behind after the emission of conversion electrons are filled with electrons from outer shells, which leads to the emission of X-rays and Auger electrons. The fluorescence yield, i.e., the emission probability of a K X-ray per hole in the K shell, for Fr is 0.971 [74]. The energies and relative intensities of the most prominent X-rays are listed in Table 3/3, p. 19. Correspondingly, the Auger electron yield is 0.029. Semiempirical energies for Auger electrons are listed by Larkin [75].

Table 3/1
Mass Excess (in MeV) of the At Isotopes as Evaluated by Wapstra et al. [30] and Calculated by Liran et al. [21]. A is the nucleon number of the isotope.

A	Wapstra et al. [30]	Liran et al. [21]	A	Wapstra et al. [30]	Liran et al. [21]
185	–	–	210	−12.075±0.018	−12.07
186	–	–	211	−11.654±0.009	−11.64
187	–	15.22[a)]	212	− 8.626±0.006	− 8.50
188	–	13.37[a)]	213	− 6.590±0.013	− 6.53
189	–	9.93[a)]	214	− 3.390±0.006	− 3.30
190	–	8.33[a)]	215	− 1.262±0.008	− 1.21
191	–	5.15[a)]	216	2.239±0.008	2.16
192	–	3.80[a)]	217	4.382±0.012	4.40
193	–	0.87[a)]	218	8.098±0.013	7.93
194	–	− 0.23[a)]	219	10.520±0.080	10.36
195	–	− 2.89[a)]	220	(14.20)	14.07
196	(−4.050)	− 3.73	221	–	16.71
197	(−6.030)	− 6.12	222	–	20.63
198	−6.670±0.310	− 6.69	223	–	23.48
199	−8.430±0.280	− 8.78	224	–	27.61
200	(−8.67)	− 9.08	225	–	30.69
201	(−10.52)	−10.85	226	–	35.03
202	(−10.52)	−10.88	227	–	38.34
203	(−11.97)	−12.29	228	–	42.91
204	(−11.97)	−12.07	229	–	46.45
205	(−12.96)	−13.09	230	–	51.23
206	(−12.73)	−12.66	231	–	54.99
207	13.310±0.050	−13.24	232	–	59.99
208	(−12.64)	−12.64	233	–	63.99
209	12.888±0.009	−12.74	234	–	69.20
			235	–	73.41
			236	–	78.83
			237	–	83.26
			238	–	88.88

a) Unstable against proton emission. – Values given in parantheses are obtained from systematics [30].

Each section in the following survey contains the structure and decay data for a particular At isotope in the following sequence: mass, level scheme, half-life, α decay, β decay, electron capture, γ radiation, decay scheme, and methods of production. In general, not all available data are listed; however, references to other published values are provided. The main criteria for inclusion are the certainty and age of the data and the extent and completeness of decay schemes and tables. In some cases, values obtained from systematics are included if reliable experimental data are not available. The literature was covered up to June 1984.

Fig. 3-1

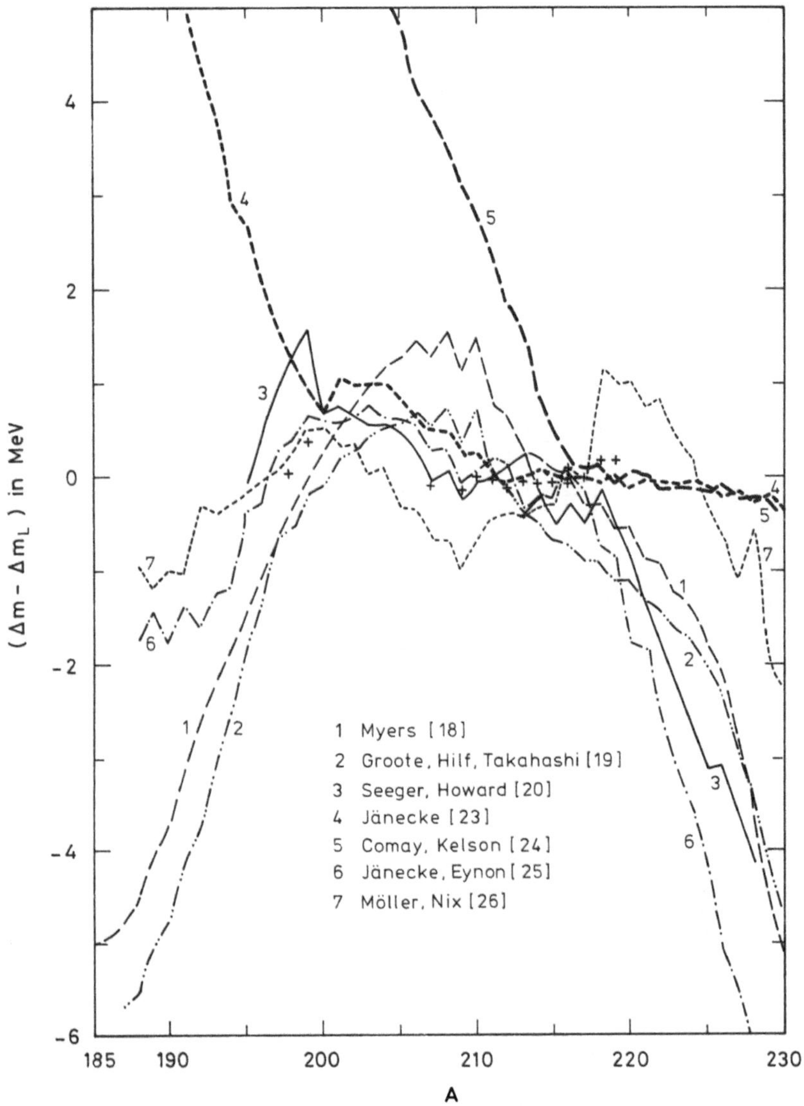

Calculated mass excess Δm for At isotopes relative to the values Δm_L obtained by Liran et al. [21]. For comparison the values obtained by Wapstra et al. [30] from experimental data are given as crosses.

The reference in which the isotope was first described is given in the heading of each section.

The level and decay schemes are usually reproduced as published. In some cases, however, small changes were made to make the schemes more self-explanatory. Dots at the upper end of the transition arrows denote observed coincidences.

References on pp. 87/94

Table 3/2
Electron Binding Energies [65] for Astatine and Neighboring Elements in the Elemental State.

shell	element		
	$_{84}$Po	$_{85}$At	$_{86}$Rn
K(1s)	93099.9 ± 1.2	95724.0 ± 1.3	98397.2 ± 1.5
L_1(2s)	16927.9 ± 1.6	17481.5 ± 1.6	18048.7 ± 1.5
$L_2(2p_{1/2})$	16238.0 ± 1.2	16777.3 ± 1.2	17329.7 ± 1.1
$L_3(2p_{3/2})$	13810.6 ± 1.2	14208.0 ± 1.2	14611.4 ± 1.2
M_1(3s)	4153.5 ± 1.5	4311.7 ± 1.5	4474.3 ± 1.5
$M_2(3p_{1/2})$	3844.3 ± 1.5	3995.8 ± 1.5	4151.5 ± 1.5
$M_3(3p_{3/2})$	3293.4 ± 1.2	3410.5 ± 1.2	3530.5 ± 1.5
$M_4(3d_{3/2})$	2793.6 ± 1.2	2901.8 ± 1.1	3012.3 ± 1.1
$M_5(3d_{5/2})$	2679.2 ± 1.0	2780.7 ± 1.1	2884.2 ± 1.1
N_1(4s)	987.5 ± 1.2	1038.2 ± 1.2	1090.5 ± 1.2
$N_2(4p_{1/2})$	850.9 ± 1.9	897.7 ± 1.8	946.2 ± 1.7
$N_3(4p_{3/2})$	715.2 ± 1.0	753.7 ± 1.0	791.2 ± 1.0
$N_4(4d_{3/2})$	495.7 ± 1.2	527.6 ± 1.2	560.4 ± 1.2
$N_5(4d_{5/2})$	469.9 ± 1.0	500.1 ± 1.0	531.1 ± 1.0
$N_6(4f_{5/2})$	184.6 ± 1.0	207.0 ± 1.0	230.1 ± 1.1
$N_7(4f_{7/2})$	178.9 ± 1.0	200.8 ± 1.5	223.6 ± 1.1

Fig. 3-2

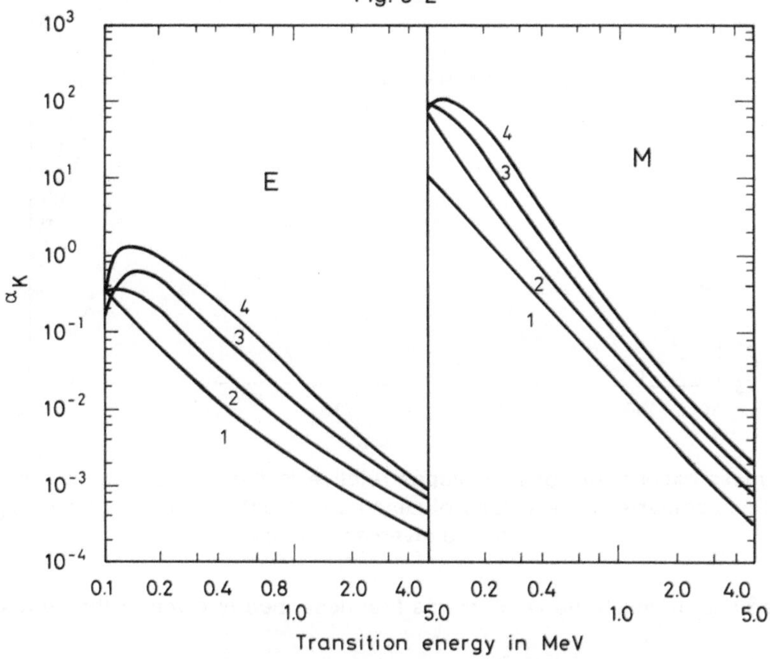

Calculated conversion coefficients for the emission of K electrons from At isotopes versus the transition energy $E_i^* - E_f^*$ [66]. The electric and magnetic multipole order are denoted by E_x and M_x, respectively.

Table 3/3
Energies and Relative Intensities of the Most Prominent X-Rays (as listed in [29], based on the values given in [76, 77]).

element	energy in keV[b)]					relative intensity[a), b)]			
	K_{α_1}	K_{α_2}	K_{β_1}	K_{β_3}	$\langle K_{\beta_2}\rangle$	K_{α_2}	K_{β_1}	K_{β_3}	$\langle K_{\beta_2}\rangle$
$_{83}$Bi	77.108	74.815	87.343	86.834	89.8	59.5	22.1	11.6	10.5
$_{84}$Po	79.290	76.862	89.80	89.25	92.4	59.7	22.1	11.6	10.8
$_{85}$At	81.52	78.95	92.30	91.72	95.0	60.0	22.1	11.6	11.0
$_{86}$Rn	83.78	81.07	94.87	94.24	97.6	60.2	22.1	11.6	11.3
$_{87}$Fr	86.10	83.23	97.47	96.81	100.3	60.5	22.1	11.6	11.5

a) relative to $I(K_{\alpha_1})=100$

b) $K_{\alpha_1}=E_B(K)-E_B(L_2)$ $K_{\beta_1}=E_B(K)-E_B(M_3)$
$K_{\alpha_2}=E_B(K)-E_B(L_3)$ $K_{\beta_3}=E_B(K)-E_B(M_2)$
$\langle K_{\beta_2}\rangle$ weighted average of $E_B(K)-E_B(N_2)$ and $E_B(K)-E_B(N_3)$

The following abbreviations are used:

Units:

ps	picosecond	min	minute	keV	10^3 electron volt (eV)
ns	nanosecond	h	hour	MeV	10^6 electron volt
µs	microsecond	d	day	b	barn ($=10^{-28}$ m^2)
ms	millisecond	a	year	mb	millibarn ($=10^{-31}$ m^2)
s	second				

Decay properties:

α	α decay
α, α_K, α_L	conversion coefficient (total, K shell, L shell)
β^-, β^+	β^- decay, β^+ decay
ε	electron capture
γ	γ ray emission
ce	conversion electrons
Q_α, Q_{β^-}, Q_{β^+}	decay energy for α decay, β^- decay, β^+ decay
S_n	neutron binding energy
S_p	proton binding energy
E_α, E_β	energy of the radiation; in the case of β decay, the maximum energy is given
I_i	absolute intensity for the radiation type i in %, i.e., I_i rays per hundred decays
I_i^R	relative intensity of the radiation type i
(syst)	value obtained from the cited systematics

The systematics of the properties of the low-lying states of the At isotopes are discussed by Schmorak [78].

References on pp. 87/94

Fig. 3-3

Section of Karlsruhe Chart of Nuclides [79], explanations on p. 22.

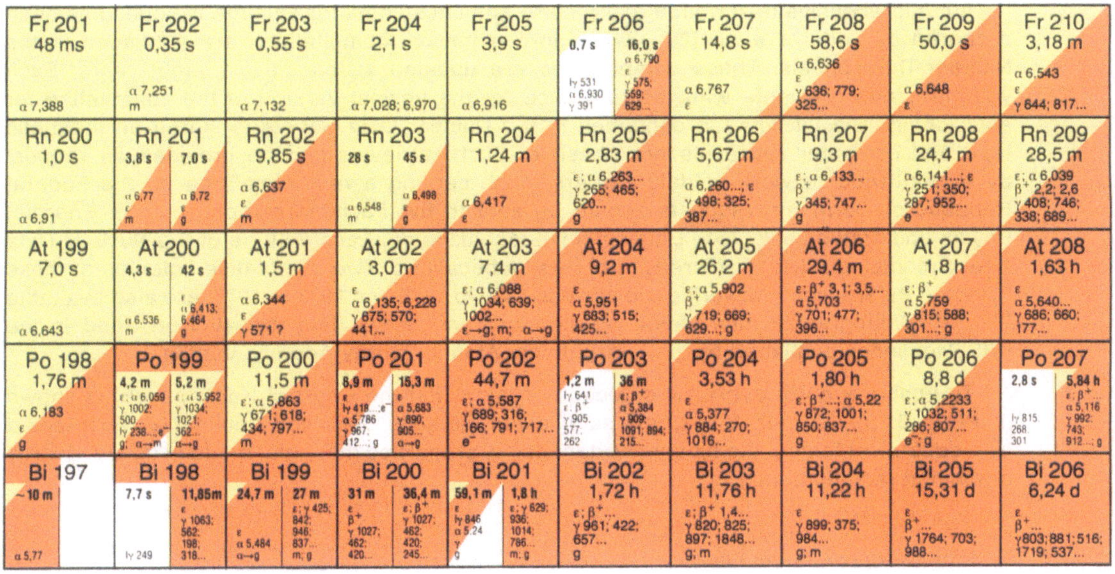

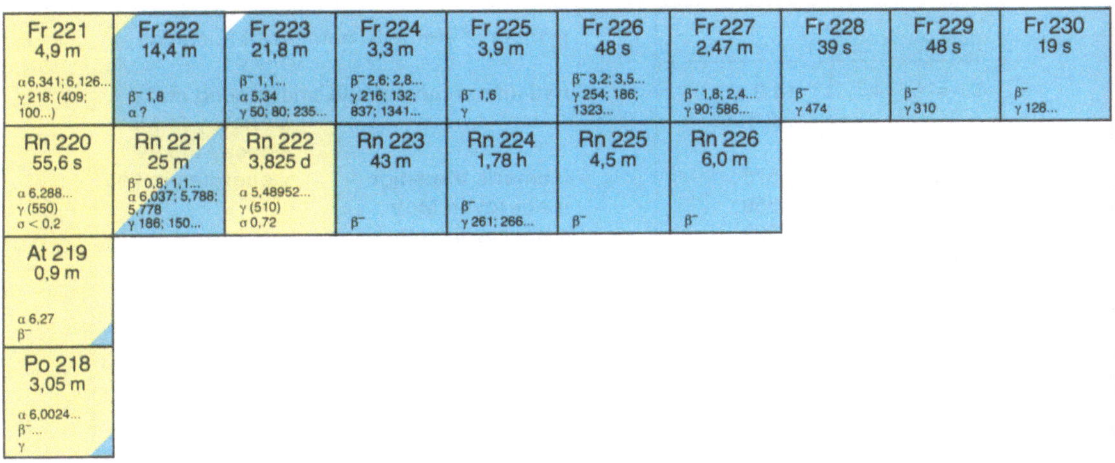

References on pp. 87/94

A comprehensive description of structure and decay data for astatine isotopes has been prepared by Lederer et al. [29]. Even more extensive compilations are published in the Nuclear Data Sheets. These compilations are updated at least every eight years. Each volume contains a table with the reference for the newest version of the compilation for an A chain. The information contained in the Nuclear Data Sheets is stored on magnetic tape and computer retrievals are possible. Such retrievals will be provided on request by a worldwide network of data centers, each serving a restricted area. In the Federal Republic of Germany, requests should be sent to the Fachinformationszentrum Energie, Physik und Mathematik, 7514 Eggenstein-Leopoldshafen. Each volume of the Nuclear Data Sheets contains a list of references of new publications. Two cumulative volumes of these "Recent References", covering the periods 1969 to 1975 and 1975 to 1977, are also available [80, 81]. Bibliographic information on structure and decay data are also provided in the INIS Atomindex published by the International Atomic Energy Agency, Vienna.

To enable a quick overview a section of the Karlsruhe Chart of Nuclides [79] is shown in **Fig. 3-3**, pp. 20/1. In this Chart each experimentally observed nuclide is represented by a square containing the most important decay properties (half-life, mode of decay, energy of emitted radiation, etc.).

The following lay-out is used:

Radionuclide

^{206}Fr		
0.7 s	16.0 s	half-life of metastable and ground state
	α 6.790	α energy in MeV
	ε	ε decay in keV
I γ 531	γ 575;	isomeric transition / γ energies in keV
α 6.930	559;	α energy in MeV
γ 391	629	γ energy in keV

(Row ^{206}Fr: symbol of element, number of nucleons)

decay data for
metastable state / ground state

Stable nuclide

^{209}Bi	number of nucleons, symbol of the element
100	abundance in natural element (in at.%)
0.014 ± 0.019	thermal neutron capture cross section (in 10^{-28} m^2) for the formation of ^{210m}Bi and ^{210}Bi, respectively

The colors designate the mode of decay

yellow	α decay
red	β^+ decay and electron capture (ε)
blue	β^- decay
white	isomeric transition
black	stable nuclide

The probability of the mode of decay is indicated by the size of colored field (wedge):

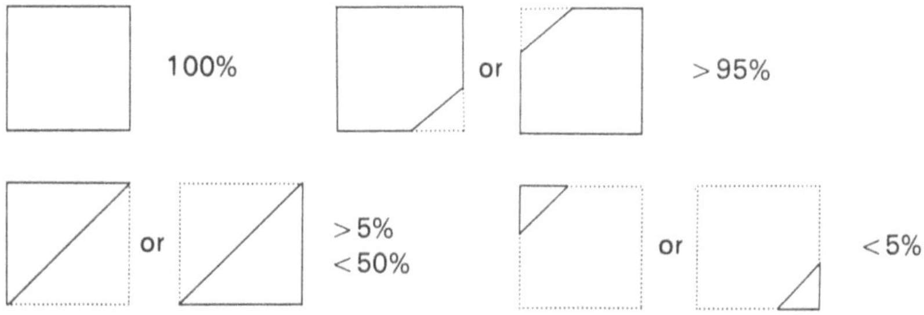

The abbreviations and symbols mentioned on p. 19 are applied with one exception: In the Chart minute is denoted to be m (and not min). In addition the following two abbreviations are used:

g formation of the ground state of the daughter nuclide favored
m formation of the metastable state of the daughter nuclide favored

Many At isotopes decay to nuclides which are themselves radioactive. Therefore, when the At isotopes are investigated, the radiation emitted in the decay of their daughter nuclides are also observed. On the other hand At isotopes are also members of the decay chains of heavier predecessors. A comprehensive survey of the chains starting with U and Pa isotopes are given in other volumes of this handbook [16, 82].

Astatine-196 [33]

mass: 195.99566 u (syst) [30]
S_p 0.28 MeV (syst) [30]

half-life: 0.3 ± 0.1 s [33]

α decay: [33]
Q_α 7202 ± 7 keV [30]
E_α 7055 ± 7 keV [33]

electron capture + β^+ decay:
not observed yet
Q_{β^+} 9160 keV (syst) [30]

production: $^{185}Re(^{20}Ne,9n)^{196}At$ [33]

Astatine-197 [33]

mass: 196.99352 u (syst) [30]
S_n 10.060 MeV (syst) [30]

S_p 0.11 MeV (syst) [30]

levels: only the ground state of ^{197}At, for which according to systematics J^π 9/2$^-$ was assumed [41], is populated in the decay of ^{201}Fr [38]

half-life: 0.4 ± 0.1 s [33]

α decay: [33], ~96% [39, 40]
Q_α 7103 ± 5 keV [30]
E_α 6959 ± 5 keV [33, 37]

References on pp. 87/94

electron capture + β^+ decay: ≈4% [39, 40]
Q_{β^+} 7200 keV (syst) [30]

decay scheme: Harmatz proposes that the α decay feeds the ground state of ^{193}Bi [41]

production: ^{185}Re(^{20}Ne,8n)^{197}At [33]

Astatine-198 [33]

mass: 197.99284 ± 0.00034 u [30]
S_n 8.70 MeV (syst) [30]
S_p 0.72 MeV (syst) [30]

half-life: 4.9 ± 0.5 s [33]

α decay: [33]
Q_α 6894 ± 50 keV [43], 6887 ± 5 keV [30]
E_α 6748 ± 5 keV [33] further reference [38]
6755 ± 5 keV [37], based on [33, 35]

electron capture + β^+ decay: not observed yet
Q_{β^+} 8410 keV (syst) [30]

production: ^{185}Re(^{20}Ne,7n)^{198}At [33]

Astatine-198m [33]

half-life: 1.5 ± 0.3 s [33]

α decay: [33]
E_α 6849 ± 5 keV [33, 37] further reference [38]

production: ^{185}Re(^{20}Ne,7n)^{198m}At [33]

Astatine-199 [33]

mass: 198.99090 ± 0.00023 u [30]
S_n 9.88 ± 0.38 MeV [30]
S_p 0.69 MeV (syst) [30]

levels: for the ground state, which is fed directly in the α decay of ^{203}Fr [61], J^π 9/2$^-$ was proposed (syst) [61]

half-life: 7.2 ± 0.5 s [33] further reference [35]

α decay: [33]
Q_α 6779.4 ± 3.1 keV [30]
E_α 6643 ± 3 keV evaluated by [61], based on [33, 35, 37] further reference [38]

electron capture + β^+ decay: not observed yet
Q_{β^+} 6580 keV (syst) [30]

decay scheme: α decay populates directly the ground state of ^{195}Bi [61]

production: ^{185}Re(^{20}Ne,6n)^{199}At [33]
^{209}Bi(^{3}He,13n)^{199}At [35]

Astatine-200 [31]

mass: 199.99069 u (syst) [30]
S_n 8.27 MeV (syst) [30]
S_p 0.90 MeV (syst) [30]

levels:

E in keV	J^π	
0.0	(5+)	[36] by analogy to ^{202}At, ^{204}At
	(7+)	[34] from ^{204}Fr decay
61.5		[34] from ^{204}Fr decay
~140	(10−)	[36] from systematics; see ^{200m}At

half-life: 42 ± 2 s [33]; 44 ± 3 s [35] — further references [31, 32]

α decay: [31], 53 ± 8% [34], 35 ± 8% (syst) [36]
Q_α 6669.6 ± 2.0 keV [30]

E_α in keV [36] [a)]	$I_\alpha^{R\ b)}$ [35 to 37]
6412 ± 2	60.4
6465 ± 2	39.0
6574 ± 5	0.6

further reference [31]

a) Evaluation based on the data given in [32 to 35, 37].
b) To obtain absolute intensities multiply by 0.35 ± 0.08 [36].

electron capture $+\beta^+$ decay: [31], 47 ± 8% [34], 65 ± 8% (syst) [36]
Q_{β^+} 8070 keV (syst) [30]

γ radiation: see ^{201}At

production: ^{185}Re(^{20}Ne,5n)^{200}At [33]
^{197}Au(^{12}C,9n)^{200}At [32]

Astatine-200 m [33]

level: the values of E* ≈ 140 keV and J^π 10^- proposed by [36] are based on systematics

half-life: 4.3 ± 0.3 s [33], 5 ± 2 s [35]

α decay: [33], 20% (syst) [36]
E_α 6536 ± 4 keV [35]

isomeric transition: 80% (syst) [36]

production: ^{185}Re(^{20}Ne,5n)^{200m}At [33]

Astatine-201 [32]

mass: 200.98871 u (syst) [30]
S_n 9.92 MeV (syst) [30]
S_p 1.07 MeV (syst) [30]

level: the level scheme is shown in **Fig.** 3-**4**, p. 26

half-life: 1.5 ± 0.1 min [32] — further references [31, 46]
1.50 ± 0.07 min

Fig. 3-4

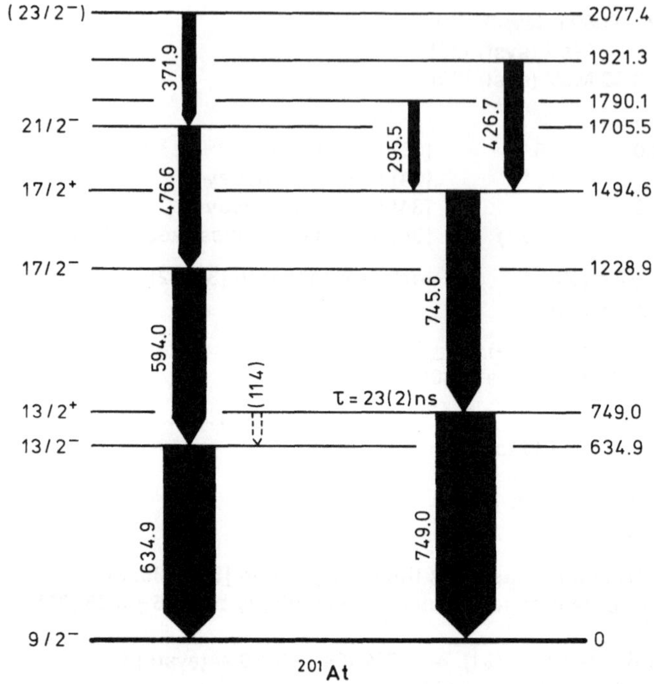

Levels of ^{201}At observed in the reaction $^{192}Pt(^{14}N,5n)^{201}At$. Transition and level energies are quoted in keV [62].

α decay: [32], 71±7% [34] — further reference [32]
Q_α 6472.0±3.1 keV [30]
E_α 6344±3 keV [49], evaluation based on [33, 34, 37] — further references [31, 32, 45]

electron capture $+\beta^+$ decay: [32], 29±7% [34] — further reference [32]
Q_{β^+} 5890 keV (syst) [30]

γ radiation: with $\varepsilon+\beta^+$
571.0 keV [46]
this γ line was tentatively assigned to ^{201}At but could also belong to ^{200g}At [61]

production: $^{169}Tm(^{32}Cl,4np)^{201}At$ [264]; σ_{max} 0.4 mb at $E(^{37}Cl)$ 182 MeV [264]
$^{185}Re(^{20}Ne,4n)^{201}At$ [33]
$^{187}Re(^{20}Ne,6n)^{201}At$ [33]
$^{197}Au(^{12}C,8n)^{201}At$ [44]

Astatine-202 [47]

mass: 201.98870 u (syst) [30]
S_n 8.08 MeV (syst) [30]
S_p 1.40 MeV (syst) [30]

levels: for the ground state J^{π} 5^+ was proposed [28]. A level at $E^* \approx 110$ keV was suggested by [33] which, however, was not confirmed by [34]. A γ ray with E_γ of 391 keV was detected by [64] and accordingly a level at E^* 391 keV was assumed

for the α particles with E_α 6226 keV a slightly lower half-life (2.6 to 3 min [32, 33, 45 to 47]) was found which was attributed to the decay of a metastable state in ^{202}At [47]. This assumption was not supported by [33, 34]

half-life: 181 ± 3 s [28]. This recommended value is based on the values given in [34, 46], further references [32, 33, 35, 47 to 49, 63]

α decay: [47], 12.0 ± 0.8% [48, 49], 15 ± 3% [34], further reference [32]
Q_α 6354.0 ± 2.0 keV [30]

E_α (in keV) [28, 37] [a)]	I_α^R [b)] [28, 37] [a)]
6135 ± 2	64 ± 2
6228 ± 2	36 ± 2

a) These recommended values are based on the data given in [32 to 35, 47 to 49].

b) To obtain absolute intensities multiply by 0.12 [28].

Fig. 3-5

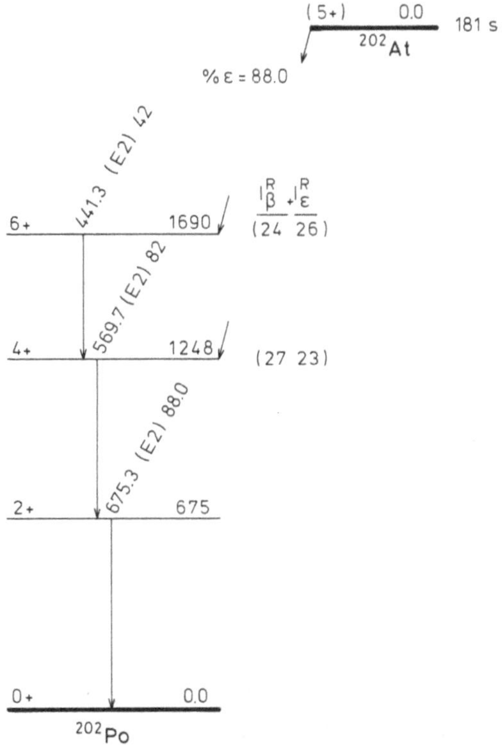

Decay scheme for the ε+β⁺ decay of ^{202}At [28]. Transition and level energies are given in keV.

References on pp. 87/94

electron capture $+\beta^+$ decay: 88.0±0.8% [48, 49], 85±3% [34]
in [28] an intensity ratio $I(\beta^+):I(\varepsilon)=51:49$ is proposed (see also **Fig.** 3-**5**, p. 27).
Q_{β^+} 7250 keV (syst) [30]

γ radiation: with α decay: the expected γ radiation with E_γ 95±3 keV was not observed in α-γ coincidences [32]

with $\varepsilon+\beta^+$ decay:

E_γ in keV	I^R_γ	
[46] a)	[46]	[28] b)
441.3±0.3	33.0±2.5	47±5
569.7±0.4	94 ±5	93±5
675.3±0.5	100	100

a) About 2 keV higher values are given in [39, 50, 63]. Two additional lines at 603 and 618 keV were observed by [78].
b) Recommended values, based on [46, 50].

decay scheme: the α decay populates the ground state and the first excited state in ^{198}Bi [28, 29]
$\varepsilon+\beta^+$ decay, see Fig. 3-5, p. 27

Fig. 3-6

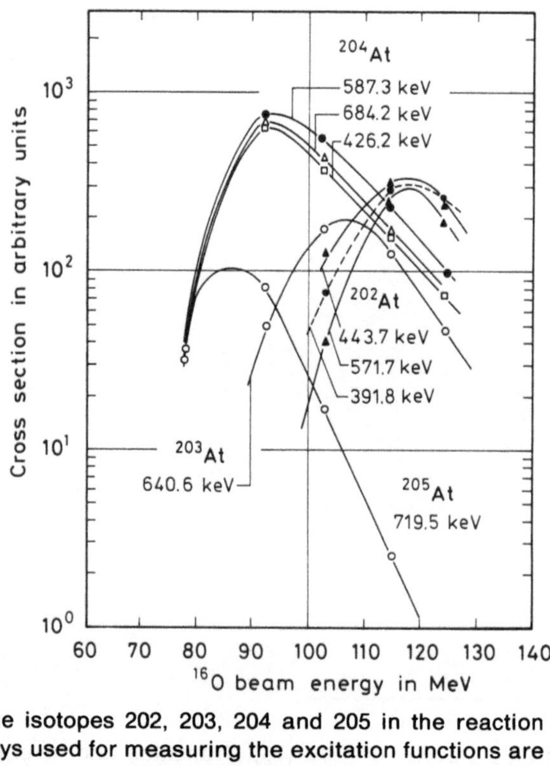

Yield of the astatine isotopes 202, 203, 204 and 205 in the reaction $^{193}Ir(^{16}O,...)$ [57]. The energies of the γ rays used for measuring the excitation functions are given with the respective curves.

production: $^{169}Tm(^{37}Cl,3np)^{202}At$ [264]
σ_{max} 0.3 mb at $E(^{37}Cl)$ 167 MeV [264]
$^{187}Re(^{20}Ne,5n)^{202}At$ [33]
$^{193}Ir(^{16}O,7n)^{202}At$ [57, 86]; see **Fig.** 3-**6**
$^{197}Au(^{12}C,7n)^{202}At$ [44, 47, 48, 253]; see **Fig.** 3-**7** and **Fig.** 3-**8**, p. 30
$^{209}Bi(^{3}He,10n)^{202}At$ [35]

Astatine-203 [54]

mass: 202.98715 (syst) [30]
S_n 9.52 MeV (syst) [30]
S_p 1.48 MeV (syst) [30]

level: see **Fig.** 3-**9**, p. 31
Morek et al. [87] found a γ line with E_γ 585.3±0.2 keV which they attributed to an isomeric transition in ^{203}At. On this basis they propose a level with $E^* \geqq 585.3$ keV with a half-life of 110±10 ms in ^{203}At. However, this level belongs most probably to ^{204}At [57]

half-life: 7.37±0.20 min [86], this recommended value is based on the data given in [32, 35, 45, 51, 52], further references [32, 46 to 49, 52, 53]

α decay: [54], 13.8±0.6% [48, 49], 31±3% [34]. The higher value is in better agreement with systematics [86]
Q_α 6210.4±1.3 [30]

Fig. 3-7

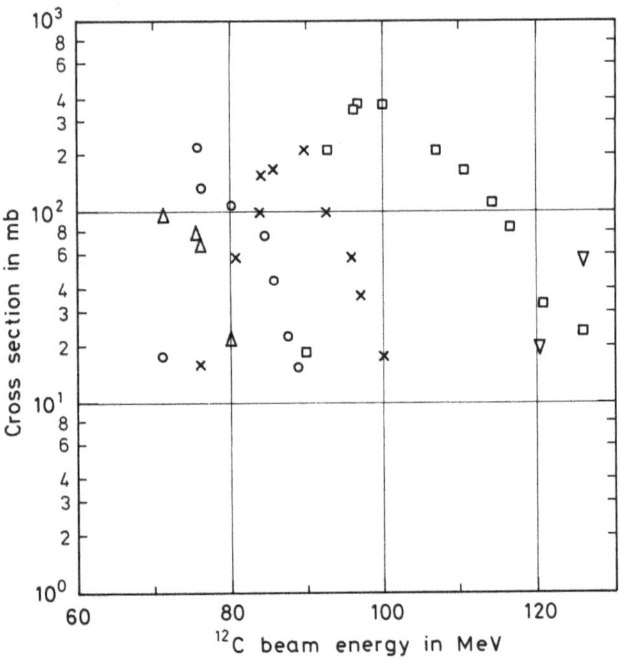

Cross sections for the formation of ^{206}At (△), ^{205}At (○), ^{204}At (×), ^{203}At (□) and ^{202}At (▽) in the reaction $^{197}Au(^{12}C,...)$ [253].

References on pp. 87/94

Fig. 3–8

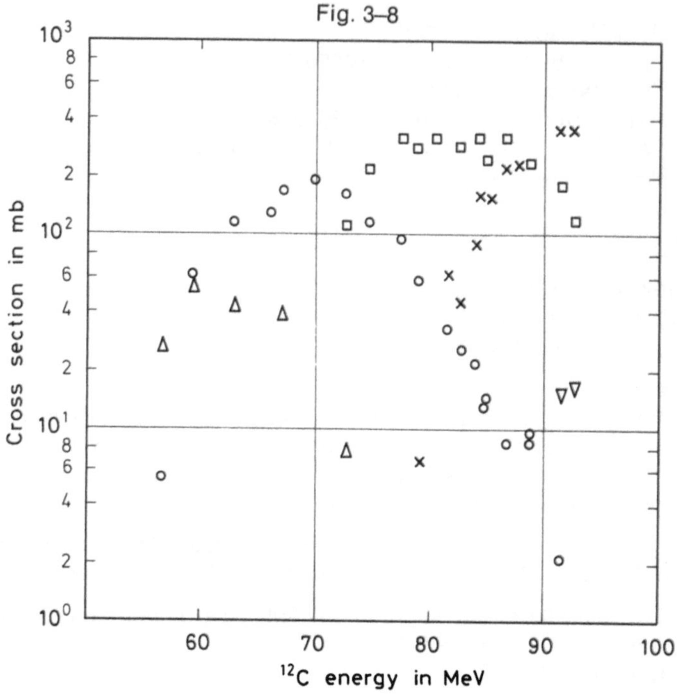

Cross sections for the formation of ^{202}At (△), ^{203}At (○), ^{204}At (□), ^{205}At (×), and ^{206}At (▽) in the reaction $^{197}Au(^{12}C,\ldots)$ [266].

E_α 6088±1 keV [37], this recommended value is based on [32 to 34, 51], further references [31, 46, 47, 52, 53]

electron capture: 86.2±0.6% [48, 49], 69±3% [34] (see α decay above)
Q_ε 5.39 MeV (syst) [30]

γ radiation: with ε, see Table 3/4, further reference [46]

decay scheme: the α decay feeds the ground state of ^{199}Bi [86]. The electron capture populates among other levels also the 1.2 min metastable state of ^{203}Po at E* 641.4 keV. A tentativ decay scheme for ^{203}At was proposed by [88]

Table 3/4
Energies and Intensities of the γ Rays Following the Electron Capture of ^{203}At [55].

energy in keV	$I_\gamma^{R\,b)}$	energy in keV	$I_\gamma^{R\,b)}$	energy in keV	$I_\gamma^{R\,b)}$
145.8±0.1	14.1±0.9	390.3±0.2	8.1±0.8	656.2±0.1	29.6±1.8
152.1±0.1	5.0±0.7	416.9±0.1	14.3±1.1	737.9±0.1	41.6±2.1
154.7±0.1	4.1±0.4	487.3±0.1	6.0±1.8	845.8±0.1	30.0±2.1
204.4±0.2	5.0±2.0 a)	531.9±0.1	18.2±1.8	880.4±0.1	40.7±2.4
206.6±0.1	9.0±0.8	608.8±0.1	20.4±1.8	1002.0±0.1	85.7±4.3
245.9±0.2	47.6±2.4	639.3±0.1	96.6±4.8	1034.0±0.1	100
361.6±0.3	23.1±1.4	641.4±0.1	53.1±3.2		

a) Disturbed by the 204.7 keV transition in ^{203}Bi. — b) For absolute intensities per 100 decays of ^{203}At multiply by 0.69±0.03 [86].

Fig. 3-9

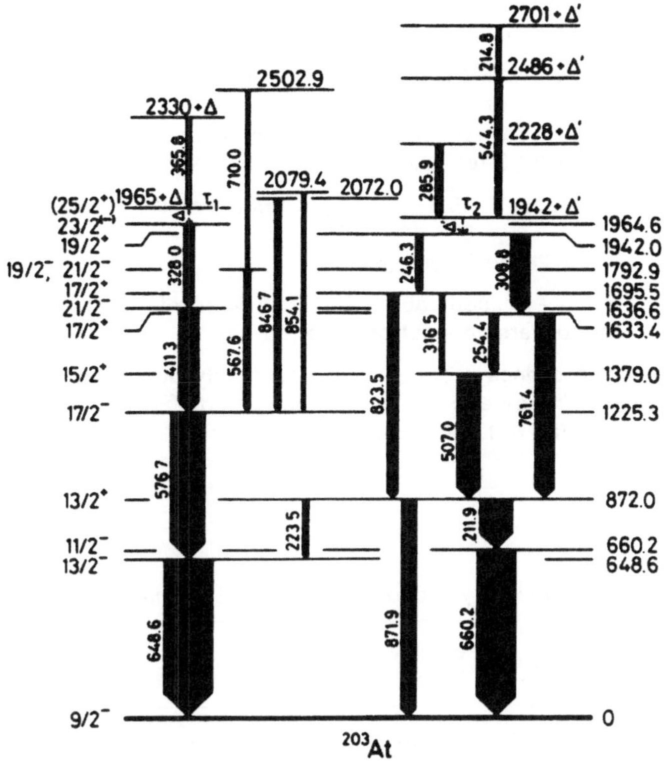

Levels of ^{203}At populated in the reaction ^{194}Pt(^{14}N,5n) [62]. The energies of the levels and γ rays are given in keV. The half-lives of the two metastable states are $\tau_1 = 17 \pm 2$ ns and $\tau_2 = 25 \pm 1$ ns.

production: ^{193}Ir(^{16}O,6n)^{203}At [57]; see Fig. 3-6, p. 28
^{197}Au(^{12}C,6n)^{203}At [44, 253, 266]; see Fig. 3-7, p. 29, and Fig. 3-8
Th(p_{600}, ...)^{203}At [34, 55]

Astatine-204 [47, 48]

mass: 203.98715 u (syst) [30]
S_n 8.07 MeV (syst) [30]
S_p 1.89 MeV (syst) [30]

levels: for the ground state J^{π} 5^+ was proposed by Schmorak [56], however, 6^+ or 7^+ could also be possible. Vakhtel et al. [91] favor also 5^+. Hornshoj et al. [34] postulate that in the decay of ^{208}Fr a 7^+ state with $E^* \approx 20$ keV is populated in ^{204}At. A metastable state at E^* 600 ± 20 keV with J^{π} 10^- and a half-life of 108 ± 10 ms is formed in heavy ion reactions with Au, Ir and Pt [57]. This metastable state decays by isomeric transition (E_{γ} 587.3 ± 0.2 keV, α 0.07) leading either to the ground or the first (≈20 keV) state [87]

half-life: 9.3 ± 0.2 min [32] further references [47 to 49]
8.9 ± 0.2 min [53]
9.2 ± 0.2 min [46], recommended by [56]

α decay: $4.4 \pm 0.2\%$ [56] evaluated on the basis of [34, 48, 49]; $3 \pm 1\%$ [90]
Q_α 6170 ± 100 keV [30]
E_α 5951 ± 2 keV [56] evaluated on the basis of [32, 33, 37, 47, 51]
5953 ± 3 keV [90]; no other groups with $I_\alpha^R > 10$ observed [90]

electron capture $+\beta^+$ decay: 95.6 ± 0.2 [56] evaluated on the basis of [34, 48, 49]
$97 \pm 1\%$ [90]
Q_{β^+} 6.28 MeV (syst) [30]

γ radiation: with $\varepsilon + \beta^+$, see Table 3/5 further references [46, 50, 56 to 60]
for conversion electrons see [56, 89]

decay scheme: with α decay: The decay presumably populates a state with J^π 7^+ and $E^* \approx 20$ keV and not the ground state of ^{200}Bi [56, 91]
with $(\varepsilon + \beta^+)$ decay, see **Fig. 3-10**; further references [56, 91]

Table 3/5
Energies and Intensities of γ Rays Following the $(\varepsilon + \beta^+)$ Decay of ^{204}At [89].

energy in keV	I_γ^R	energy in keV	I_γ^R	energy in keV	I_γ^R
12 [a]	0	489.52 ± 0.05	2.88 ± 0.18	841.06 ± 0.02	8.66 ± 0.32
24	0.5	m_0c^2 [d]	26.3 ± 0.8	844.66 ± 0.04	1.47 ± 0.12
74 [a]	0	516.32 ± 0.01	95.4 ± 3.0	858.18 ± 0.10	1.10 ± 0.14
Po K_{α_1}	86 [b]	522.47 ± 0.09	0.66 ± 0.12	867.80 ± 0.04	1.98 ± 0.14
Po K_{β_2}	4.0 ± 0.4	527.88 ± 0.06	0.76 ± 0.10	899.39 ± 0.06	2.62 ± 0.16
152 [c]	0.5	539.22 ± 0.13	0.53 ± 0.12	908.49 ± 0.07	0.98 ± 0.10
209.46 ± 0.30	0.25 ± 0.03	558.30 ± 0.10	0.43 ± 0.09	920.72 ± 0.14	0.46 ± 0.09
222 [c]	0.5	588.30 ± 0.02	7.26 ± 0.26	926.34 ± 0.06	1.41 ± 0.12
261.44 ± 0.03	2.45 ± 0.19	596.66 ± 0.25	0.38 ± 0.09	935.50 ± 0.25	0.37 ± 0.11
297.72 ± 0.12	0.33 ± 0.08	609.14 ± 0.03	26.0 ± 0.8	950.25 ± 0.07	1.06 ± 0.13
327.69 ± 0.06	3.51 ± 0.25	621.20 ± 0.09	6.0 ± 0.09	953.05 ± 0.10	0.55 ± 0.11
335.21 ± 0.03	3.73 ± 0.26	650.66 ± 0.06	1.74 ± 0.14	1003.45 ± 0.25	0.59 ± 0.17
338.37 ± 0.15	0.44 ± 0.09	656.60 ± 0.20	0.70 ± 0.11	1044.32 ± 0.09	0.61 ± 0.08
350.00 ± 0.15	0.68 ± 0.14	662.72 ± 0.04	1.46 ± 0.13	1101.45 ± 0.15	0.33 ± 0.08
351.70 ± 0.10	3.08 ± 0.20	672.40 ± 0.60	0.21 ± 0.12	1137.74 ± 0.30	0.20 ± 0.09
361.17 ± 0.27	0.50 ± 0.10	684.34 ± 0.01	100.0 ± 3.0	1162.23 ± 0.18	0.39 ± 0.10
414.62 ± 0.05	1.00 ± 0.09	696.73 ± 0.09	0.64 ± 0.11		
426.24 ± 0.01	71.3 ± 2.1	712.26 ± 0.11	0.35 ± 0.08		
479.78 ± 0.15	0.69 ± 0.13	749.45 ± 0.15	0.68 ± 0.18		
485.50 ± 0.25	0.51 ± 0.10	761.65 ± 0.05	5.0 ± 0.8		

[a] Existence implied by γ-γ coincidence relations. — [b] Value uncertain. — [c] Seen only in γ-γ coincidence spectra. — [d] Annihilation radiation.

Fig. 3-10

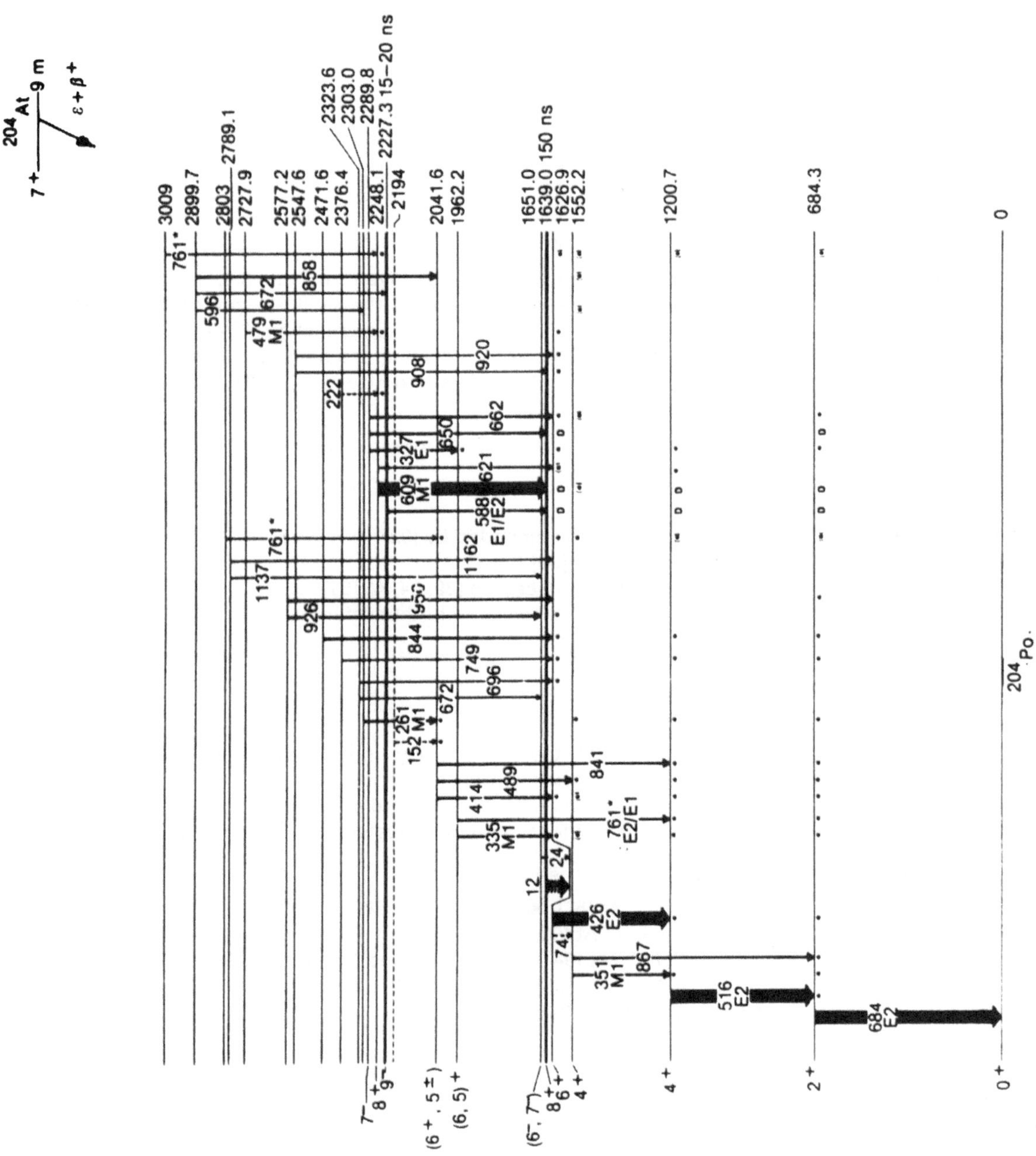

Proposed decay scheme for ^{204}At [89]. The observed prompt γ-γ coincidence relations are indicated by the dots and delayed coincidences by D. The energies are in keV.

References on pp. 87/94

production: $^{193}Ir(^{16}O,5n)^{204}At$ [57, 89]; see Fig. 3-6, p. 28
$^{197}Au(^{12}C,5n)^{204}At$ [32, 44, 47, 48, 94, 253, 266]; see Fig. 3-7, p. 29, and Fig. 3-8, p. 30
$^{209}Bi(\alpha,9n)^{204}At$ [53]
$U(p_{660},\ldots)^{204}At$ [90]

Astatine-205 [31]

mass: 204.98608 u (syst) [30]
S_n 9.07 MeV (syst) [30]
S_p 2.00 MeV (syst) [30]

levels: the levels proposed for ^{205}At are given in Table 3/6. The listed values are obtained by the investigation of the $\varepsilon+\beta^+$ decay of ^{205}Rn [98], the isomeric transitions [94] and the nuclear reaction $^{197}Au\ (^{12}C,4n)^{205}At$ [96]. The latter two investigations yield both the high spin states but they differ completly with each other. This is noticed in [96, 97] but no explanation was given

half-life: 26.2±0.5 min [32], further references [47, 93]
25.0±0.5 min [51]
27.2±0.6 min [46]

α decay: 10±2% [34, 55] recommended by [92], 18.4±1.6% [44, 48]
Q_α 6020.6±2.0 keV [30]
E_α 5902±2 keV [37] based on [32, 34, 46, 51]
further references [31, 47, 52, 93]

electron capture: 87±2% [34, 55]

β^+ decay: 3±2% [34, 55]
Q_{β^+} 4.61 MeV (syst) [30]

γ radiation: in Table 3/7, p. 36, the prominent γ lines following the $\varepsilon+\beta^+$ decay of ^{205}At are given, further reference [95]
conversion electrons were investigated in [55, 85]
the isomeric transitions were investigated in [94, 96, 97]; see Table 3/6

decay scheme: see **Fig. 3-11**, p. 37, see also [92]

production: $^{193}Ir(^{16}O,4n)^{205}At$ [57]; see Fig. 3-6, p. 28
$^{197}Au(^{14}N,p5n)^{205}At$ [93, 253]
$^{197}Au(^{12}C,4n)^{205}At$ [94, 96, 253, 266]; see Fig. 3-7, p. 29, and Fig. 3-8, p. 30
$^{209}Bi(^{3}He,7n)^{205}At$ [253]; see **Fig. 3-12**, p. 38
$^{232}Th(p_{660},\ldots)^{205}At$ [55]

Table 3/6
Levels Proposed for ^{205}At.

level		transition		intensity I_{γ}^{R} observed in		
energy in keV	J^{π}	E_{γ} in keV	final level in keV [a]	^{205}Rn decay [98]	^{205m}At decay [94]	^{197}Au + ^{12}C [e] [96, 97]
264.9	(5/2,7/2$^-$)	264.9	g.s.	100		
566.0	(11/2$^-$)	566.0	g.s.		56	
620.0	(5/2,7/2$^-$)	620.2	g.s.	25		
		354.9	264.9	3.7		
638	11/2$^-$	637.9	g.s.			s
664	13/2$^-$	664.1	g.s.		100	vs
729.6	(5/2,7/2$^-$)	729.6	g.s.	20		
		464.5	264.9	25		
939.9	(5/2,7/2$^-$)	675.0	264.9	20		
969	(13/2$^+$)	331.4	638			m
1035	(15/2$^-$)	469	566		55	
		372	664		67	
1132	15/2$^-$	493.8	638			m
		467.9	664			s
1230	17/2$^-$	566.2	664			vs
		(98)	1132			vw
1301	(13/2$^-$)	638	664		42	
1367	(17/2$^-$)	332	1035		93	
1441	(15/2$^+$)	471.4	969			m
1492 [b]	(21/2$^-$)	191	1301		28	
		125	1367		31	
1563	21/2$^+$	332.8	1230			vs
1756	(17/2$^+$)	786.2	969			w
		315.0	1441			m
1862	(19/2$^+$)	729.8	1132			s
1877	(17/2$^+$)	436.2	1441			m
1935	23/2$^{(-)}$	372.3	1563			s
2053	(21/2$^+$)	191.6	1862			m
		176.1	1877			w
2062 [c]	25/2$^{(+)}$	126.4	1935			m
		9	2053			vw
2339	25/2$^{(-)}$	403.4	1935			m
? [d]	(29/2$^+$)	? [d]	2339			vw
2784	27/2$^{(+)}$	722.1	2062			m

[a] g.s. = ground state. — [b] Half-life = 110 ns [94]. — [c] Half-life = 92 ns [96, 97]. — [d] Half-life ≈ 3 μs [96, 97]; energy of level not known, [96] suggest a position between E* 2339 and 2784 keV. — [e] No numerical values given; the ratings s (strong), m (medium) and w (weak) eventually with the prefix v (very) were taken from a graph of the level scheme in [96].

References on pp. 87/94

Table 3/7
Prominent γ Rays Following the Electron Capture of ^{205}At. (All rays with total relative intensities ≧2 are given; in [55] 29 and in [84] 111 more rays with less intensities are listed.)

energy in keV	I^R_γ		$I^R_{\gamma+ce}$	
[84]	[55]	[84]	[55]	[84]
76.8 K_{α_2}		86±6		
79.29 K_{α_1}		146±8		
89.6 K_{β_1}		52±3		
92.4 K_{β_2}		16.5 ±1.0		
143.166±0.017	2.9±0.3	~2.69±0.19	7±2	7.3±0.7
154.198±0.012	8.8±0.6	8.4 ±0.6	40±6	36.2±3.5
161.030±0.017	4.6±0.4	3.39±0.21	79±8	61±6
202.60 ±0.20	1.2±0.3	1.07±0.13	1.2	2.8±0.34
311.090±0.025	12.7±0.9	13.5 ±0.7	18±1	19.5±1.3
312.50 ±0.20	1.5±0.3	2.10±0.14	1.5	3.05±0.24
360.91 ±0.07	2.8±0.3	3.80±0.25	3.8±0.4	4.53±0.31
364.60 ±0.09	1.9±0.4	2.75±0.22	1.9	3.58±0.30
384.61 ±0.14	3.0±0.3	3.98±0.20	3.0	4.75±0.30
448.61 ±0.07	5.3±0.5	5.51±0.29	5.8±0.5	6.30±0.36
511		14.3 ±0.9		
516.04 ±0.12	2.7±0.5	4.24±0.30	3.0±0.5	4.4 ±0.4
520.44 ±0.06	13.1±1.0	14.4 ±0.6	13.4±1.0	14.8 ±0.7
528.90 ±0.13	1.3±0.4	2.33±0.14	1.3	2.37±0.15
553.94 ±0.07	1.6±0.4	2.01±0.14	1.6	2.16±0.16
617.80 ±0.07	6.9±0.6	7.16±0.32	6.9±0.6	7.23±0.33
628.88 ±0.07	17±2	18.3 ±1.3	17.5±2.0	18.7 ±1.4
659.63 ±0.06	7.4±0.6	7.46±0.33	7.4	7.60±0.34
669.410±0.037	30±2	28.1 ±1.2	31±2	28.6 ±1.2
672.85 ±0.05	11.0±1.0	10.5 ±0.5	11.3±1.0	10.8 ±0.5
719.30 ±0.04	100	100	101	101.46
756.82 ±0.18	1.6±0.4	2.06±0.13	1.6	2.08±0.14
782.80 ±0.12	5.9±0.6	6.41±0.28	5.9	6.51±0.29
789.20 ±0.16	4.1±0.8	4.16±0.22	4.1	4.22±0.23
819.49 ±0.10		2.14±0.17		2.18±0.18
872.4 ±0.5	8±4	10.2 ±1.5	8	10.5 ±1.5
976.00 ±0.12	2.5±0.5	2.97±0.12	2.5	3.02±0.13
1013.70 ±0.14		2.05±0.09		2.06±0.09
1026.2 ±0.3 [a)]	3.0±0.6			
1031.69 ±0.08	7.6±0.7	6.5 ±0.6	7.7±0.7	6.6 ±0.7
1171.04 ±0.08	2.1±0.4	2.28±0.12	2.1	2.29±0.12
1307.60 ±0.08	3.1±0.5	3.48±0.22	3.1	3.51±0.22
1324.95 ±0.08	4.2±0.5	4.14±0.24	4.2	4.15±0.24
1398.3 ±0.3	2.2±0.5	1.23±0.09	2.2	1.24±0.09
1475.36 ±0.09	2.2±0.5	2.75±0.13		2.76±0.13
1479.16 ±0.10	2.2±0.5	2.89±0.19		2.90±0.19
2050.46 ±0.20	2.3±0.5	2.38±0.11		2.38±0.12

a) From [55].

Fig. 3-11

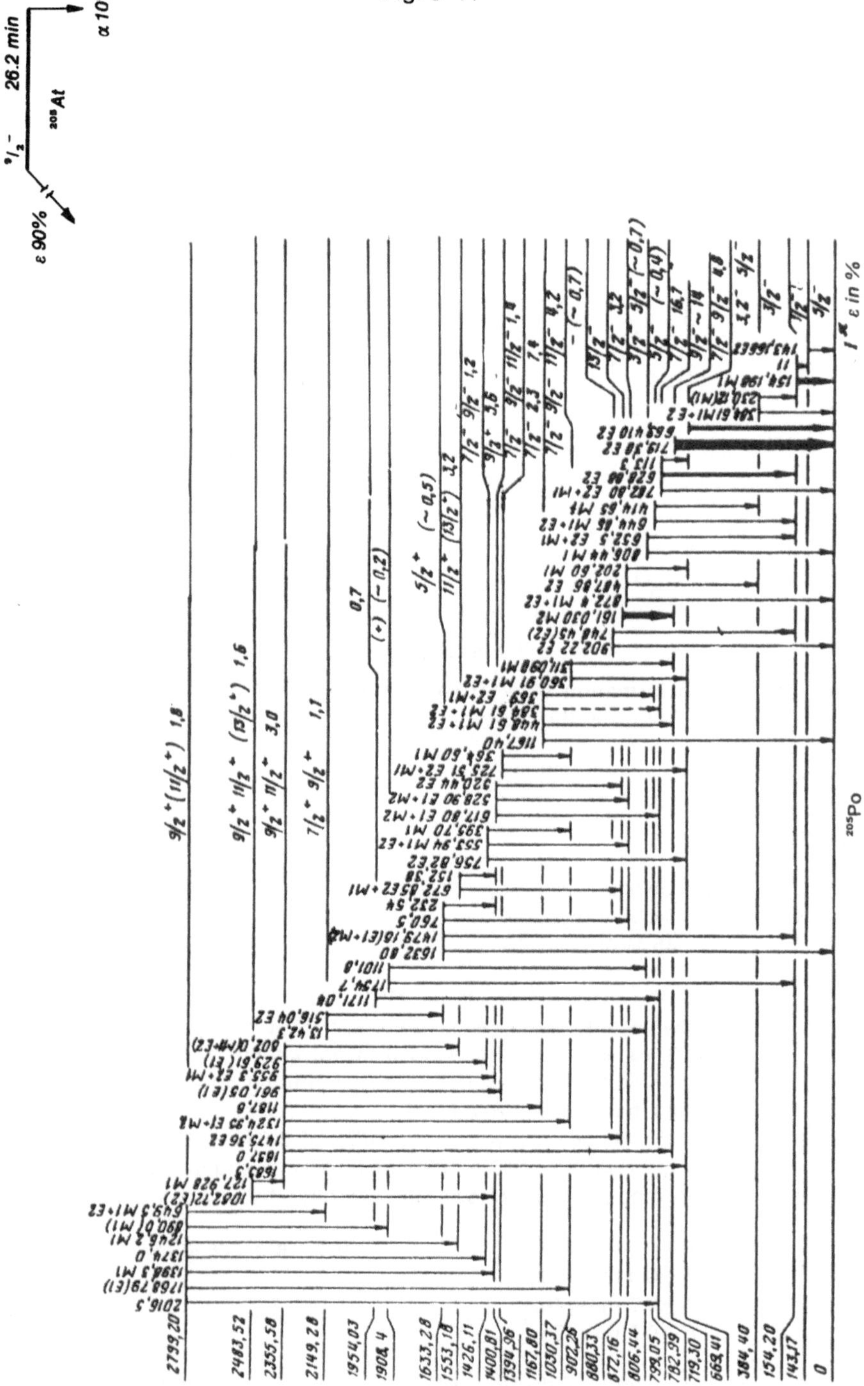

Decay scheme for the electron capture of ^{205}At [85]. The energies are given in keV.

References on pp. 87/94

Fig. 3-12

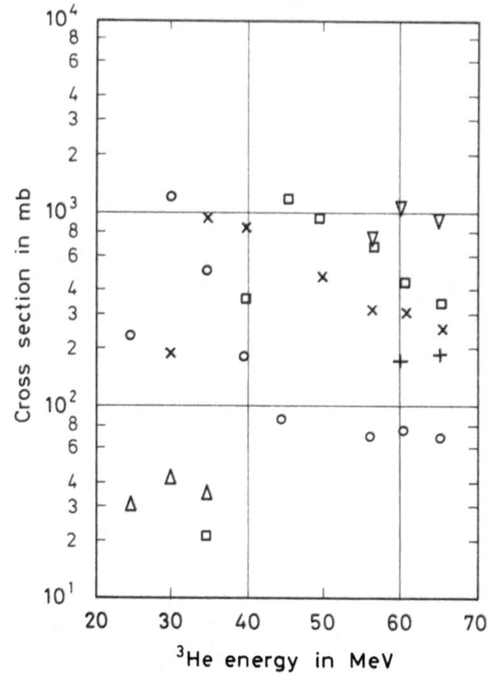

Cross sections for the formation of ^{205}At (+), ^{206}At (▽), ^{207}At (□), ^{208}At (×), ^{209}At (○) and ^{210}At (△) in the reaction ^{209}Bi(^{3}He, xn)$^{212-x}$At [253].

Astatine-206 [47, 103]

mass: 205.98633 u (syst) [30]
S_n 7.84 MeV (syst) [30]
S_p 2.45 MeV (syst) [30]

level: J^π of the ground state is 5^+ [101]. Using the results of [98] for the ^{206}Rn decay a tentative level scheme for ^{206}At was constructed by [101] (see **Fig. 3-13**, pp. 40/1). Because it is not known if the transition lead to the ground state or an excited state, eventually to all level energies a small value x has to be added [101]
in 1951 an activity with a half-life of 2.6 h was assigned to ^{206}At [31], Stoner measured 2.9 h [99]. In later investigations these could not be confirmed [103], therefore, this assigment was apparently erroneous

half-life: 29.5±0.6 min [48]
29.3±0.4 min [102]
31.4±0.7 min [46]
32.8±0.5 min [53]
further references [32, 51]

α decay: [103], 0.96±0.09% [29, 101] based on [44, 48, 102], further reference [51]
Q_α 5884.1±2.0 keV [30]

E_α in keV	I_α^R	further references [32, 47, 51, 100, 103]
[90]	[90]	
5774 ± 4	0.9 ± 0.3	
5767 ± 3	2.4 ± 0.4	
5734 ± 3	1.2 ± 0.3	
5703.0 ± 2.0 [a]	100	

[a] Value adopted from [37].

electron capture: 82% [102], $\varepsilon+\beta^+$ 99.04 ± 0.09% [29, 101] based on [44, 48, 102]

β^+ decay: 17% [102]
Q_{β^+} 5460 keV (syst) [30]
E_{β^+} 3092 ± 150 keV [102]

γ radiation: with α decay: E_γ 68 keV, e/γ 5.3 ± 1.3 [32]
with $\varepsilon+\beta^+$ decay, see Table 3/8, p. 44, further references [46, 50, 58, 104]

decay scheme: for α decay: it is in general assumed that the α decay populates predominantly the 68 keV 5^+ state in ^{202}Bi [101]. However, based on their experimental results, a different decay scheme was proposed in [91] (see **Fig.** 3-**14**) for $\varepsilon+\beta^+$ decay, see **Fig.** 3-**15**, p. 42. This decay scheme is mainly based on the results of Lingeman [102]. However, the levels with high excitation energies are considered in [105] as rather uncertain

production: Pt(^{14}N, ...)^{206}At [44, 48]
^{197}Au(^{12}C,3n)^{206}At [32, 44, 47, 50, 94, 253, 266]; see Fig. 3-7, p. 29, and Fig. 3-8, p. 30
^{197}Au(^{14}N,5n)^{206}Rn $\xrightarrow{\varepsilon+\beta^+}$ ^{206}At [32]
^{209}Bi(^{3}He,6n)^{206}At [102, 253]; see Fig. 3-12
^{209}Bi(α,7n)^{206}At [31, 51, 55, 99]
Th(p_{660}, ...)^{206}At [105]
U(p_{660}, ...)^{206}At [90]

Fig. 3-14

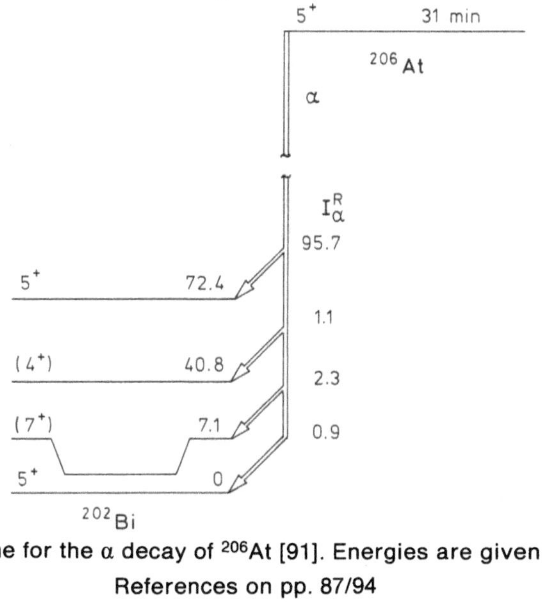

Decay scheme for the α decay of ^{206}At [91]. Energies are given in keV.

Fig. 3-13

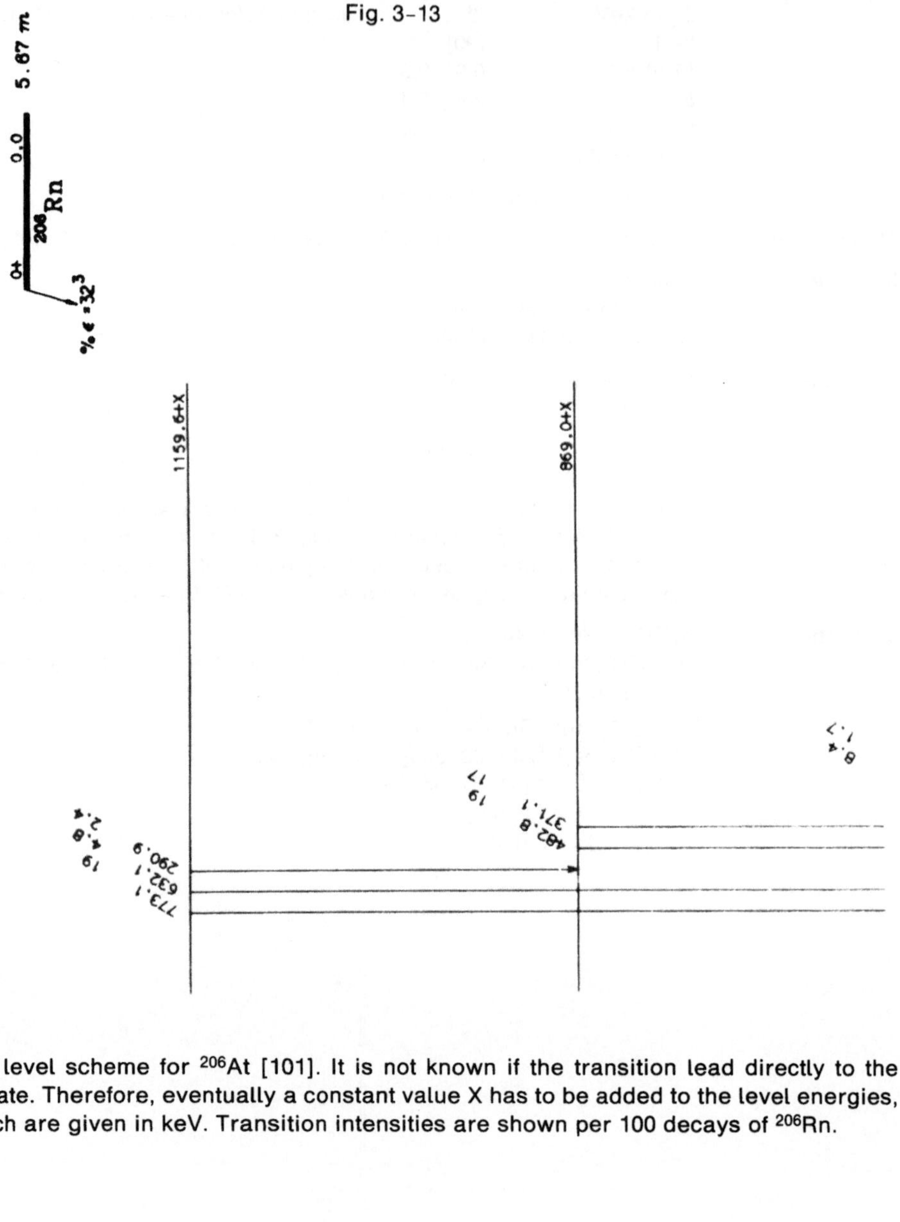

Tentative level scheme for ^{206}At [101]. It is not known if the transition lead directly to the ground state. Therefore, eventually a constant value X has to be added to the level energies, which are given in keV. Transition intensities are shown per 100 decays of ^{206}Rn.

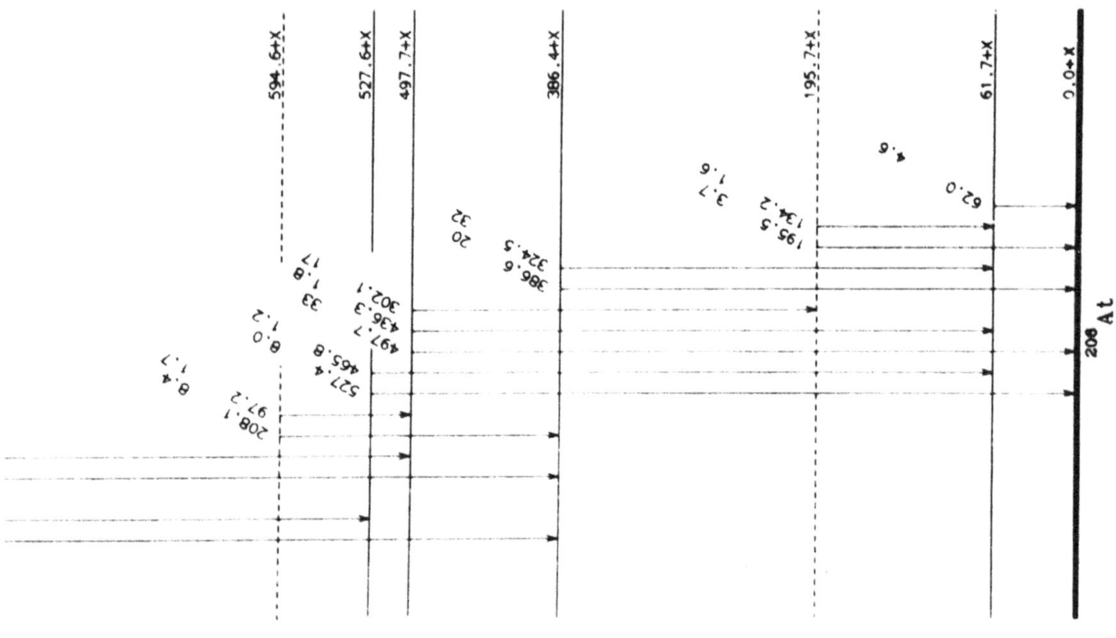

References on pp. 87/94

Fig. 3-15

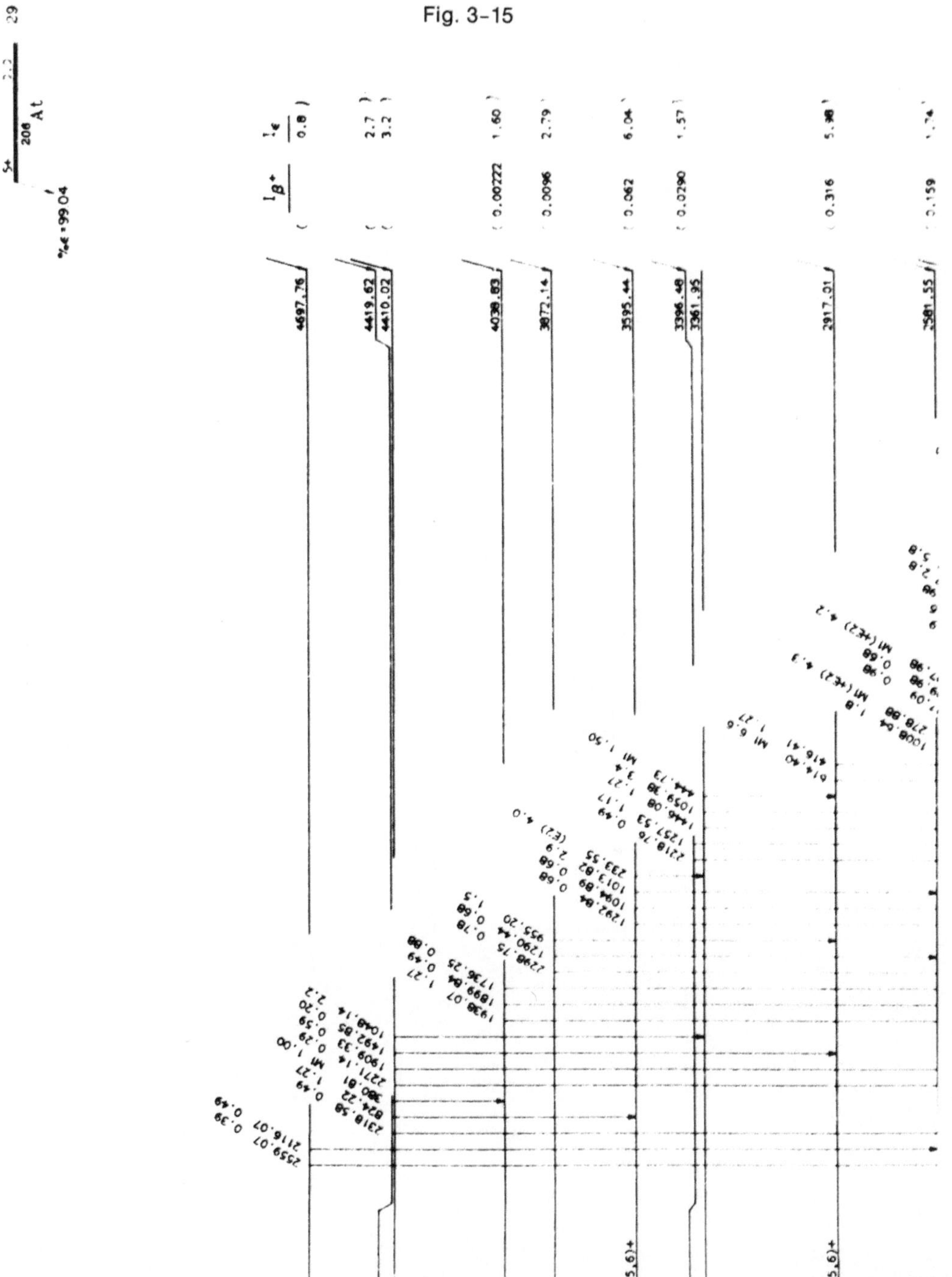

Decay scheme for the $\varepsilon+\beta^+$ decay of ^{206}At [101], which is mainly based on the result of [102]. The energies are given in keV, the $\varepsilon+\beta^+$ transition intensities are shown per 100 decays of ^{206}At.

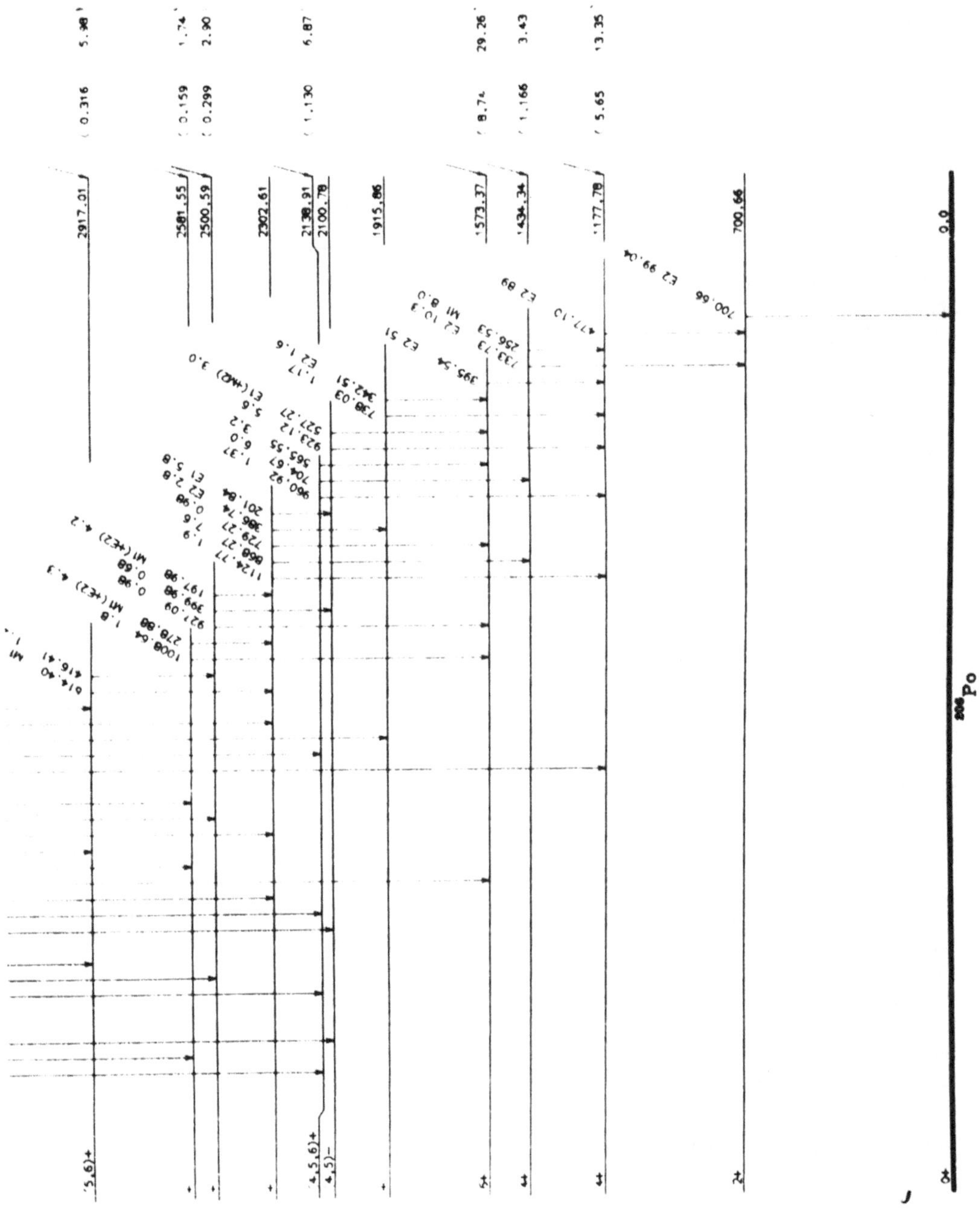

References on pp. 87/94

Table 3/8

Energies and Relative Intensities of Prominent γ Rays and Conversion Electrons Following the $\varepsilon+\beta^+$ Decay of ^{206}At. (All γ lines with $I_\gamma^R > 1$ are given; 33 γ lines with $E_\gamma \leqq 2559$ keV were omitted. The values in parentheses, corresponding to the last digits of the number, give the uncertainty of the data.)

E_γ in keV	I_γ^R		I_{ce} [a)]	
[105]	[105]	[102]	[105]	[102]
61.766(19)	1.34(14)		0.3	
256.438(10)	4.20(26)	4.5(4)	2.54(20); L0.45(9)	3.0(3)
269.468(20)	1.28(14)	1.3(1)	0.68(6)	0.15(5)
275.06(3)	1.66(16)	2.1(2)	0.87(8)	0.17(5)
279.03(4)	1.92(14)	2.7(3)	0.35(3)	1.4(1)
386.894(19)	2.9(4)	2.7(3)	0.85(7)	0.10(5)
395.504(11)	44.4(8)	49.3(29)	1.67(15)	1.8(2)
444.96(9)	1.37(7)	1.3(1)	0.2(1)	—
477.094(16)	87.7(24)	88(4)	2.32(25)	2.2(2)
527.498(27)	2.98(12)	3.0(3)	0.27(3)	0.02(1)
614.58(4)	6.14(19)	6.3(6)	0.369(20)	0.4(1)
617.80(20) [b)]	1.22(12)	—	—	—
700.548(23)	100(3)	100(3)	1.14	1.14
704.65(15)	5.2(10)	6.1(6)	~0.15	—
729.18(8)	2.20(8)	1.0(1)	0.034(20)	—
733.54(4)	7.89(21)	10.4(7)	0.115(18)	—
737.71(18)	1.07(7)	1.2(1)	~0.03	—
794+796	1.86(12)	1.2(1)	—	—
824.16(9)	1.16(6)	1.3(1)	—	—
838.01(15)	2.6(3)	—	~0.06	—
864.30(11)	1.76(10)	—	~0.02	—
868.26(4)	8.1(4)	7.8(8)	~0.06	—
922.95(4)	5.30(21)	5.7(6)	~0.07	—
927.08(16)	1.00(13)	1.0(1)	~0.02	—
939.27(5)	2.06(8)	2.0(2)	~0.025	—
955.38(5)	2.00(20)	1.5(2)	~0.025	—
976.47(6)	1.12(6)	1.4(1)	~0.06	—
1008.64(28)	—	1.8(2)	—	—
1013.86(5)	2.45(20)	3.0(3)	0.13(4)	—
1048.18(8)	1.88(8)	2.3(2)	~0.04	—
1059.59(7)	2.80(10)	3.5(4)	0.13(4)	—
1124.58(7)	1.85(10)	1.9(2)	~0.02	—
1198.00(7)	1.04(6)	1.5(2)	—	—
1258.10(8)	1.10(7)	1.2(1)	—	—
1637.32(12)	0.96(7)	1.2(1)	—	—
1937.91(11)	1.15(5)	1.3(1)	—	—

a) Intensities for the K line listed unless otherwise specified. — b) Partially due to the decay of ^{207}At.

Astatine-207 [31, 110]

mass: 206.98571 ± 0.00005 u [30]
S_n 8.65 MeV (syst) [30]
S_p 2.41 MeV (syst) [30]

levels: the levels populated in the decay of ^{207}Rn [98] and in the reaction $^{204}Pb(^6Li,3n\gamma)$ [97, 214, 215] are given in Table 3/9

half-life: 1.80 ± 0.05 h [44]
1.78 ± 0.08 h [106]
1.80 ± 0.08 h [93]
further references [31, 47, 93, 103, 106, 111]

Table 3/9
Levels of ^{207}At Populated in the Decay of ^{207}Rn [98] and in the Reaction $^{204}Pb(^6Li,3n\gamma)$ [97, 214, 215].

E* in keV	J^π	observed in ^{207}Rn decay	observed in $^{204}Pb(^6Li,3n\gamma)$
0	9/2⁻	+	+
344.53 ± 0.04	(7/2)⁻	+	
643	11/2⁻		+
674.01 ± 0.04	(5/2, 7/2)⁻	+	
686	13/2⁻		+
747.18 ± 0.05	7/2⁻	+	
973.27 ± 0.08		+	
976.08 ± 0.10		+	
1018.55 ± 0.06	(5/2, 7/2)⁻	+	
1041.71 ± 0.06		+	
1055	13/2⁻		+
1085	15/2⁻		+
1108.1 ± 0.3		+	
1119.95 ± 0.08		+	
1197.90 ± 0.10		+	
1224.74 ± 0.16		+	
1234			+
1284.14 ± 0.15		+	
1351.20 ± 0.15		+	
1495			+
1534.79 ± 0.16		+	
1539.36 ± 1.3		+	
1553.5 ± 0.2		+	
1799.4 ± 0.2		+	
1823.3 ± 0.3		+	
1841.09 ± 0.17		+	
1898			+
1966.56 ± 0.17		+	
2038.6 ± 0.2		+	
2117			+
2149.55 ± 0.17		+	

References on pp. 87/94

α decay: ≈10% [31, 110]
Q_α 5871.6±3.1 keV [30]
E_α 5759±2 keV [37] based on [32, 58]
further references [44, 47, 52, 93, 99, 100, 106, 109]

electron capture +β^+ decay: ~90% [31, 110]

γ radiation: with $\varepsilon+\beta^+$, see Table 3/10
further references [104, 108, 111, 112]

Table 3/10
Prominent γ Rays Following the $\varepsilon+\beta^+$ Decay of ^{207}At [213]. (All lines with $I_\gamma^R>10$ are listed. There are about 150 additional lines with E_γ up to 3458 keV which are omitted. The results given in [108] agree well with the data obtained by [213]. The intensity of the K conversion electrons are also listed).

E_γ in keV	I_γ^R	$I_{ce(K)}^R$	$I_{\gamma+ce}^R$
167.900±0.020	22.5±1.2	47 ±5	81 ±6
191.256±0.008	11.8±0.7	16.6 ±1.7	34.5±2.4
221.270±0.020	26.7±1.5	24.0 ±2.6	57.0±3.5
236.477±0.015	21.7±1.3	17.4 ±1.8	43.4±2.3
300.648±0.013	287 ±14	393 ±19	805 ±42
324.408±0.020	17.8±1.2	5.9 ±0.6	24.7±1.4
357.153±0.015	60.2±3.5	15.6 ±1.0	81 ±4
392.94 ±0.06	17.9±1.4	3.9 ±0.7	22.4±1.6
411.10 ±0.04	13.3±0.9	<2.5	16.2±1.1
456.75 ±0.02	40.5±2.6	6.0 ±0.6	48.0±2.8
459.69 ±0.03	37.9±2.5	1.11±0.12	39.5±2.7
467.116±0.013	160 ±10	5.8 ±0.7	168 ±10
503.40 ±0.13	11.0±3.0	0.88±0.13	12.1±3.1
511	59 ±7	annihilated radiation	59 ±7
520.78 ±0.09	19.0±1.3	1.77±0.20	21.2±1.4
529.790±0.025	77.5	7.1 ±0.8	86 ±5
583.340±0.030	49.0±2.9	2.49±0.28	52 ±3
588.333±0.023	432 ±23	9.3 ±1.0	444 ±23
617.20 ±0.04	40.5±2.4	0.70±0.11	41.4±2.4
626.77 ±0.04	43.0±2.5	0.57±0.10	43.8±2.5
637.270±0.020	56 ±5	3.1 ±0.4	60 ±5
641.00 ±0.07	14.0±1.7	0.56±0.10	14.7±1.7
648.095±0.020	96 ±7	5.2 ±0.7	103 ±8
658.40 ±0.15	144 ±16	4.3 ±0.7	150 ±17
670.41 ±0.07	84 ±8	0.79±0.21	85 ±8
675.154±0.023	152 ±10	0.62±0.11	153 ±10
686.0 ±1.0	~45		
693.33 ±0.06	57.7±3.3	0.75±0.10	58.7±3.5
721.14 ±0.04	135 ±11	6.2 ±0.9	143 ±12
755.08 ±0.09	11.1±0.9		

Table 3/10 [continued]

E_γ in keV	I^R_γ	$I^R_{ce(K)}$	$I^R_{\gamma+ce}$
765.03 ±0.10	12.8±0.9		
768.30 ±0.30	11.2±0.7		
814.407±0.027	1000 ±50	8.5 ±0.4	1010 ±50
820.50 ±0.15	12.0±1.6		
833.06 ±0.10	10.6±1.5		
862.46 ±0.05	16.0±1.1	0.42±0.04	16.4±1.1
880.92 ±0.04	24.5±1.8	0.27±0.05	24.9±1.8
907.08 ±0.03	149 ±9	3.1 ±0.5	153 ±9
948.37 ±0.10	15.0±1.2		
1021.67 ±0.12	19.6±1.2	0.13±0.04	19.8±1.3
1054.22 ±0.04	24.0±1.7	0.34±0.04	24.4±1.8
1077.68 ±0.03	44.4±2.7		
1115.196±0.024	108 ±6	0.55±0.07	109 ±7
1131.72 ±0.06	10.2±0.8		
1171.62 ±0.04	27.9±1.7	0.106±0.015	28.0±1.7
1174.60 ±0.08	10.3±0.8		
1188.260±0.030	83.7±2.3	0.20±0.05	39.0±2.3
1193.44 ±0.07	10.6±0.8		
1225.620±0.030	26.4±1.8	< 0.28	26.6±1.9
1242.62 ±0.07	17.5±1.5	0.180±0.027	17.7±1.6
1245.46 ±0.05	13.4±1.1		
1263.71 ±0.04	12.1±0.9	0.18±0.06	12.3±0.9
1283.08 ±0.04	27.4±1.6		
1396.19 ±0.04	31.8±1.8	~ 0.06	31.9±1.8
1409.86 ±0.05	25.9±1.5	0.16±0.04	26.1±1.5
1413.15 ±0.05	22.4±1.3	0.16±0.04	22.6±1.4
1506.97 ±0.09	14.3±0.9		
1510.89 ±0.08	11.7±0.8		
1548.21 ±0.08	26.1±1.6		
1641.82 ±0.06	21.1±1.2		
1676.50 ±0.10	68 ±4		
1712.60 ±0.09	29.8±1.8		
1716.39 ±0.10	20.9±1.2		
1730.76 ±0.06	85 ±5		
1745.32 ±0.07	15.6±0.9		
1772.77 ±0.07	15.0±0.9		
1781.67 ±0.07	12.1±0.8		
1786.57 ±0.07	19.3±1.1		
1805.25 ±0.06	16.4±1.2		
1854.54 ±0.09	12.1±0.9		
2016.25 ±0.10	16.0±1.1		
2075.27 ±0.07	11.0±1.1		
2342.65 ±0.10	16.3±1.2		
2712.50 ±0.15	27.4±1.7		

References on pp. 87/94

Fig. 3-16

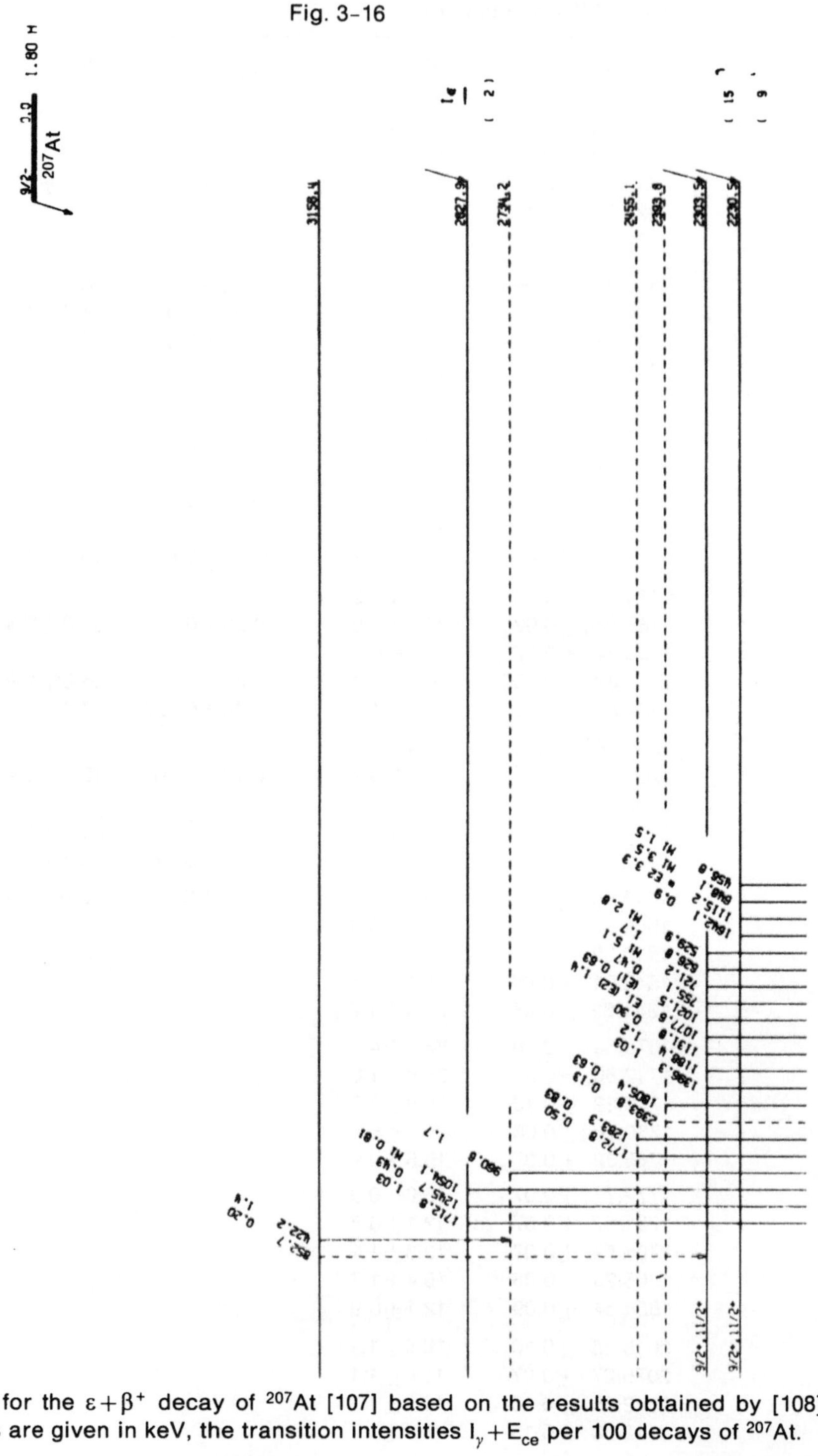

Decay scheme for the $\varepsilon+\beta^+$ decay of ^{207}At [107] based on the results obtained by [108]. The energies are given in keV, the transition intensities $I_\gamma+E_{ce}$ per 100 decays of ^{207}At.

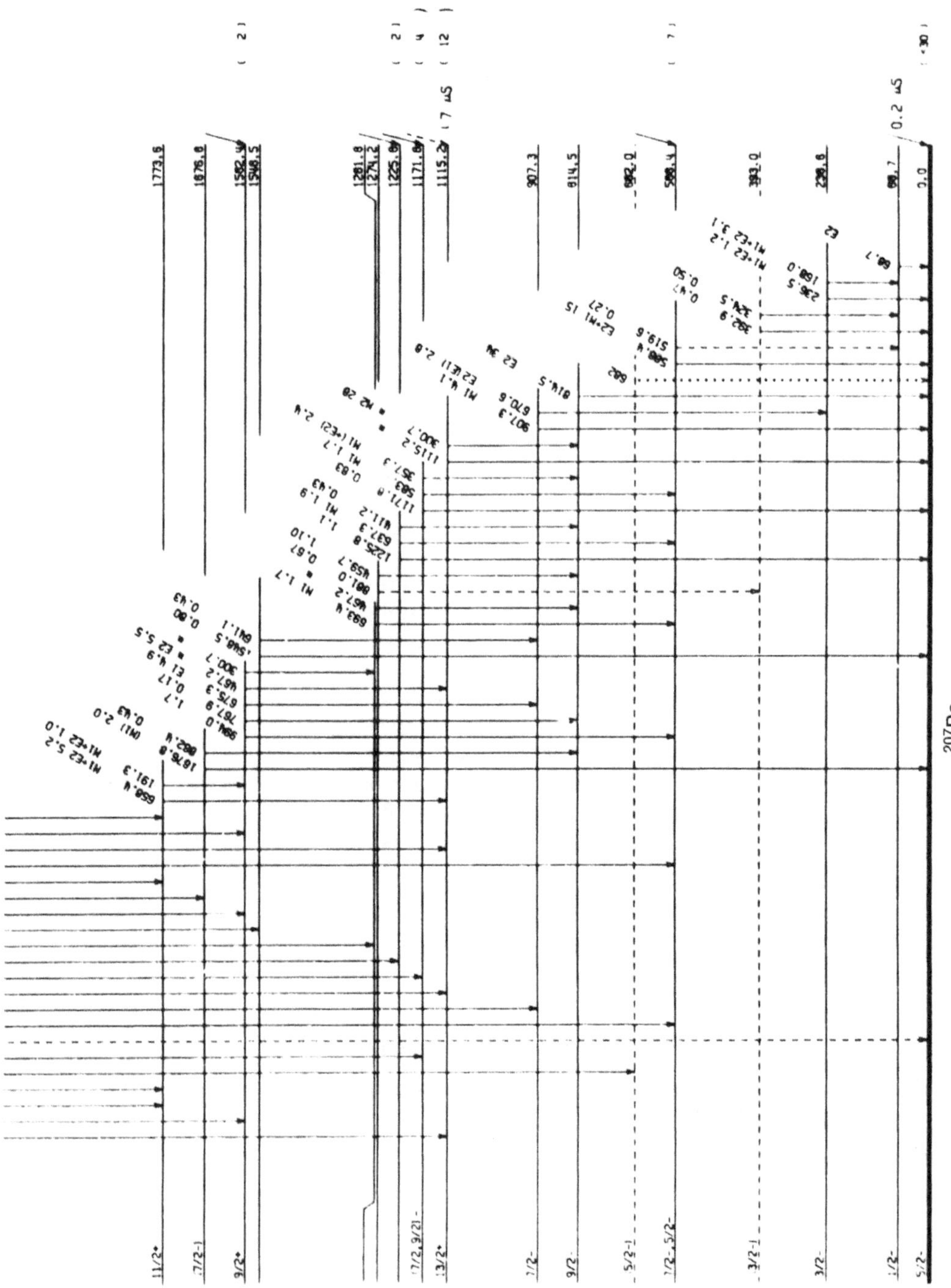

References on pp. 87/94

Fig. 3-17

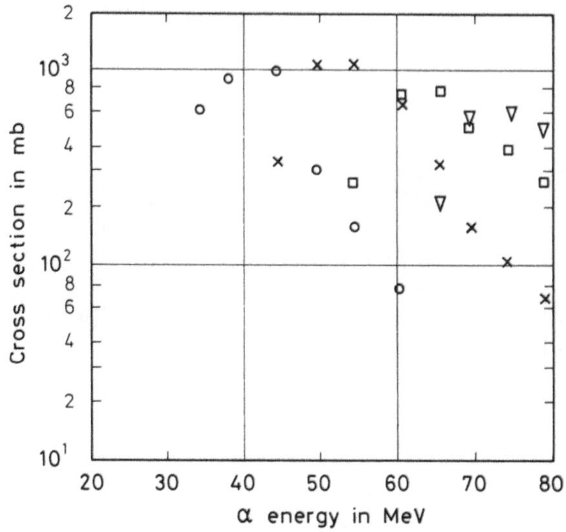

Cross sections for the formation of ^{207}At (▽), ^{208}At (□), ^{209}At (×), and ^{210}At (○) in the reaction ^{209}Bi(α, xn)$^{213-x}$At [253].

decay scheme: the α decay populates the ground state of ^{203}Bi [29]
ε+β^+ decay: see **Fig.** 3-**16**, p. 48; further reference [216]

production:
^{197}Au(^{12}C,2n)^{207}At [32, 44]
^{197}Au(^{14}N,4n)^{207}Rn $\xrightarrow{\varepsilon+\beta^+}$ ^{207}At [32]
^{209}Bi(^{3}He,5n)^{207}At [253]; see Fig. 3-12, p. 38
^{209}Bi(α,6n)^{207}At [31, 110, 253]; see **Fig.** 3-**17**
^{209}Bi(^{19}F,...)^{207}At [262]; the cross section at E(^{19}F) 186 MeV is 10 mb [262]
^{209}Bi(^{40}Ar, ...)^{207}At [262]; see **Fig.** 3-**18**
^{209}Bi(^{40}Ca, ...)^{207}At [262]; see **Fig.** 3-**19**, p. 52
Th(p_{660}, ...)^{207}At [108, 109]

Astatine-208 [113]

mass:
207.98643 u (syst) [30]
S_n 7.40 MeV (syst) [30]
S_p 2.78 MeV (syst) [30]

levels: see **Fig.** 3-**20**, p. 53; further references [29, 116]

half-life:
1.63 ± 0.03 h [53] further references [31, 44, 47, 99, 113]
1.6 h [115]

α decay:
0.55% [113]
Q_α 5750 ± 3.1 keV [30]

further references [37, 51, 113, 115, 117]

E_α in keV [90]	I_α^R [90]
5641 ± 3	100
5626 ± 4	2.2 ± 0.2
5586 ± 2	0.9 ± 0.1

Fig. 3-18

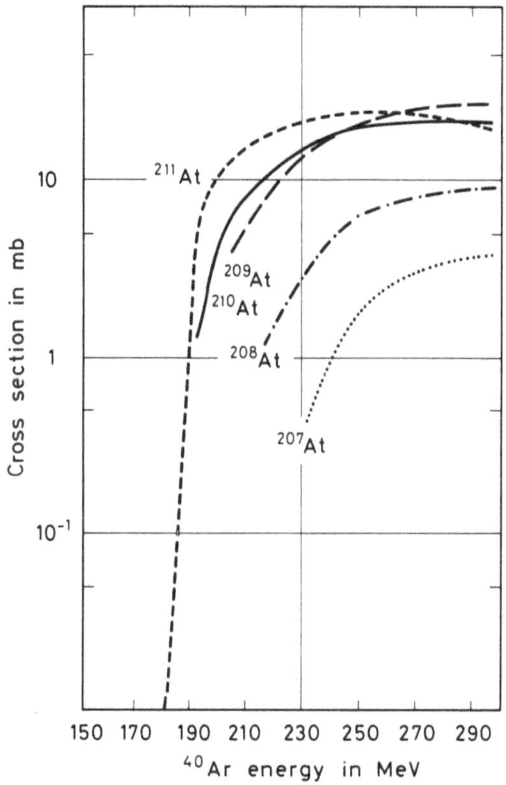

Cross sections for the formation of the At isotopes 207 to 211 in the reaction ^{209}Bi(^{40}Ar,...) [262].

electron capture and β^+ decay: 99.45% [113]
E_{β^+} 4.30 MeV (syst) [30]

γ radiation: with α decay: 120±10 keV; γ/α (3±3)×10^{-4} [99]
with ε+β^+ decay: see Table 3/11, p. 57; further references [114, 218, 219, 221]

decay scheme: the 5641 keV α populate the ground state of ^{206}Bi [29, 91, 116]
the decay scheme for ε+β^+ decay is given in **Fig.** 3-**21**, pp. 54/6; further references [29, 116, 218, 219]
the γ line with E_γ 177 keV has two components with $T_{1/2}$ 380 and 4 ns, respectively [114]

production: ^{209}Bi(^{3}He,4n)^{208}At [208, 253]; see Fig. 3-12, p. 38
^{209}Bi(α,5n)^{208}At [53, 253]; see Fig. 3-17
^{209}Bi(^{19}F, ...)^{209}At [262]
the cross section at E(^{19}F) 186 MeV is 16 mb [262]
^{209}Bi(^{20}Ne, ...)^{208}At [262]

Fig. 3-19

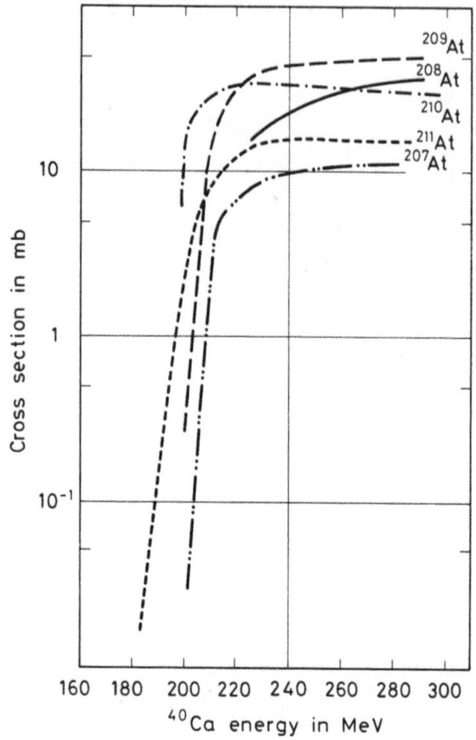

Cross sections for the formation of the At isotopes 207 to 211 in the reaction $^{209}Bi(^{40}Ca,...)$ [262].

the cross section at $E(^{20}Ne)$ 201 MeV is 38 mb [262]
$^{209}Bi(^{40}Ar, ...)^{208}At$ [262]; see Fig. 3-18, p. 51
$^{209}Bi(^{40}Ca, ...)^{208}At$ [262]; see Fig. 3-19
$^{209}Bi(^{56}Fe, ...)^{208}At$ [262]; see **Fig.** 3-**22**, p. 59
$^{209}Bi(^{63}Cu, ...)^{208}At$ [262]; see **Fig.** 3-**23**, p. 60
$^{232}Th(p_{350}, ...)^{212}Fr \xrightarrow{\alpha} {}^{208}At$ [113]
$^{238}U(p_{660}, ...)^{208}At$ [90]

Fig. 3-20

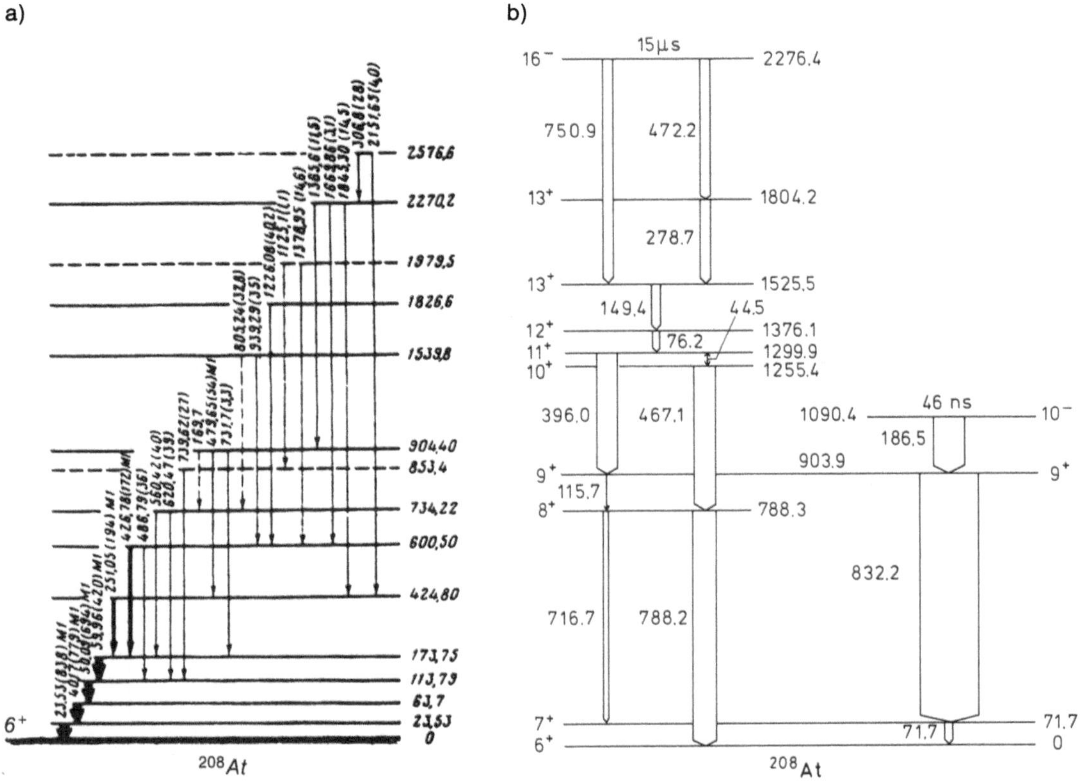

Levels of ^{208}At populated (a) in the electron capture of ^{208}Rn [217] and (b) in the reaction ^{209}Bi(α, 5n) [236]. The energies are given in keV.

Astatine-209 [31]

mass:	208.986165 ± 0.000010 u [30] S_n 8.32 MeV (syst) [30] S_p 2702 ± 8 keV [30]	
levels:	see Table 3/12, p. 57	further references [223, 224]
half-life:	5.41 ± 0.05 h [226] 5.42 ± 0.06 h [109]	further references [31, 47]

Fig. 3-21

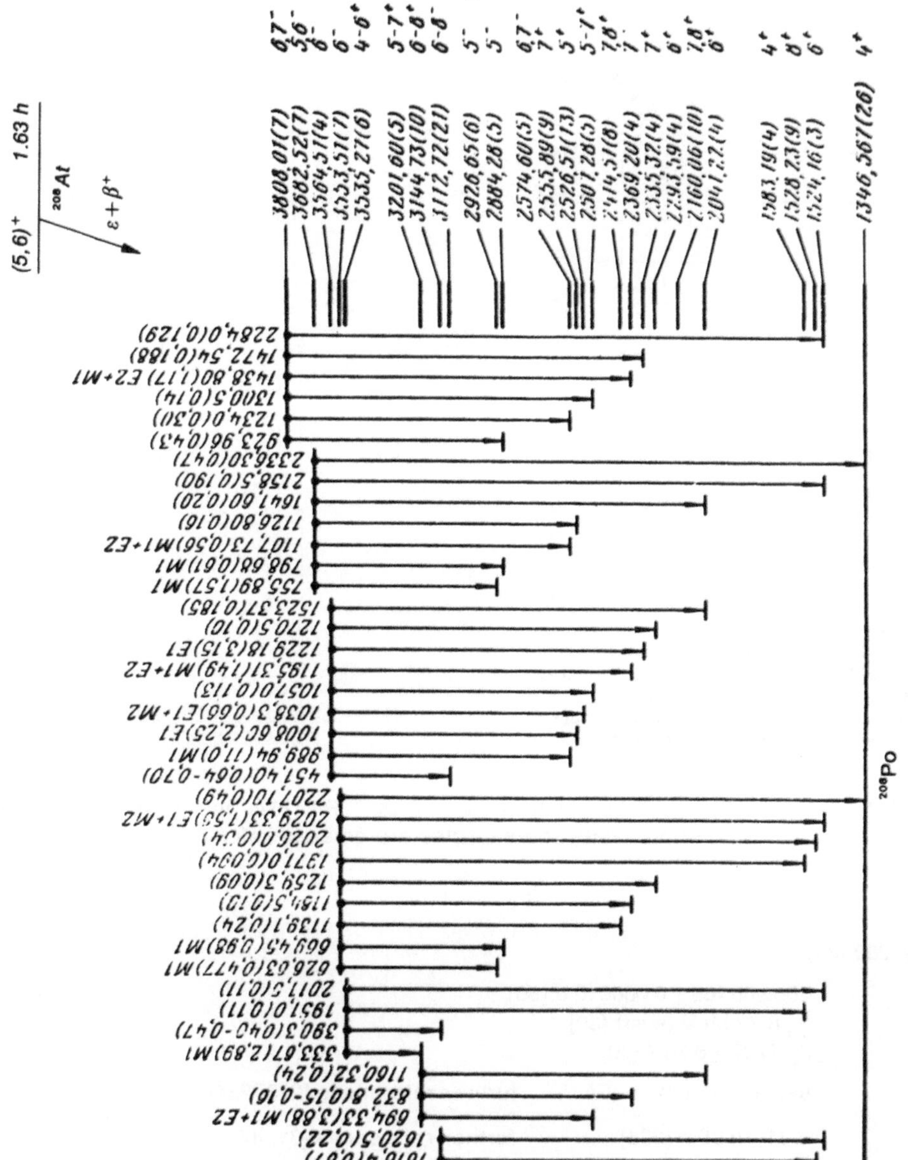

Decay scheme for the $\varepsilon+\beta^+$ decay of ^{208}At [222]. Part 1, 2 and 3 show the levels with $E^*>4$ MeV, $4<E^*>3.15$ MeV and $E^*<3.15$ MeV, respectively, populated in the decay. The energies in the drawing are given in keV. The values in the brackets on the right which correspond to the last digits in the level energies, give the uncertainty of these data. The relative transition intensities are given in brackets together with the transition energies.

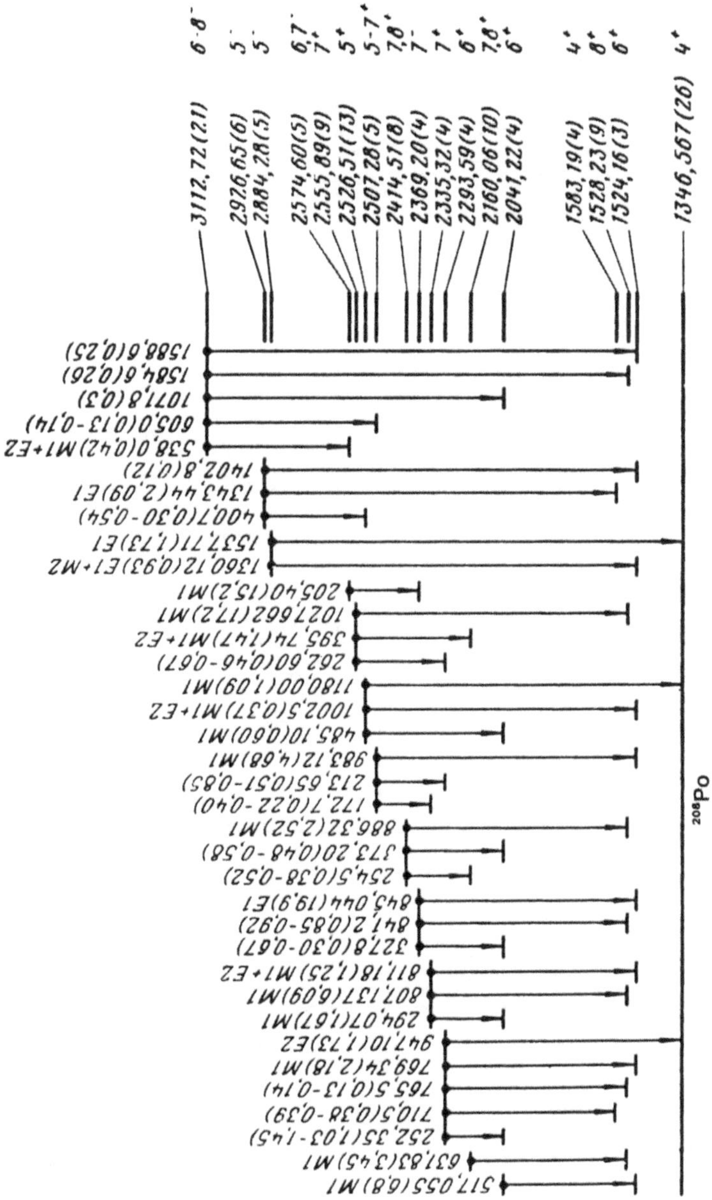

References on pp. 87/94

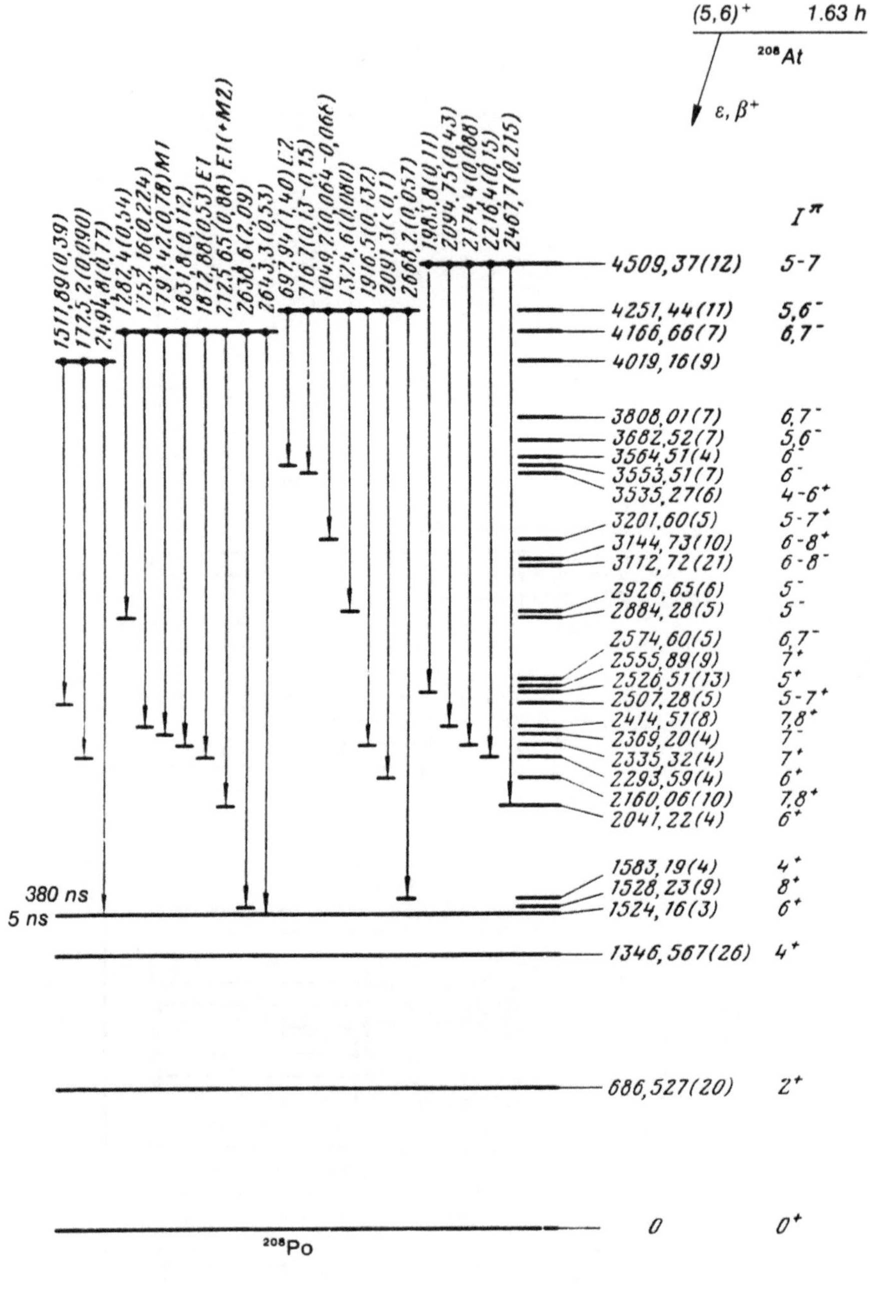
(5,6)⁺ 1.63 h
208At
ε, β⁺
Iπ
4509,37(12) 5-7
4251,44(11) 5,6⁻
4166,66(7) 6,7⁻
4019,16(9)
3808,01(7) 6,7⁻
3682,52(7) 5,6⁻
3564,51(4) 6⁻
3553,51(7) 6⁻
3535,27(6) 4-6⁺
3201,60(5) 5-7⁺
3144,73(10) 6-8⁺
3112,72(21) 6-8⁻
2926,65(6) 5⁻
2884,28(5) 5⁻
2574,60(5) 6,7⁻
2555,89(9) 7⁺
2526,51(13) 5⁺
2507,28(5) 5-7⁺
2414,51(8) 7,8⁺
2369,20(4) 7⁻
2335,32(4) 7⁺
2293,59(4) 6⁺
2160,06(10) 7,8⁺
2041,22(4) 6⁺
1583,19(4) 4⁺
1528,23(9) 8⁺
1524,16(3) 6⁺
1346,567(26) 4⁺
686,527(20) 2⁺
0 0⁺
380 ns
5 ns
208Po

Table 3/11
Prominent γ Rays Following the $\varepsilon+\beta^+$ Decay of ^{208}At [220]. (All γ lines with $I^R>20$ are listed, in [220] about 240 additional lines with E_γ up to 3223 keV are given. The agreement with the data given in [221] is good whereas the data in [114] differ up to 9 keV. The energies are stated in keV; 1 unit in I^R_γ corresponds to 0.098% decays of ^{208}At [220].)

E_γ	I^R_γ	$I^R_{ce(K)}$	$I^R_{\gamma+ce}$
177.595 ± 0.017	498 ± 21	102 ± 5	840 ± 40
205.40 ± 0.03	64.8 ± 2.7	74 ± 4	155 ± 8
333.67 ± 0.03	21.5 ± 1.2	6.5 ± 0.4	29.5 ± 1.6
517.055 ± 0.020	62 ± 6	6.2 ± 0.4	69 ± 7
631.83 ± 0.04	33.0 ± 1.4	1.84 ± 0.11	35.2 ± 1.5
660.040 ± 0.017	910 ± 40	11.8 ± 0.6	930 ± 40
686.527 ± 0.020	1000	11.9	1016
694.33 ± 0.04	38.4 ± 2.3	0.99 ± 0.10	39.6 ± 2.4
769.34 ± 0.05	21.3 ± 1.2	0.70 ± 0.07	22.2 ± 1.3
807.137 ± 0.025	60.0 ± 2.5	1.80 ± 0.11	62.1 ± 2.6
845.044 ± 0.020	202 ± 9	0.58 ± 0.06	203 ± 10
886.32 ± 0.05	25.0 ± 1.4	0.55 ± 0.08	25.7 ± 1.5
896.66 ± 0.04	55.0 ± 2.3	0.41 ± 0.04	55.4 ± 2.3
983.12 ± 0.04	46.8 ± 2.2	0.75 ± 0.07	47.8 ± 2.3
989.91 ± 0.03	110 ± 8	2.01 ± 0.14	112 ± 8
1008.60 ± 0.04	23.0 ± 2.4	0.057 ± 0.011	23.0 ± 2.4
1027.662 ± 0.024	172 ± 7	2.60 ± 0.21	175 ± 8
1229.18 ± 0.03	32.0 ± 2.4	0.045 ± 0.007	32.1 ± 2.6
1343.44 ± 0.05	21.3 ± 0.9	0.029 ± 0.005	21.3 ± 0.9
2638.6 ± 0.3	21.3 ± 1.5		21.3 ± 1.5

Table 3/12
Levels of ^{209}At Populated in the Decay of ^{209}Rn [128] and in the Nuclear Reactions ^{209}Bi + α [126] and ^{206}Pb + ^{6}Li [127]. (Most of the data given below are taken from the compilation of Martin [131] which is based on the results given in the references mentioned above. The investigation of the reaction Bi + ^{3}He by Adam et al. [225] was also taken into account.)

E^* a) in keV	$J\pi$ e)	$T_{1/2}$	observed in: Rn decay [128]	Pb + Li [127]	Bi + ^{4}He [126]	Bi + ^{3}He [225]
0.0	9/2⁻		+	+	+	+
408.34 ± 0.03	5/2⁻		+	+		+
577.1 ± 0.1 b)	11/2⁻			+	+	+
640 d)						+
725.1 ± 0.1 b)	13/2⁻	≦0.5 ns [127]		+	+	+
745.78 ± 0.04	7/2⁻		+			+

References on pp. 87/94

Table 3/12 [continued]

E* [a) in keV	Jπ [e)	$T_{1/2}$	observed in Rn decay [128]	Pb+Li [127]	Bi+^{4}He [126]	Bi+^{3}He [225]
789.0 ±0.7 [c)				+		+
794.69±0.06	(5/2,7/2)$^-$		+			
1081.15±0.05	(5/2,7/2)$^-$		+			
1093.24±0.11	(5/2)$^-$		+			
1097.65±0.05	(5/2,7/2)$^-$		+			
1212.0 ±0.7 [c)			+			
1240.9 ±0.7 [c)	(13/2$^+$,13/2$^-$)			+		+
1270.5 [d)	(15/2$^-$)					+
1321.6 ±0.1 [b)	17/2$^-$			+	+	+
1394.37±0.08	(5/2,7/2)$^-$		+			
1427.7 ±0.2	21/2$^-$	25±1 ns [126] 29±2 ns [127]		+	+	+
1771.5 ±0.8 [c)	(15/2$^+$,15/2$^-$)			+		+
1851.8 ±0.2 [b)	23/2$^-$			+	+	+
1953.45±0.06	(3/2$^+$,5/2$^+$,7/2$^+$)		+			
2135.64±0.06	(3/2,5/2)$^+$		+			
2414.86±0.08	(3/2,5/2)$^+$		+			
2429.2 ±0.3 [b)	29/2$^+$	680± 75 ns [127] 875±100 ns [126]		+	+	+
2516.53±0.14	(3/2$^+$,5/2$^+$,7/2$^+$)		+			
2522.37±0.19			+			
2569.1 ±0.3			+			
2581.3 ±0.3			+			
2689.8 ±0.3			+			
2712.75±0.10			+			
2821.8 ±0.3			+			
3140.59±0.20			+			
3172.2 ±0.3			+			
3388.45±0.21			+			
3544.29±0.17			+			
3551.3 ±0.3			+			
3626.9 ±0.4			+			
3753.6 ±0.3			+			

[a) Energies are from Rn decay unless otherwise noted. — [b) From [126]. — [c) From [127]. — [d) From [225]. — [e) Mainly based on [126, 225].

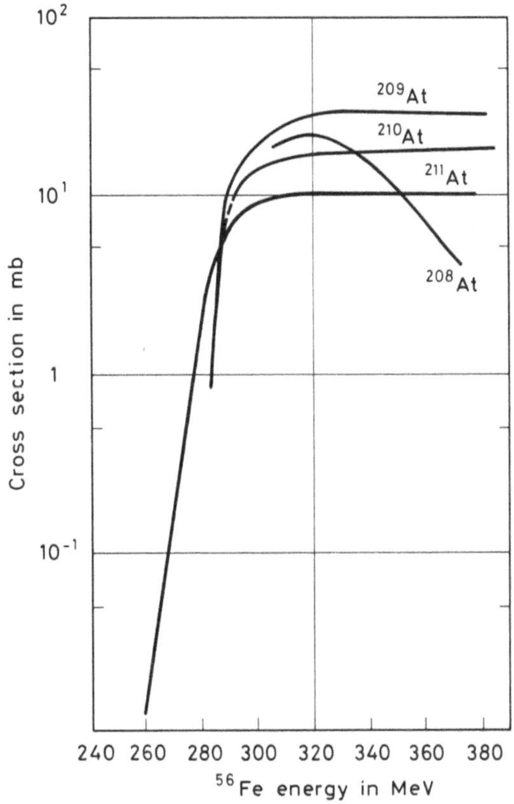

Cross sections for the formation of the At isotopes 208 to 211 in the reaction $^{209}Bi(^{56}Fe,\ldots)$ [262].

α decay: ~5% [31], 4.1 ± 0.5% [109]
Q_α 5757.2 ± 2.0 keV [30]
E_α 5647 ± 2 keV [37] based on [99, 100, 109, 226]
further references [31, 118, 122]
α particles with E_α of 5116 ± 2 keV and I_α^R of 0.10 ± 0.05 were observed in [109]; however, were not confirmed in [131]

electron capture + β^+ decay: ~95% [31], 95.9 ± 0.5% [109]
Q_{β^+} 3486 ± 8 keV [30]

γ radiation: with ε + β^+ decay see Table 3/13, p. 60; further references [99, 119, 124]

decay scheme: the 5647 keV α particles populate the ground state of ^{205}Bi [29, 131]
the decay scheme for the ε + β^+ decay is shown in **Fig.** 3-**24**, pp. 62/3; further reference [29]

References on pp. 87/94

Fig. 3-23

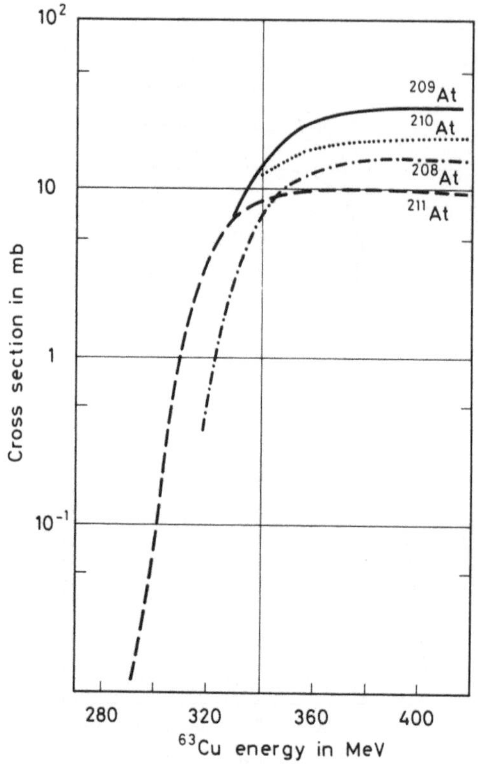

Cross sections for the formation of the At isotopes 208 to 211 in the reaction $^{209}Bi(^{63}Cu,...)$ [262].

Table 3/13
Prominent γ Rays Following the $\varepsilon+\beta^+$ Decay of ^{209}At as Given in the Compilation by Martin [131]. (All lines with $I_\gamma^R \geqq 1$ are listed; 76 additional lines with E_γ up to 2654.4 keV are omitted. α is the total conversion coefficient; for values of α(K), α(L), etc., see [131]).

E_γ [a)] in keV	I_γ^R [b)]	α	E_γ [a)] in keV	I_γ^R [b)]	α
90.8±0.1	2.02±0.22	10.97	790.2±0.1	69.8 ±1.9	0.00420
104.2±0.1	2.6 ±0.4	9.8 ±0.8	863.9±0.1	2.27±0.09	0.039 ±0.0012
195.0±0.1	24.8 ±1.1	1.55±0.07	903.0±0.1	4.01±0.11	0.00328
233.6±0.1	1.05±0.07	0.96±0.05	1103.4±0.1	5.94±0.18	0.0097 ±0.0011
239.1±0.1	13.6 ±0.5	0.0537	1147.6±0.1	1.50±0.10	0.00571

545.0±0.1	100	0.0265	1170.6±0.1	3.4 ±0.1	0.00550
551.0±0.1	5.4 ±0.2	0.0259	1175.3±0.1	2.1 ±0.1	0.00546
552.5±0.2	1.7 ±0.2	0.096±0.007	1217.2±0.1	1.22±0.06	
666.1±0.1	2.05±0.07	0.01698	1262.6±0.1	2.08±0.07	0.00181
781.9±0.1	91.8 ±2.4	0.01213	1581.6±0.1	1.97±0.07	

a) Based on the results given in [120, 121, 123, 125]. — b) Based on the results given in [121, 123]. For absolute intensities per 100 decays of ^{209}At multiply by 0.9098±0.0048 [131].

Fig. 3-25

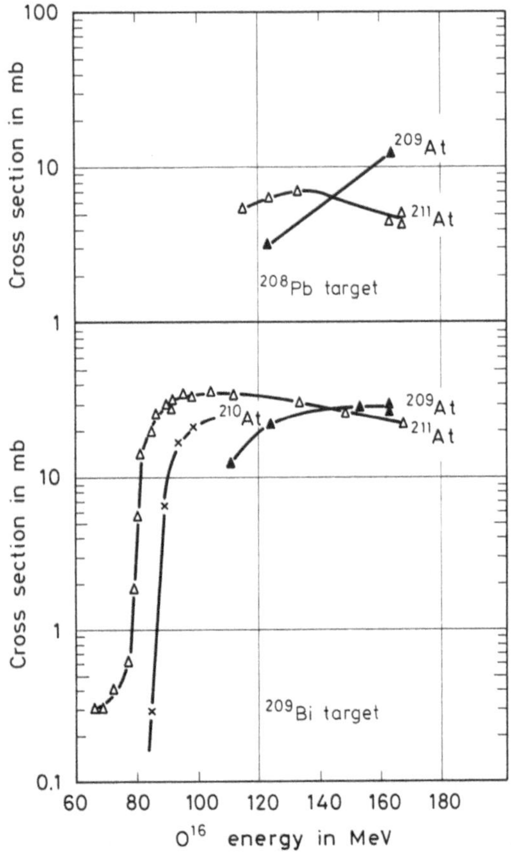

Cross sections for the formation of ^{209}At (▲) ^{210}At (×) and ^{211}At (△) in the reactions ^{208}Pb(^{16}O,...) and ^{209}Bi(^{16}O,...), respectively [122]. The data for ^{209}Bi(^{16}O,...)^{210}At are given in [262].

production: ^{208}Pb(^{16}O, ...)^{209}At [122]; see **Fig.** 3-**25**
^{209}Bi(^{3}He,3n)^{209}At [253]; see Fig. 3-12, p. 38
^{209}Bi(α,4n)^{209}At [121, 122, 162, 253]; **Fig.** 3-**26**, p. 64, and Fig. 3-17, p. 50
^{209}Bi(^{12}C, ...)^{209}At [262]; see **Fig.** 3-**27**, p. 64; see also [263]

References on pp. 87/94

Fig. 3-24

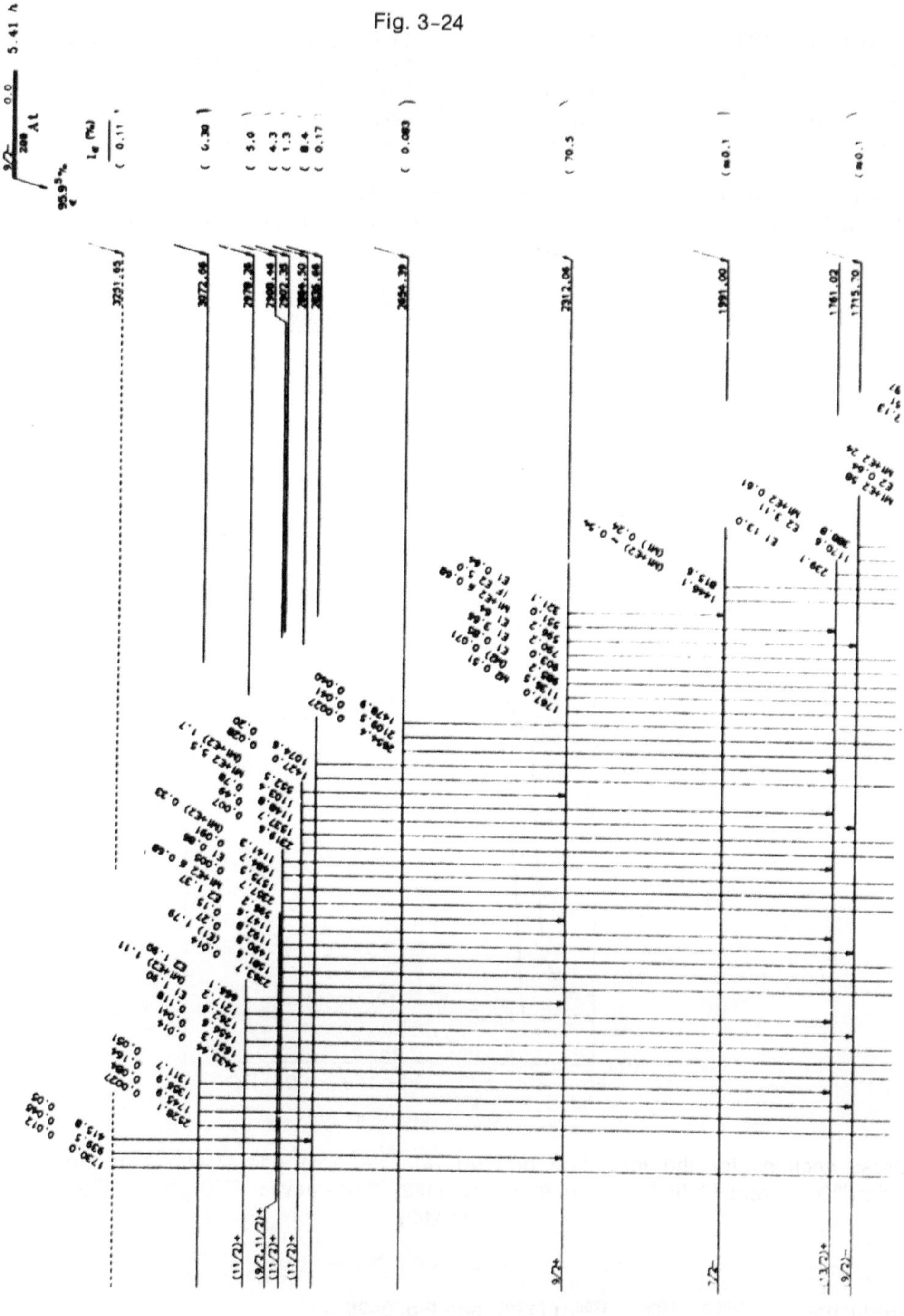

Decay scheme of ^{209}At [131]. The energies are given in keV, the intensities are $I_{\gamma+ce}$ per 100 decays of ^{209}At.

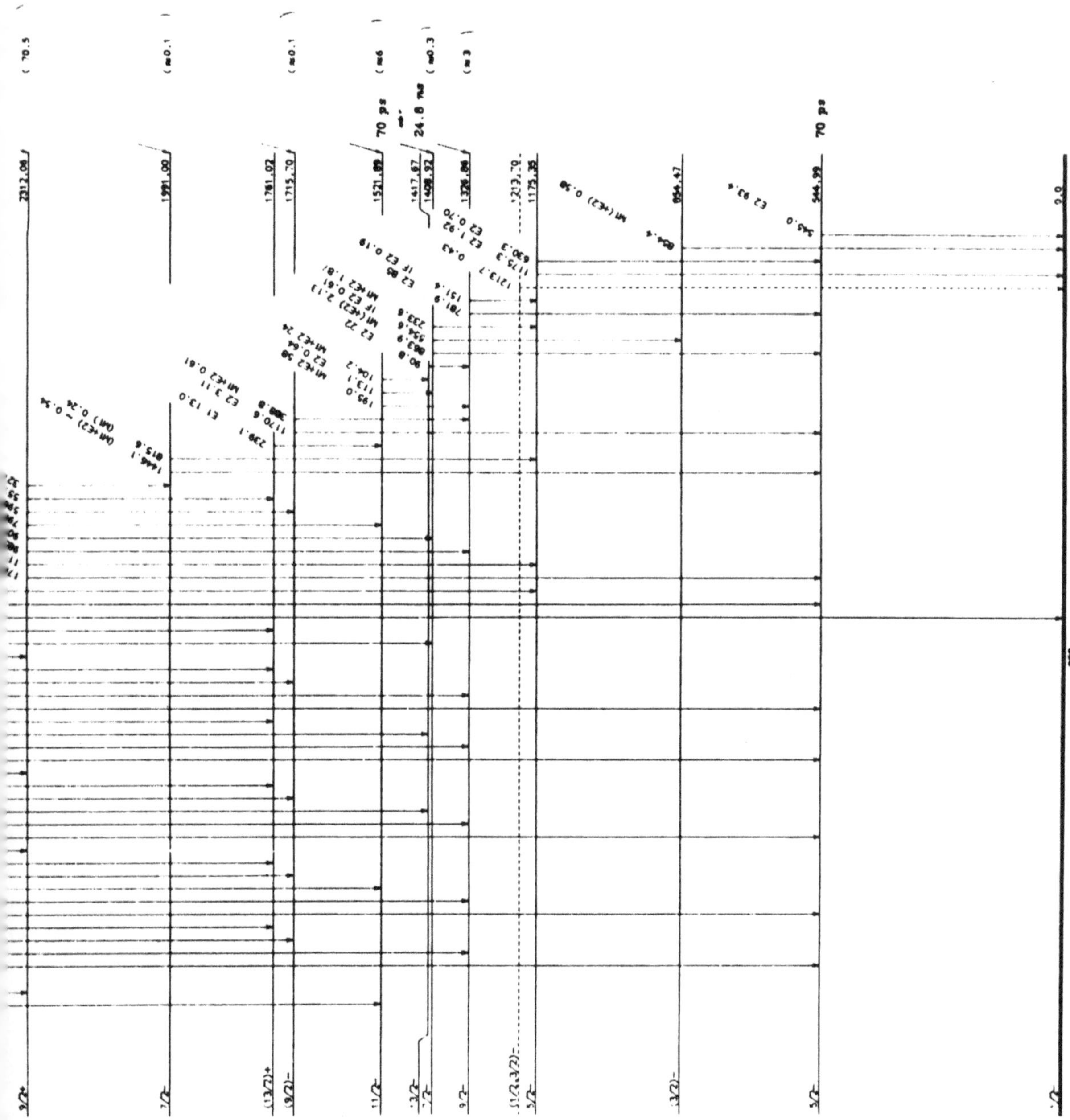

References on pp. 87/94

Fig. 3-26

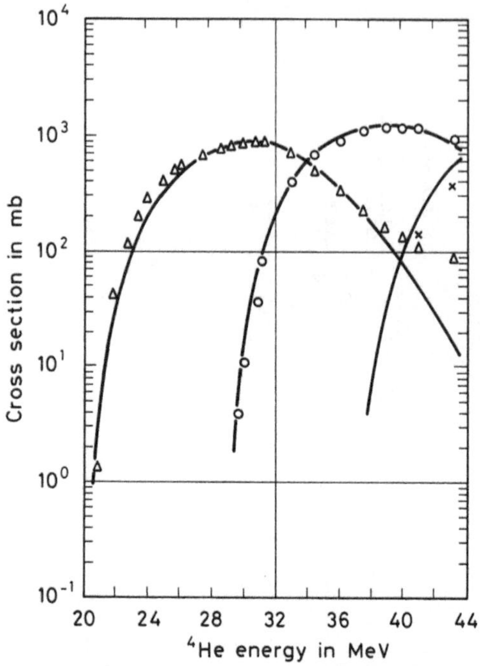

Cross sections for the formation of ^{209}At (×), ^{210}At (o) and ^{211}At (△) in the reaction ^{209}Bi(α, xn)$^{213-x}$At) [162].

Fig. 3-27

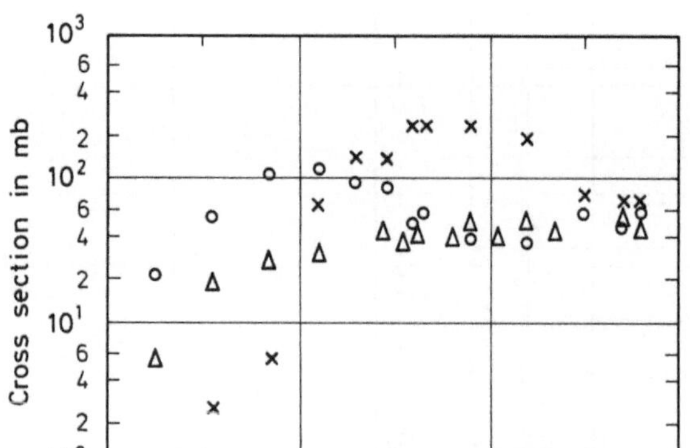

Cross sections for the formation of ^{209}At (×), ^{210}At (o) and ^{211}At (△) in the reaction ^{209}Bi(^{12}C,...) [262].

$^{209}Bi(^{16}O, \ldots)^{209}At$ [122]; see Fig. 3-25, p. 61
$^{209}Bi(^{19}F, \ldots)^{209}At$ [262]; the cross section at $E(^{19}F)$ 186 MeV is 45 mb [262]
$^{209}Bi(^{20}Ne, \ldots)^{209}At$ [262]; the cross section at $E(^{20}Ne)$ 201 MeV is 90 mb [262]
$^{209}Bi(^{40}Ar, \ldots)^{209}At$ [262]; see Fig. 3-18, p. 51
$^{209}Bi(^{40}Ca, \ldots)^{209}At$ [262]; see Fig. 3-19, p. 52
$^{209}Bi(^{56}Fe, \ldots)^{209}At$ [262]; see Fig. 3-22, p. 59
$^{209}Bi(^{63}Cu, \ldots)^{209}At$ [262]; see Fig. 3-23, p. 60
$^{232}Th(p, \ldots)^{209}At$ [256]; the cross section at E_p 660 MeV is 14.5 mb [256]
^{232}Th(bremsstrahlung, ...)^{209}At [259]; the cross sections at 250, 300 and 350 MeV maximum bremsstrahlung energy are 66, 87 and 280 μb, respectively [259]
^{238}U(bremsstrahlung, ...)^{209}At [259]; the cross section at 300 MeV maximum bremsstrahlung energy is 15.8 μb [259]

Astatine-210 [132]

mass: 209.987143 ± 0.000013 u [30] which corresponds to a mass excess Δm of −11976 ± 12 keV [30]. A new value was measured by [210]: Δm −11963 ± 14 keV
S_n 7160 ± 13 keV [30]
S_p 2892 ± 11 keV [30]

levels: the levels adopted in the compilation of Harmatz [142] are listed in Table 3/14; further reference [223]

Table 3/14
Levels of ^{210}At Adopted by [142]. (The evaluation is based on the results obtained in the investigation of the decay of ^{210}Rn [140, 146] and ^{214}Fr [141, 145] and the nuclear reaction $^{209}Bi(\alpha,3n\gamma)$ [139, 143, 144, 229].)

E* in keV	J^π	$T_{1/2}$	observed in ^{210}Rn decay	^{214}Fr decay	$^{209}Bi+\alpha$
0.0	5+	8.1 h	+	+	+
72.7 ± 0.1	4+		+	+	−
496.2 ± 0.1	4+		+	+	−
507.4 ± 1.0	6+		−	+	+
530.9 ± 0.1	3+		+	−	−
576.4 ± 1.0	7+		−	−	+
594 ± 5			−	+	−
721.3 ± 0.1	3+		+	−	−
769.0 ± 0.1	2+		+	−	−
837 ± 8			−	+	−
854 ± 5			−	+	−
970 ± 5			−	+	−
984.8 ± 0.2	(3+)		+	−	−
1036.8 ± 0.2	(3)+		+	+	−
1052.9 ± 0.2	2+		+	−	−
1129.1 ± 0.2	(2)+		+	−	−
1222.3 ± 1.5			−	+	+

Table 3/14 [continued]

E* in keV	J^π	$T_{1/2}$	observed in ^{210}Rn decay	^{214}Fr decay	^{209}Bi+α
1251.9±1.5	9+		−	−	+
1292.4±0.2	2+		+	−	−
1363.2±1.7	11+	27±3 ns [139] 25±5 ns [230] 30±5 ns [231] IT 100% [142]	−	−	+
1402.9±1.8	(8+)		−	−	+
1488.7±0.2	1+		+	−	−
1495.0±1.7	(10+)		−	−	+
1525.6±0.2	1+		+	−	−
1688.5±1.8	(10−)	15±2 ns [144, 229] 16 ns [143] IT 100% [142]	−	−	+
1739.9±1.8			−	−	+
1905.2±1.8	12+		−	−	+
1967.1±0.3	1−, 2−		+	−	−
1969.9±2.0			−	−	+
2042.9±1.8	13+		−	−	+
2281.2±0.3	0−, 1−		+	−	−
2467.1±2.1			−	−	+
2549.6±2.0	15−	580±5 ns [143] 740±8 ns [231] 750±10 ns [130, 230] IT 100% [142]	−	−	+
2572.0±2.0	(14−)		−	−	+
2683.7±2.1			−	−	+
2783.4±2.1			−	−	+
3107.2±2.2	16−		−	−	+
3112.0±2.3			−	−	+
3542.4±2.3	17−		−	−	+
3551.8±2.2			−	−	+
3655.3±2.1	16−		−	−	+
3774.5±2.5			−	−	+
4027.7±2.3	19+	4.0±1.7 μs [143]	−	−	+
4078.0±2.5			−	−	+

half-life: 7.9±0.4 h [109]; further references [31, 132]

α decay: 0.17% [133], 0.175±0.020% [109]
Q_α 5631.4±1.5 keV [30]

E_α in keV [90]	$I_\alpha^{R\,b)}$ [90]
5524.3 ± 1.3 a)	100
5465.5 ± 1.5	23.6 ± 1.0
5456.0 ± 2.0	1.3 ± 0.2
5443.0 ± 1.5	93 ± 5
5387.0 ± 2.0	15.0 ± 1.0
5361.1 ± 1.3	91 ± 5
5242 ± 3	2.8 ± 0.4
5175 ± 4	0.7 ± 0.2

further references [37, 99, 109, 133, 134, 226]

a) Value adopted from [37]. — b) To obtain absolute intensities per 100 decays of ^{210}At multiply by $(5.3 \pm 0.6) \times 10^{-4}$ [142].

electron capture $+\beta^+$ decay: 99.82% [133], 99.93% [109]
Q_{β^+} 3986 ± 12 keV [30]

γ radiation: with α decay

E_γ in keV [134]	$I_\gamma^{R\,a)}$ [134]
83 ± 1	766 ± 153
106 ± 1	272 ± 54
140 ± 1	100
167 ± 2	174 ± 35

a) To obtain absolute intensities per 100 decays of ^{210}At multiply by $(1.61 \pm 0.19) \times 10^{-5}$ [142].

with $\varepsilon+\beta^+$ decay see Table 3/15; further references [63, 104, 119, 137, 147, 148, 227, 228]

decay scheme: α decay, see **Fig.** 3-**28**, p. 68; further references [29, 142]
$\varepsilon+\beta^+$ decay, see **Fig.** 3-**29**, p. 70/1; further references [29, 140, 141]

production: ^{209}Bi(^{3}He,2n)^{210}At [253]; see Fig. 3-12, p. 38
^{209}Bi(α,3n)^{210}At [132, 162, 253]; see Fig. 3-17, p. 50, and 3-26, p. 64
cross sections are also given in [132] and thick target yields in [268, 269]
^{209}Bi(^{12}C, ...)^{210}At [262]; see Fig. 3-27, p. 64, see also [263]
^{209}Bi(^{16}O, ...)^{210}At [262]; see Fig. 3-25, p. 61

Table 3/15
Prominent γ Rays Following the $\varepsilon+\beta^+$ Decay of ^{210}At as Adopted by [142]. (All lines with $I_\gamma^R \geqq 1$ are listed; 57 additional lines with E_γ between 46.5 and 2386.8 keV are omitted. α is the total conversion coefficient; for values of α(K), α(L), etc., see [142].)

E_γ a)	I_γ^{R} a), b)	α	E_γ a)	I_γ^{R} a) b)	α
245.3 ± 0.1	80 ± 4	0.24	1181.4 ± 0.1	100.0 ± 2.5	0.0054
527.6 ± 0.1	1.15 ± 0.04	0.0093	1436.7 ± 0.1	29.2 ± 1.3	0.00145
817.2 ± 0.2	1.72 ± 0.05	0.037	1483.3 ± 0.1	46.8 ± 2.0	0.0014
852.7 ± 0.2	1.39 ± 0.05	0.033	1599.5 ± 0.1	13.5 ± 0.6	0.001
955.8 ± 0.1	1.81 ± 0.06	0.025	2254.0 ± 0.2	1.53 ± 0.05	

a) The evaluation is mainly based on [135, 136, 138]. — b) To obtain absolute intensities per 100 decays of ^{210}At multiply by 0.993 ± 0.025 [142].

References on pp. 87/94

Fig. 3-28

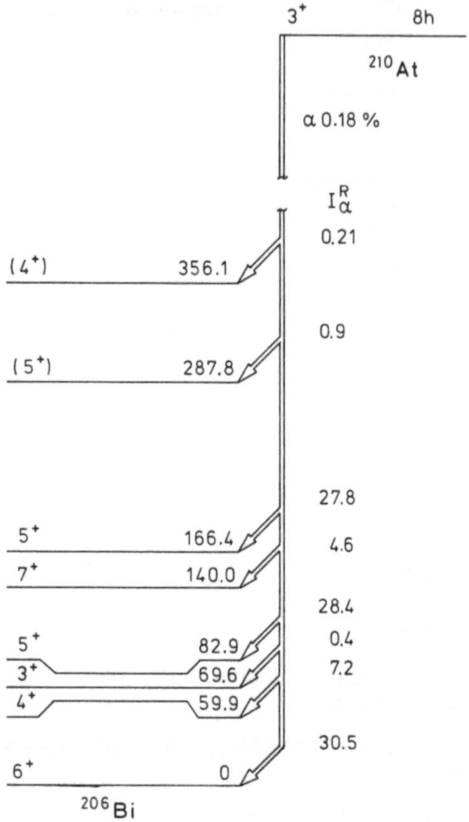

Decay scheme for the α decay of ^{210}At [91]. The energies are given in keV.

$^{209}Bi(^{19}F, \ldots)^{210}At$ [262]; see **Fig.** 3-**30**, p. 72
$^{209}Bi(^{20}Ne, \ldots)^{210}At$ [262]; the cross section at $E(^{20}Ne)$ 201 MeV is 26 mb [262]
$^{209}Bi(^{40}Ar, \ldots)^{210}At$ [262]; see Fig. 3-18, p. 51
$^{209}Bi(^{40}Ca, \ldots)^{210}At$ [262]; see Fig. 3-19, p. 52
$^{209}Bi(^{56}Fe, \ldots)^{210}At$ [262]; see Fig. 3-22, p. 59
$^{209}Bi(^{63}Cu, \ldots)^{210}At$ [262]; see Fig. 3-23, p. 60
$^{232}Th(p, \ldots)^{210}At$ [258]; the cross sections at E_p 200, 250, 300 and 660 MeV are 11, 13, 15 and 18 mb, respectively [258, 260]
$^{232}Th(bremsstrahlung, \ldots)^{210}At$ [259]; the cross sections at 250, 300 and 350 MeV maximum bremsstrahlung energies are 74, 98 and 230 μb, respectively [290]
$^{238}U(p, \ldots)^{210}At$ [90]
$^{238}U(bremsstrahlung, \ldots)^{210}At$ [259]; the cross section at 300 MeV maximum bremsstrahlung energy is 17.9 μb [259]

Astatine-211 [150]

mass: 210.987490 ± 0.000009 u [30]
S_n 7748 ± 14 keV [30]
S_p 2979 ± 8 keV [30]

levels: the spin of the ground state was measured with the atomic beam method to be 9/2 [152]. The levels adopted by Martin [164] are listed in Table 3/16

half-life: 7.214±0.007 h [149] further references [44, 132, 150, 151, 153, 162]
7.23±0.02 h [165]

α decay: [150], 41.94±0.50% [161], 41.8±0.2% [109]; further references [154, 165]
Q_α 5980.5±2.0 keV [30]

E_α in keV	I_α^R		further references [31, 37, 133, 232, 234]
[109]	[109]	[161]	
5866±2	100 [a)]	100	
5210.0±1.5	0.013±0.002	0.0086±0.0019	
5141±2	0.004±0.002	0.0023±0.0008	

a) The absolute intensity is according to [165] 41.7 per 100 decays of ^{211}At.

electron capture $+\beta^+$ decay: 58.06±0.50% [161], 58.2±0.2% [109]; further references [154, 165]
Q_{β^+} 791±8 keV [30]

Table 3/16
Levels of ^{211}At Adopted by Martin [164]. (The evaluation was based on the results obtained in the investigation of the decay of ^{211}Ru [155, 156] and the nuclear reactions ^{204}Hg, ^{208}Pb, ^{209}Bi(HI,xnγ) [157], ^{208}Pb(^{7}Li,4nγ) [163] and ^{209}Bi(α,2nγ) [158, 159, 166, 223].)

E* in keV	J^π	$T_{1/2}$	observed in Rn decay	HI reaction	^{208}Pb+Li	^{209}Bi+α
0.0	9/2−	7.214 h	+	+	+	+
674.11±0.07	7/2−, 9/2−		+	−	−	+
865.99±0.07	5/2−, 7/2−, 9/2−		+	−	−	+
947.44±0.07	5/2−, 7/2−		+	−	−	+
1067.1 ±0.5	(13/2)−		−	+	+	+
1116.21±0.08	3/2−, 5/2−	0.57±0.04 ns	+	−	+	+
1123.2	(9/2 to 13/2)−		−	−	+	+
1270.3	(15/2, 17/2)−	11±3 ns	−	−	+	+
1320.6 ±0.7	(17/2)−		−	+	+	+
1355.0	(9/2, 11/2, 13/2)		−	−	−	+
1416.6 ±0.9	(21/2)−	34±5 ns	−	+	+	+
1800.80±0.08	3/2−, 5/2−		+	−	−	−
1927.8 ±1.0	(23/2)−		−	+	+	+
1992.57±0.09	(5/2, 7/2)−		+	−	−	−
2062.87±0.09	(1/2, 3/2)−		+	−	−	−
2108.72±0.10	(3/2, 5/2)−		+	−	−	−
2128.71±0.09	(5/2, 7/2)−		+	−	−	−
2479.19±0.08	1/2+, 3/2+		+	−	−	−
2617.2 ±1.1	(25/2+, 27/2−)		−	+	+	+
2641.4 ±1.1	(29/2+)	54±3 ns	−	+	+	+
2655.16±0.10	1/2, 3/2		+	−	−	−
4177.4 ±1.5	(31/2, 33/2)	≦10 ns	−	+	−	−
4381.1 ±1.6	(33/2, 35/2)		−	+	−	−
4816.2 ±1.7	(39/2, 41/2)	4.2±0.4 ns	−	+	−	−

References on pp. 87/94

Fig. 3-29

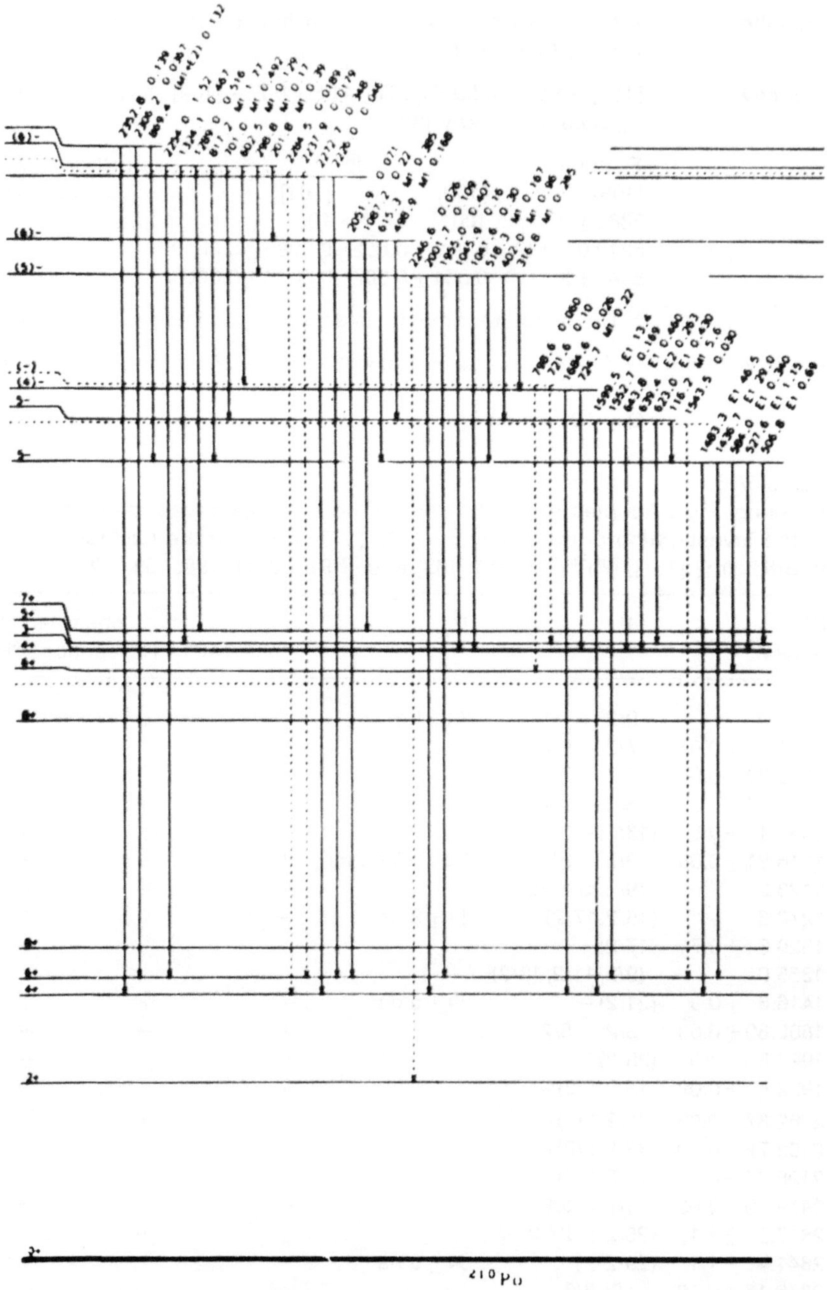

Decay scheme for the $\varepsilon+\beta^+$ decay of ^{210}At [142]. This scheme is mainly based on the results of [135]. The energies are given in keV. The half-lives of the Po levels were determined by [139, 143, 144].

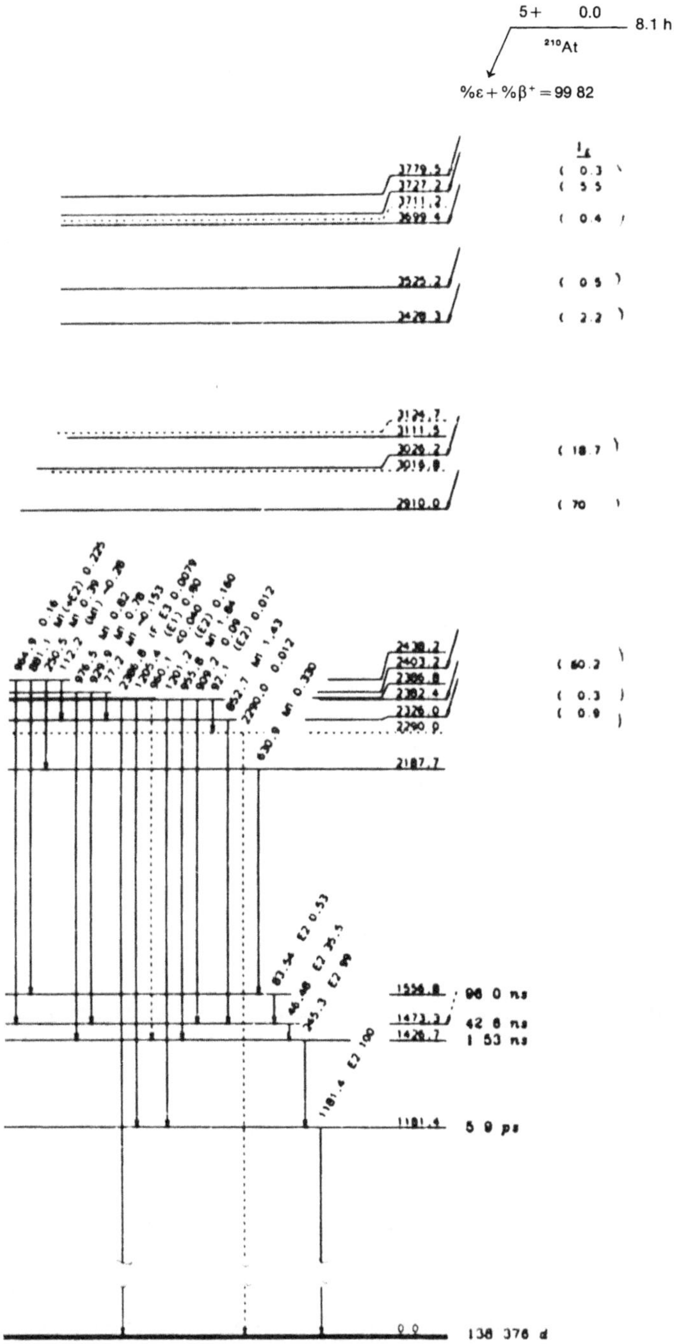

References on pp. 87/94

Fig. 3-30

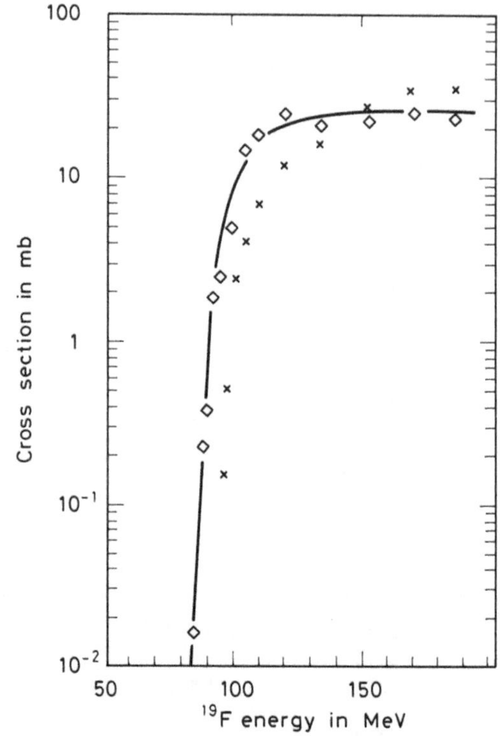

Cross sections for the formation of ^{210}At (×) and ^{211}At (△) in the reaction ^{209}Bi(^{19}F, ...) [262].

γ radiation: with α decay:

E_γ in keV	$I_\gamma^{R\,a)}$	α	further reference [119]
[161]	[161]	[164]	
669.6 ± 0.2	1.1 ± 0.2	0.053	
742.7 ± 0.5	0.3 ± 0.1	0.042	

a) To obtain absolute intensities per 100 decays of ^{211}At multiply by $(3.11 \pm 0.11) \times 10^{-3}$ [161].

with $\varepsilon + \beta^+$ decay

E_γ in keV		I_γ	further reference [119]
[161]	[165]	[161]	
687.0 ± 0.1	686.7 ± 0.5	0.245 ± 0.012	

decay scheme: the transition with E_α 5866 keV populates the ground state of ^{207}Bi [29, 164]. In the electron capture 58% of the transitions populate the ground state and only 0.26% populate the first excited state (E* 687 keV) of ^{211}Po [29, 164]

production: ^{205}Tl(^{19}F, ...)^{211}At [262]; the cross section at E(^{19}F) 178 MeV is 0.092 mb [262]
^{205}Tl(^{20}Ne, ...)^{211}At [262]; cross sections of 0.15 up to 1.46 mb were observed in the energy range 193 ≦ E(^{20}Ne) ≦ 239 MeV [262]
^{205}Tl(^{40}Ar, ...)^{211}At [262]; the cross section at E(^{40}Ar) 218 MeV is 1.06 mb [262]

$^{208}Pb(^{16}O, \ldots)^{211}At$ [122]; see Fig. 3-25, p. 61
$^{209}Bi(\alpha,2n)^{211}At$ [132, 162, 253]; see 3-26, p. 64
cross sections are also given in [132] and thick target yields in [268 to 270]
$^{209}Bi(^{12}C, \ldots)^{211}At$ [262]; see Fig. 3-27, p. 64, see also [263]
$^{209}Bi(^{14}N, \ldots)^{211}At$ [262]; see **Fig. 3-31**
$^{209}Bi(^{16}O, \ldots)^{211}At$ [122]; see Fig. 3-25, p. 61, see also [262]
$^{209}Bi(^{19}F, \ldots)^{211}At$ [262]; see Fig. 3-30
$^{209}Bi(^{20}Ne, \ldots)^{211}At$ [262]; the cross sections vary between 24 and 38 mb in the energy range $111 \leqq E(^{20}Ne) \leqq 203$ MeV [262]
$^{209}Bi(^{40}Ar, \ldots)^{211}At$ [262, 265]; see Fig. 3-18, p. 51
$^{209}Bi(^{40}Ca, \ldots)^{211}At$ [262]; see Fig. 3-19, p. 52
$^{209}Bi(^{56}Fe, \ldots)^{211}At$ [262]; see Fig. 3-22, p. 59
$^{209}Bi(^{63}Cu, \ldots)^{211}At$ [262]; see Fig. 3-23, p. 60
$^{209}Bi(^{84}Kr, \ldots)^{211}At$ [254]; in the energy range $432 \leqq E(^{84}Kr) \leqq 490$ MeV cross sections of 2.8 up to 15.7 mb were observed

Astatine-212g [168, 169]

mass: 211.990741 ± 0.000006 u [30]
S_n 5044 ± 9 keV [30]
S_p 3470 ± 5 keV [30]

levels: the levels of ^{212}At populated in nuclear reactions are shown in **Fig. 3-32**, p. 74. Only the ground state is populated in the α decay of ^{216}Fr [173]

Fig. 3-31

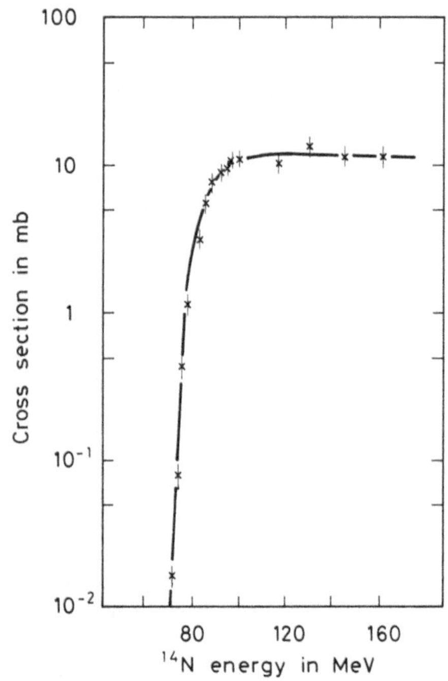

Cross sections for the reaction $^{209}Bi(^{14}N, \ldots)^{211}At$ [262].

Fig. 3-32

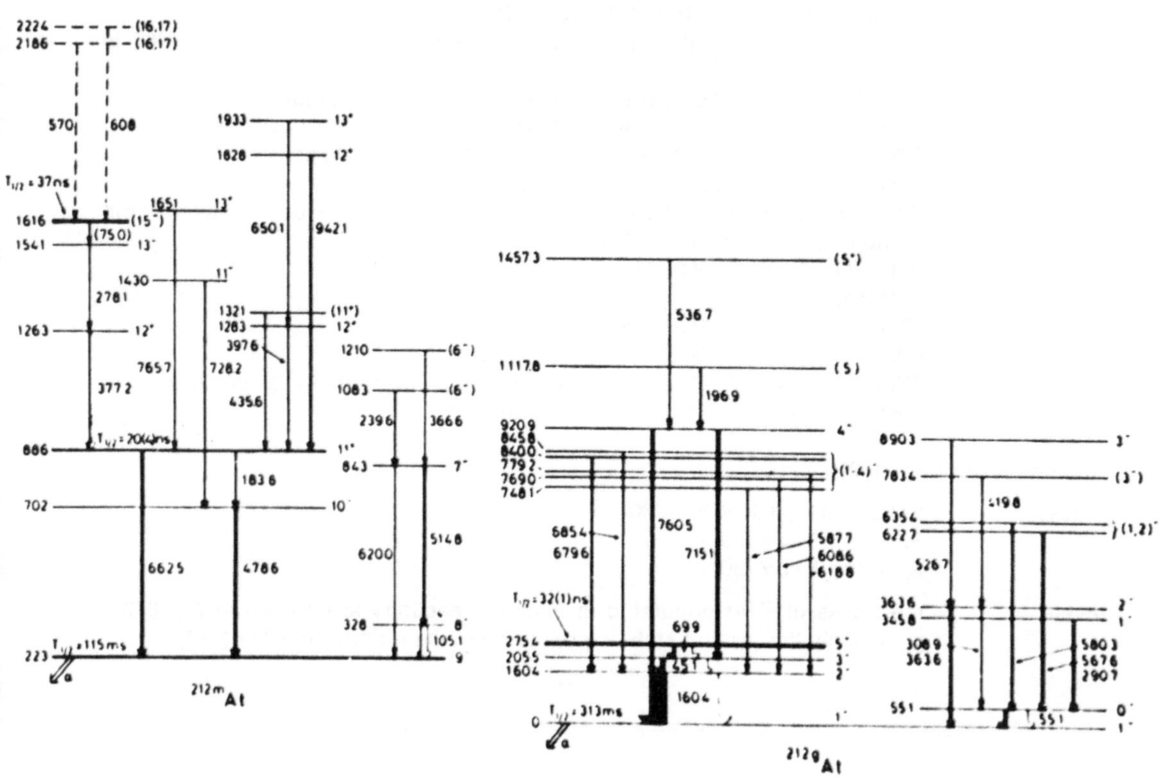

Levels of ^{212}At populated in the reactions ^{209}Bi(α,nγ) [233] and ^{208}Pb(^{7}Li,3nγ) [174]. The energies are given in keV.

half-life: 0.315 ± 0.003 s [170] further references [167 to 169]
0.313 ± 0.003 s [171]

α decay: [169], 100% [173]
Q_α 7828.9 ± 2.0 keV [30]

E_α in keV [170]	E_α in keV [175]	I_α [170]	I_α [175]
7679 ± 9	7679	80.9 ± 0.8	83.5
7616 ± 9	7618	17.0 ± 0.5	15.1
7170 ± 12	7177	0.26 ± 0.06	0.15
7076 ± 13	7088	0.63 ± 0.06	0.60
7045 ± 10	7058	0.50 ± 0.08	0.39
6799 ± 10	6805	0.26 ± 0.03	0.073
6777 ± 11	6765	0.04 ± 0.03	0.046
6754 ± 10		0.12 ± 0.03	
6730 ± 13		0.06 ± 0.02	
6668 ± 9		0.05 ± 0.02	
6612 ± 9	6629	0.15 ± 0.01	0.14

electron capture $+\beta^+$ decay: $<4\times10^{-2}\%$ [173]
Q_{β^+} 1756±4 keV [30]

β^- decay: $<4\times10^{-2}\%$ [173]
Q_{β^-} 41±8 keV [30]

γ radiation: E_γ 63 keV [169]

decay scheme: see **Fig.** 3-**33**

production: $^{209}Bi(\alpha,n)^{212}At$ [169, 171]; see **Fig.** 3-**34**, p. 76; cross sections are also given in [261]
$^{208}Pb(^7Li,3n)^{212}At$ [235]

Astatine-212m [167, 169]

levels: see Fig. 3-32

half-life: 0.122±0.001 s [170] further references [167, 169]
0.115±0.002 s [171]

Fig. 3-33

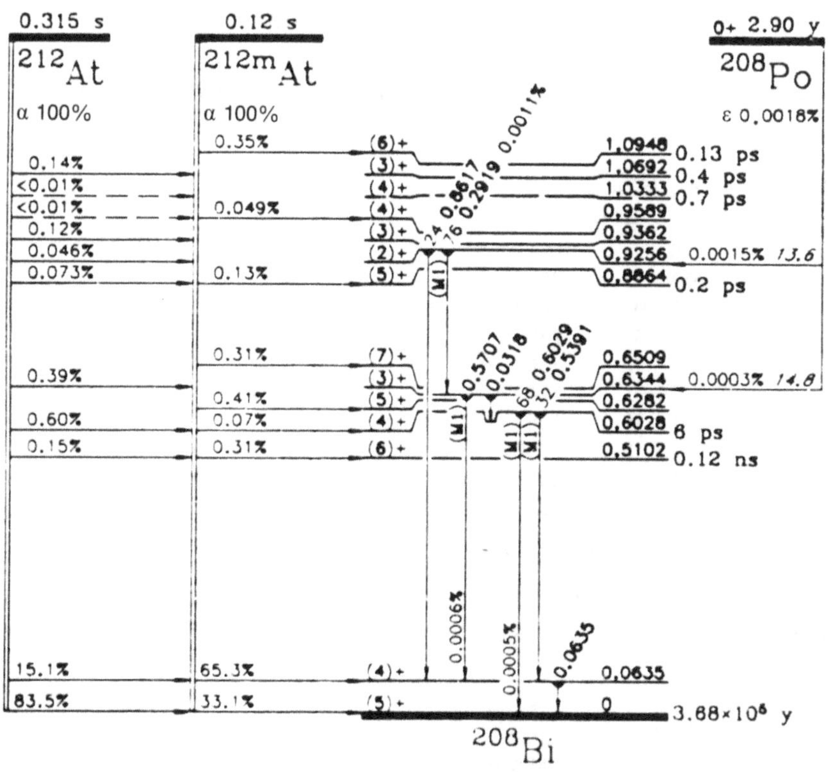

Decay scheme of ^{212}At and ^{212m}At [29]. The energies are given in MeV. The absolute γ intensities refer to the decay of ^{208}Po.

References on pp. 87/94

Fig. 3-34

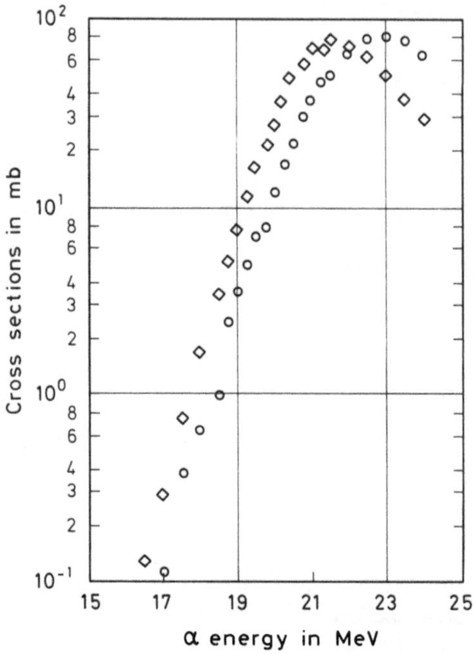

Cross sections for the formation of ^{212g}At (△) and ^{212m}At (○) in the reaction ^{209}Bi(α,n) [171].

α decay: [167, 169]

E_α in keV		I_α	
[170]	[175]	[170]	[175]
7900 ± 9	7897	29.9 ± 0.3	33.1
7837 ± 9	7837	67.3 ± 1.0	65.3
7391 ± 9	7397	0.65 ± 0.07	0.31
7300 ± 12	7309	0.11 ± 0.03	0.07
7276 ± 9	7282	0.53 ± 0.04	0.41
7253 ± 10	7261	0.61 ± 0.11	0.31
7016 ± 13	7032	0.26 ± 0.13	0.13
6954 ± 11	6954	0.16 ± 0.10	0.049
6816 ± 11	6821	0.19 ± 0.12	0.35

isomeric transition: < 1% [169]

decay scheme: see Fig. 3-33, p. 75

production: ^{209}Bi(α,n)^{212m}At [169, 171]; see Fig. 3-34; cross sections are also given in [262]

Astatine-213 [176, 178]

mass: 212.992926 ± 0.000014 u [30]
S_n 6036 ± 13 keV [30]
S_p 3498 ± 12 keV [30]

Fig. 3-35

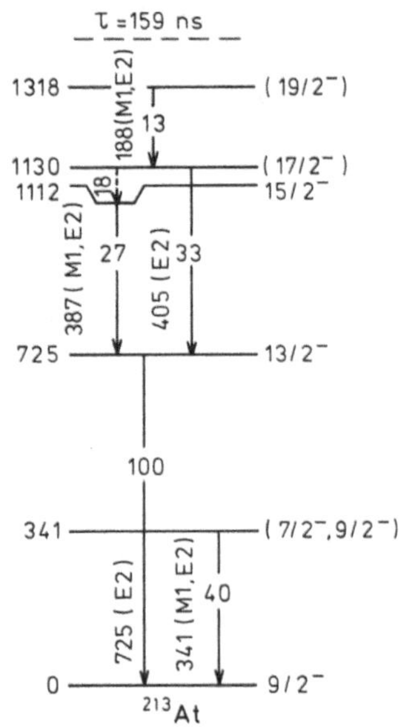

Levels of ^{213}At populated in the reaction ^{208}Pb(^{7}Li,2nγ) [235]. The energies are given in keV.

levels: see **Fig. 3-35**; further references [97, 237]

half-life: 0.11 ± 0.02 μs [177, 211] further references [176, 178]

α decay: [176]
Q_α 9254 ± 12 keV [30]
E_α 9080 ± 12 keV [177]; further references [176, 234]
9060 ± 20 keV [178]

electron capture: $<2.5\times10^{-13}$% (syst) [238]
Q_ε 75 ± 13 keV [30]

decay scheme: the α decay populates the ground state of ^{209}Bi [29]

production: ^{208}Pb(^{16,18}O, ...)^{221}Ac $\xrightarrow{2\alpha}$ ^{213}At [177] [a)]
^{208}Pb(^{19}F, ...)^{221}Ac $\xrightarrow{2\alpha}$ ^{213}At [177] [a)]
^{209}Bi(^{7}Li,p2n)^{213}At [235]
^{209}Bi(^{22}Ne, ...)^{225}Pa $\xrightarrow{3\alpha}$ ^{213}At [177] [a)]
^{233}U(p,5n)^{229}Np $\xrightarrow{4\alpha}$ ^{213}At [178]

[a)] For relative yields see [177] or the values given in this handbook [16, 255].

Astatine-214 [179]

mass: 213.996362 ± 0.000007 u [30]
S_n 4871 ± 13 keV [30]
S_p 4015 ± 7 keV [30]

References on pp. 87/94

levels: see **Fig. 3-36**; further reference [25]
in addition to the levels given in Fig. 3-20, p. 53, a level at 59 keV with $T_{1/2}$ = 265 ns was observed by [239]

half-life: ground state 558 ± 10 ns [239]
59 keV level 265 ± 30 ns [239]
233 keV level 760 ± 15ns [239]

α decay: [179], 100% [240]
Q_α 8987 ± 4 keV [30]

E_α in keV [239] I_α^R [239] further references [179 to 182, 192, 232]

Fig. 3-36

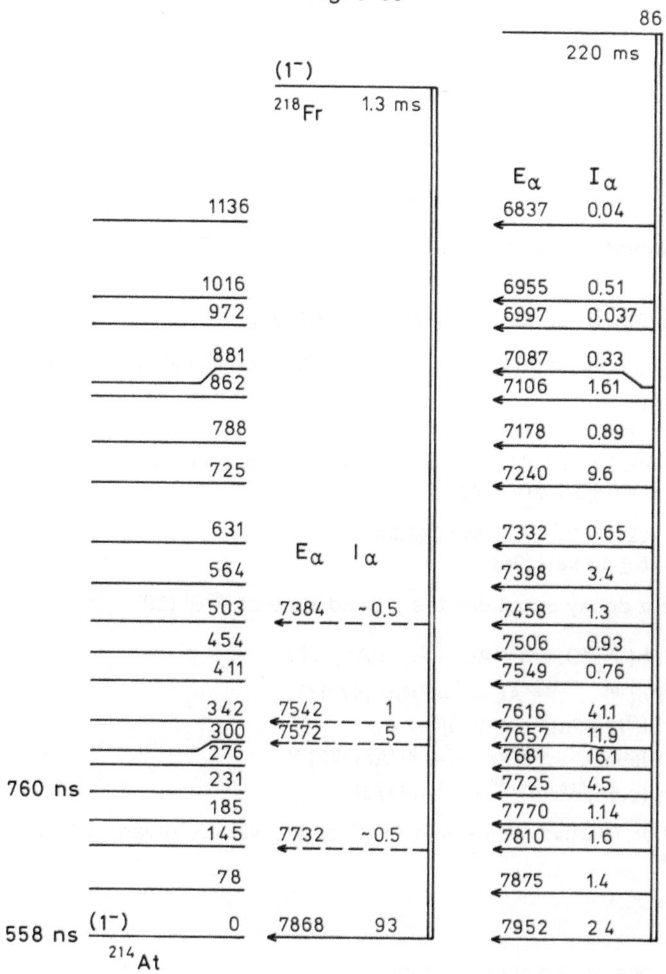

Levels of ^{214}At populated in the α decay of ^{218}Fr and ^{218m}Fr [239]. The energies are given in keV.

558 ns ground state

8270 ± 5	0.32 ± 0.03
8480 ± 6	0.58 ± 0.04
8507 ± 7	0.15 ± 0.04
8819 ± 4 [a]	98.95 ± 0.06

265 ns, 59 keV isomeric state

8877 ± 8	100

760 ns, 233 keV isomeric state

8390 ± 6	0.18 ± 0.03
8620 ± 5	0.64 ± 0.05
8782 ± 5	99.18 ± 0.06

[a] From [37].

electron capture: Q_ε 1091 ± 4 keV [30]

β^- decay: Q_{β^-} 939 ± 11 keV [30]

decay scheme: see **Fig. 3-37**, p. 80; further reference [29]

production: $Th(p_{150}, \ldots)^{226}Pa \xrightarrow{3\alpha} {}^{214}At$ [179, 180]

Astatine-215 [184]

mass: 214.998646 ± 0.000008 u [30]
S_n 5944 ± 8 keV [30]
S_p 4071 ± 7 keV [30]

levels: see **Fig. 3-38**, p. 81

half-life: 0.10 ± 0.02 ms [180, 183]; further reference [177]

α decay: [184]
Q_α 8178 ± 4 keV [30]
E_α 8026 ± 4 keV [37] based on [192]
8010 ± 10 keV [232]; further references [177, 180, 183, 184, 186]

electron capture: Q_ε −721 ± 7 keV [30]

β^- decay: Q_{β^-} −82 ± 11 keV [30]

γ radiation: a weak γ line with E_γ ~404 keV accompanying the α decay was observed by [242]

decay scheme: the α particles populate the ground state of ^{211}Bi [29]

production: $^{205}Tl(^{22}Ne,4n)^{223}Pa \xrightarrow{2\alpha} {}^{215}At$ [177]

$^{208}Pb(^{11}B,\alpha)^{215}At$ [185]

$^{208}Pb(^{19}F,4n)^{223}Pa \xrightarrow{2\alpha} {}^{215}At$ [177]

$^{209}Bi(^{12}C,\alpha 2n)^{215}At$ [186]

$^{209}Bi(^{20}Ne,\alpha 2n)^{223}Pa \xrightarrow{2\alpha} {}^{215}At$ [177]

References on pp. 87/94

Fig. 3-37

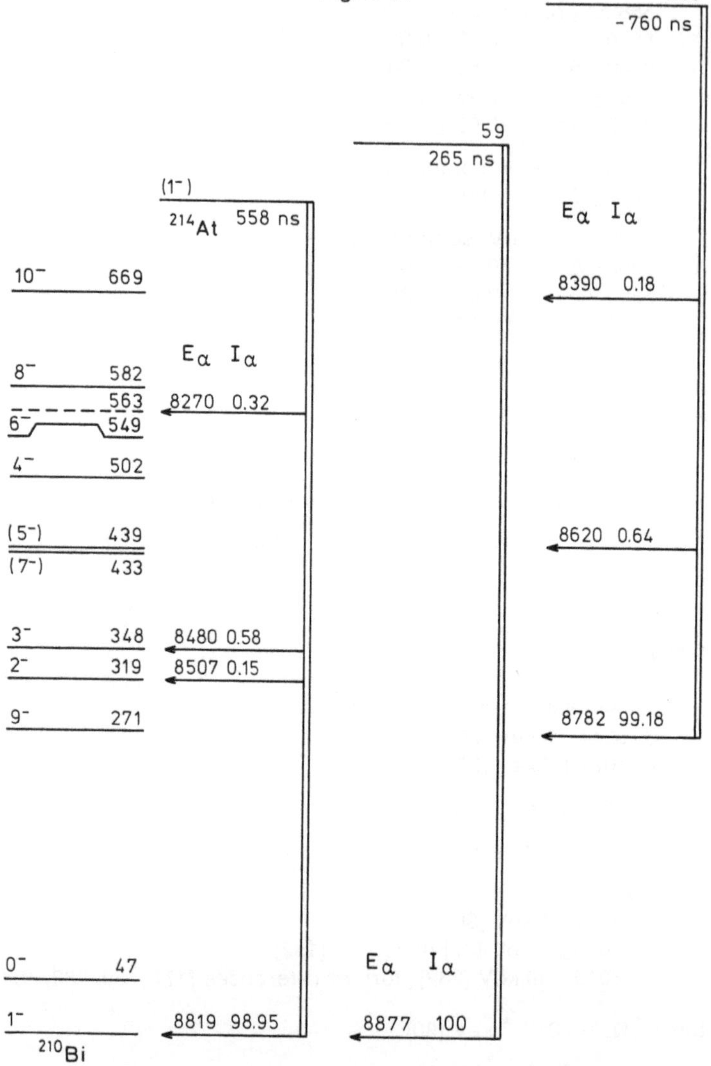

Decay scheme for the α decay of ^{214}At (ground state and the two metastable states) [239]. The energies are given in keV.

Astatine-216 [187]

mass: 216.002401 ± 0.00007 u [30]
S_n 4573 ± 9 keV [30]
S_p 4512 ± 7 keV [30]

levels: see **Fig.** 3-**39**, p. 82
in [193] a metastable state (E* and $T_{1/2}$ unknown) was proposed which decays by α emission with E_α 7960 keV. The intensity of these α particles is 0.028% of the intensity of the 7800 keV α particles emitted in the decay of ^{216}Fr

Fig. 3-38

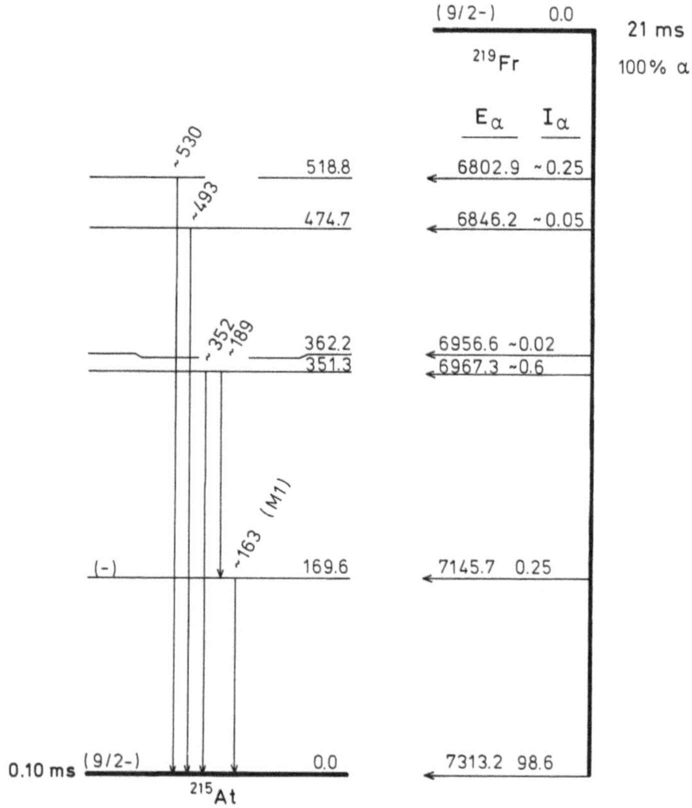

Levels of ^{215}At populated in the α decay of ^{219}Fr [191, 241]. The energies are given in keV.

half-life: 0.30 ± 0.03 ms [179, 180]

α decay: [187]

Q_α 7947.0 ± 3.0 keV [30]

further references [180, 183, 187, 189, 232]

E_α in keV [243] [a]	I_α [243] [a]
7238	0.06
7315	0.085
7393	0.23
7482	(0.1)
7565 } 7595 }	0.2
7697	2.1
7800 ± 3	97

[a] Based on [181, 188, 192, 193].

electron capture: Q_ε 468 ± 5 keV [30]

References on pp. 87/94

Fig. 3-39

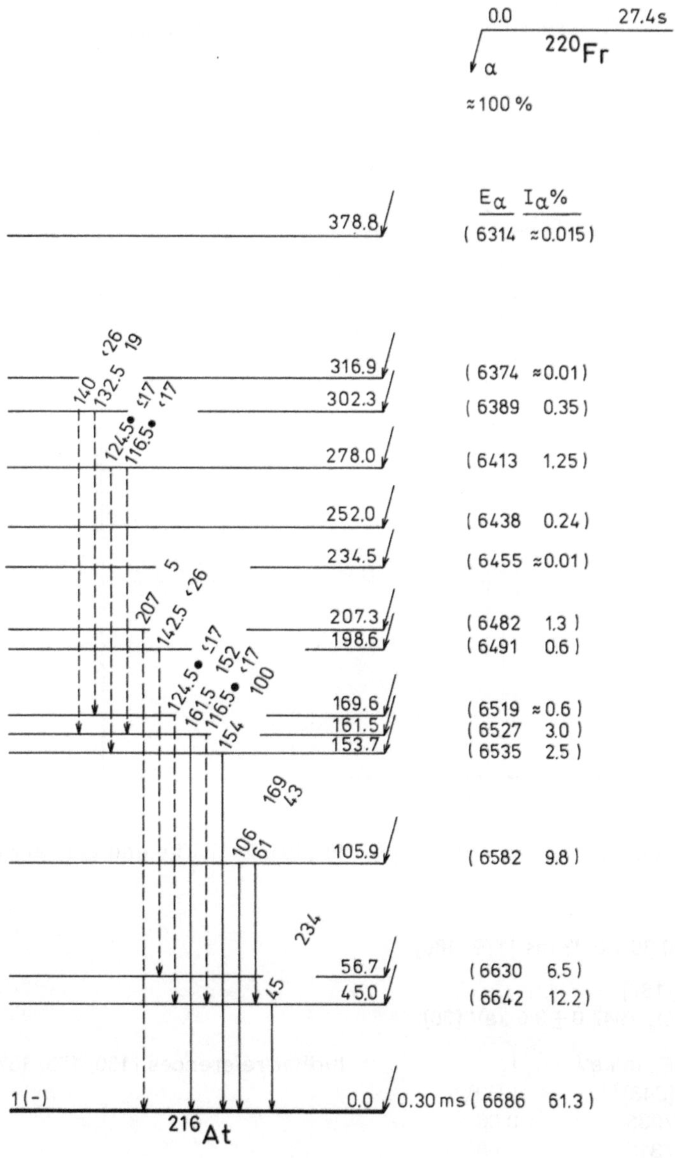

Levels of ^{216}At populated in the α decay of ^{220}Fr [243]. The energies are given in keV.

β⁻ decay: Q_{β^-} 1991 ± 11 keV [30]

decay scheme: see **Fig. 3-40**

production: descendant of ^{228}Pa [29]

Fig. 3-40

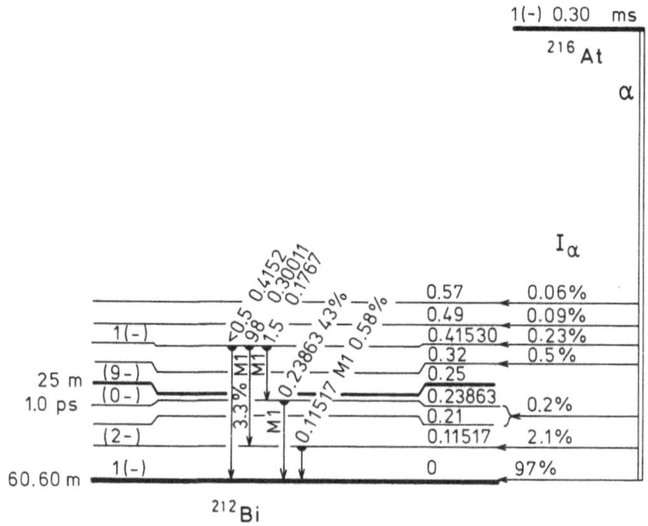

Decay scheme for the α decay of ^{216}At [29]. The energies are given in MeV, the γ intensities refer to the ^{216}Pb decay.

Astatine-217 [195, 197]

mass: 217.004704 ± 0.000012 u [30]
S_n 5926 ± 12 keV [30]
S_p 4676 ± 12 keV [30]

levels: see **Fig.** 3-**41**, p. 84/5

half-life: 32.3 ± 0.4 ms [194] further references [195 to 197]
111 ± 7 ms [203]

α decay: [195, 197], 99.9$^+$% [203]
Q_α 7200.4 ± 2.0 keV [30]

further references [192, 195, 197 to 204, 232, 246, 247]

E_α in keV [203]	I_α [203]
6484 ± 5	0.02 ± 0.01
6610 ± 7	0.010 ± 0.005
6813 ± 5	0.06 ± 0.02
7067 ± 2 [a)]	99.9 ± 0.1

a) 7069.0 ± 1.5 keV are suggested in the evaluation by Rytz [37].

β⁻ decay: 0.012 ± 0.004% [203] further reference [199]
Q_{β^-} 733 ± 11 keV [30]

electron capture: Q_ε −1580 keV (syst) [30]

References on pp. 87/94

Fig. 3-41

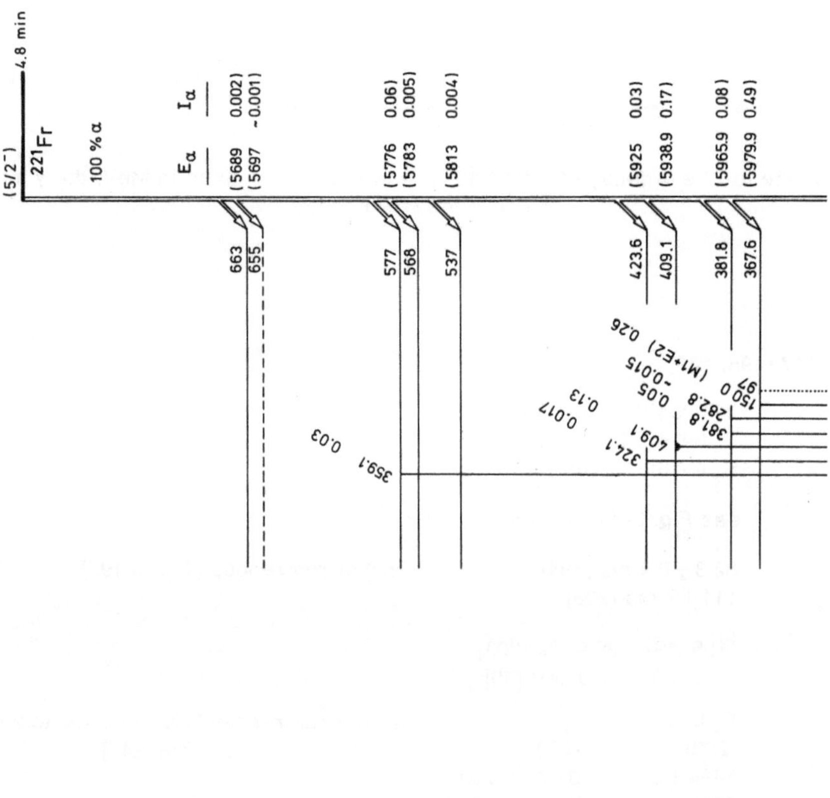
(5/2⁻)
221Fr
4.8 min
100 % α
Eα Iα
(5689 0.002)
(5697 ~0.001)
(5776 0.06)
(5783 0.005)
(5813 0.004)
(5925 0.03)
(5938.9 0.17)
(5965.9 0.08)
(5979.9 0.49)
663
655
577
568
537
423.6
409.1
381.8
367.6
359.1 0.03
324.1 0.017
409.1 0.13
381.8 0.05
282.8 ~0.015
150 (M1+E2) 0.26
97
0

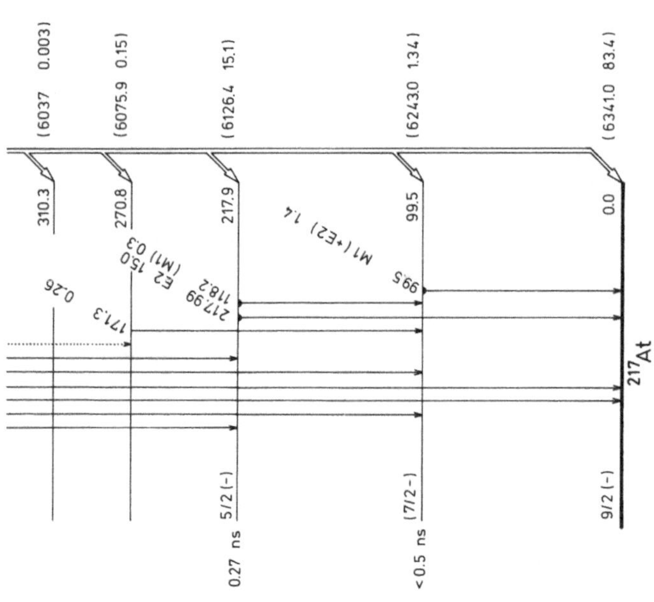

Levels of ^{217}At populated in the α decay of ^{221}Fr [244]. The energies are given in keV, the transition intensities $I_\gamma + I_{ce}$ are for 100 decays of ^{221}Fr. The half-lives for the excited levels were measured by [205].

References on pp. 87/94

Fig. 3-42

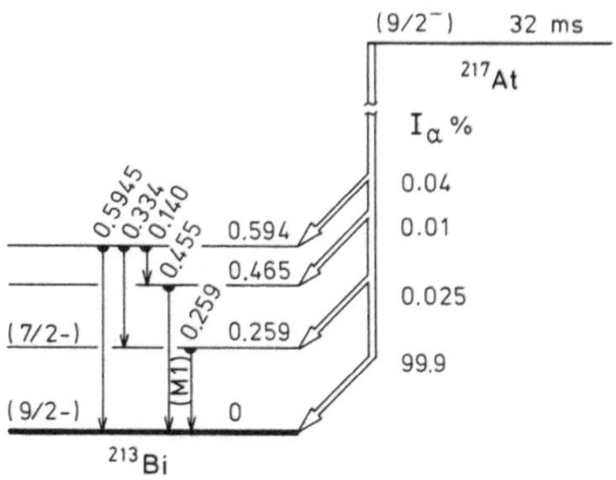

Decay scheme for the α decay of ^{217}At [29]. The energies are given in MeV.

γ radiation with α decay:

E_γ in keV			I_γ	further references
[199]	[203]	[248]	[248]	[206, 246, 247]
140 (?)				
166 (?)				
	218			
260	259	258.5	0.239±0.020	
335 (?)	334			
375 (?)				
485				
595	592	593.1	0.0507±0.0025	

decay scheme: see **Fig.** 3-**42**; further reference [266]

production: descendant of ^{225}Ac [195, 197]

Astatine-218 [208]

mass: 218.008694±0.000014 u [30]
S_n 4355±17 keV [30]
S_p 5.15 MeV (syst) [30]

half-life: 1.12±0.11 μs [245]; further references [207, 208, 249]

α decay: [208], 99⁺% [207]
Q_α 6883±5 keV [30]

E_α in keV	I_α	further references [208, 252]
[37]	[37]	
6757±5	3.6	
6694±5	90	
6653±5	6.4	

β^- decay: 0.1% [207]
Q_{β^-} 2887±13 keV [30]

electron capture: Q_ε −256±13 keV [30]

decay scheme: the emission of 6757 keV α particles leads to the ground state of ^{214}Bi [29, 250]

production: descendant of ^{218}Po [207, 208]

Astatine-219 [209]

mass: 219.01130 ± 0.00009 u [30]
S_n 5640 ± 80 keV [30]
S_p 5120 ± 80 keV [30]

half-life: 0.9 ± 0.1 min [209]

α decay: 97% [209]
Q_α 6390 ± 50 keV [30]
E_α 6275 ± 50 keV [209] as adjusted by [251]

β^- decay: 3% [209]
Q_{β^-} 1700 ± 80 keV [30]

production: descendant of ^{227}Ac [209]

References:

[1] T. Mayer-Kuckuk (Kernphysik: Eine Einführung, Teubner, Stuttgart 1984). – [2] G. Friedlaender, J.W. Kennedy, J.M. Miller (Nuclear and Radiochemistry, 3rd Ed., Wiley, New York 1981). – [3] L. Valentin (Subatomic Physics: Nuclei and Particles, Vol. 1/2, North-Holland, Amsterdam 1981). – [4] G.R. Choppin, J. Rydberg (Nuclear Chemistry, Pergamon, Oxford 1980). – [5] W.M. Gibson (The Physics of Nuclear Reactions, Pergamon, Oxford 1980).

[6] K.H. Lieser (Einführung in die Kernchemie, 2nd Ed., Verlag Chem., Weinheim/Bergstr. 1980). – [7] E. Segrè (Nuclei and Particles, 2nd Ed., Benjamin, Reading 1977). – [8] V.G. Soloviev (Theory of Complex Nuclei, Pergamon, Oxford 1976). – [9] M.A. Preston, R.K. Baduri (Structure of the Nucleus, Addison-Wesley, Reading 1975). – [10] A. Bohr, B.R. Mottelson (Nuclear Structure, Vol. 1/2, Benjamin, New York 1969/75).

[11] A. de Shalit, H. Feshbach (Theoretical Nuclear Physics, Vol. 1, Wiley, New York 1974). – [12] J.M. Eisenberg, W. Greiner (Nuclear Theory, Vol. 1/2/3, North-Holland, Amsterdam 1970/72). – [13] G.F. Bertsch (The Practioners Shell Model, North-Holland, Amsterdam 1972). – [14] W. Donner (Einführung in die Theorie der Kernspektren, Bibl. Inst., Mannheim 1971). – [15] E.K. Hyde (Properties of the Atomic Nuclei in: Gmelin Handbuch "Transurane" A 1, I, 1973, pp. 19/178).

[16] H. Münzel (Pa-Isotope und ihre Eigenschaften in: Gmelin Handbuch "Protactinium" Erg.-Bd. 1, 1977, pp. 28/142). – [17] S. Maripuu (At. Data Nucl. Data Tables **17** [1976] 411/608). – [18] W.D. Myers (At. Data Nucl. Data Tables **17** [1976] 411/7, 476/608). – [19] H. van Groote, E.R. Hilf, K. Takahashi (At. Data Nucl. Data Tables **17** [1976] 418/27, 476/608). – [20] P.A. Seeger, W.M. Howard (At. Data Nucl. Data Tables **17** [1976] 428/30, 476/608).

[21] S. Liran, N. Zeldes (At. Data Nucl. Data Tables **17** [1976] 431/41, 476/608). – [22] M. Bauer (At. Data Nucl. Data Tables **17** [1976] 442/9, 476/608). – [23] J. Jaenecke (At. Data Nucl. Data Tables **17** [1976] 455/62, 476/608). – [24] E. Comay, I. Kelson (At. Data Nucl. Data Tables **17** [1976] 463/6, Tables pp. 476/608). – [25] J. Jaenecke, B.P. Eynon (At. Data Nucl. Data Tables **17** [1976] 467/73, 476/608).

[26] P. Möller, J.R. Nix (At. Data Nucl. Data Tables **26** [1981] 165/96). – [27] W.D. Myers (Phys. Letters B **30** [1969/70] 451/4). – [28] M.R. Schmorak (Nucl. Data Sheets **25** [1978]

675/721). — [29] C.M. Lederer, V.S. Shirley (Table of Isotopes, 7th Ed., Wiley, New York 1978). — [30] A.H. Wapstra, K. Bos (At. Data Nucl. Data Tables **19** [1977] 175/297, **20** [1977] 1/126).

[31] G.W. Barton Jr., A. Ghiorso, I. Perlman (Phys. Rev. [2] **82** [1951] 13/9). — [32] R.W. Hoff, F. Asaro, I. Perlman (J. Inorg. Nucl. Chem. **25** [1963] 1303/19). — [33] W. Treytl, K. Valli (Nucl. Phys. A **97** [1967] 405/16). — [34] P. Hornshoj, P.G. Hansen, B. Jonson (Nucl. Phys. A **230** [1974] 380/92). — [35] G. Bastin, C.F. Liang (CSNSM-RA-1975 [1975] 1/125, 35/5; INIS Atomindex **7** [1976] No. 250748).

[36] M.R. Schmorak (Nucl. Data Sheets **26** [1979] 81/143). — [37] A. Rytz (At. Data Nucl. Data Tables **23** [1979] 507/33). — [38] G.T. Ewan, E. Hagberg, B. Jonson, S. Mattson, P. Tidemand-Petersson (Z. Physik A **296** [1980] 223/8). — [39] M.R. Schmorak (Nucl. Data Sheets **31** [1980] 283/380). — [40] K. Takahashi, M. Yamada, T. Kondoh (At. Data Nucl. Data Tables **12** [1973] 101/42).

[41] B. Harmatz (Nucl. Data Sheets **34** [1981] 101/82). — [42] R.L. Auble (Nucl. Data Sheets **40** [1983] 301/83). — [43] A.H. Wapstra, G. Audi, K. Bos (private communication 1982 from [42]). — [44] T.D. Thomas, G.E. Gordon, R.M. Latimer, G.T. Seaborg (Phys. Rev. [2] **126** [1962] 1805/10). — [45] P. Hornshoj, K. Wilsky, P.G. Hansen, A. Lindahl, O.B. Nielsen (Nucl. Phys. A **163** [1971] 277/88).

[46] J.M. Dairiki (Diss. Univ. California 1970, pp. 1/159; UCRL-20412 [1970] 1/159; N.S.A. **25** [1971] No. 25730). — [47] W. Forsling, T. Alväger, L.W. Holm, O. Melin, J. Uhler, B. Aström (Arkiv Fysik **19** [1961] 83/98). — [48] R.M. Latimer, G.E. Gordon, T.D. Thomas (J. Inorg. Nucl. Chem. **17** [1961] 1/5). — [49] M.R. Schmorak (Nucl. Data Sheets **25** [1978] 193/234). — [50] T. Morek, R. Broda, V. Walus, C. Droste, W. Neubert, S. Chojnacki (JINR-P6-4868 [1969] 1/28; N.S.A. **24** [1970] No. 28772).

[51] N.A. Golovkov, R.B. Ivanow, Y.V. Norseev, So-Ki Kvan, V.A. Khalkin, V.G. Chumin (Intern. Conf. Nucl. Struct., Dubna 1968, p. 54). — [52] W.E. Burcham, B.C. Haywood (Proc. Phys. Soc. [London] A **69** [1956] 862/5). — [53] P.E. Thoresen, F. Asaro, I. Perlman (J. Inorg. Nucl. Chem. **26** [1964] 1341/7). — [54] J.F. Miller, J.G. Hamilton, T.M. Purnam, H.R. Haymond, G.B. Rossi (Phys. Rev. [2] **80** [1950] 486). — [55] B. Jonson, M. Alpsten, A. Appelqvist, G. Astner (Nucl. Phys. A **174** [1971] 225/50).

[56] M.R. Schmorak (Nucl. Data Sheets **27** [1979] 581/635). — [57] P. Gippner, K.-H. Kaun, W. Neubert, W. Schulze, F. Stary (Nucl. Phys. A **237** [1975] 142/8). — [58] R. Broda, S. Chojnacki, C. Droste, T. Morek, V. Walus (JINR-E6-5197 [1970] 1/11; N.S.A. **24** [1970] No. 52559). — [59] U. Hagemann, W. Neubert, W. Schulze (Nucl. Phys. A **175** [1971] 428/32). — [60] A. Hashizume, Y. Tendow, T. Katou, H. Kumagai (IPCR [Inst. Phys. Chem. Res.] Cyclotron Progr. Rept. **7** [1973] 79).

[61] J. Halperin (Nucl. Data Sheets **24** [1978] 57/116). — [62] K. Dybdal, T. Chapuran, D.B. Fossan, W.F. Piel Jr., D. Horn, E.K. Warburton (Phys. Rev. [3] C **28** [1983] 1171/80; Bull. Am. Phys. Soc. **26** [1981] 621). — [63] O. Hausser, T.K. Alexander, J.R. Beene, E.D. Earle, A.B. McDonald, F.C. Khanna, I.S. Towner (Nucl. Phys. A **273** [1976] 253/61). — [64] B.G. Ritchie, K.S. Toth, H.K. Carter, R.L. Mlekodaj, E.H. Spejewski (Phys. Rev. [3] C **23** [1981] 2342/4). — [65] K.D. Servier (At. Data Nucl. Data Tables **24** [1979] 323/71).

[66] F. Roesel, H.M. Fries, K. Alder, H.C. Pauli (At. Data Nucl. Data Tables **21** [1978] 291/514). — [67] L.A. Sliv, I.M. Band (Gamma Rays, Izd. Nauka, Moscow 1961), K. Siegbahn (Alpha-Beta- and Gamma-Ray Spectroscopy, Vol. 2, North-Holland, Amsterdam 1965, pp. 1639/72). — [68] L.A. Sliv, I.M. Band (Coefficients of Internal Conversion of Gamma Radiation, Pt. 1: K-Shell, Pt. 2: L-Shell, Izd. Nauka, Moscow 1958/65). — [69] R.S. Hager, E.C.

Seltzer (Nucl. Data Tables A **4** [1968] 1/235, 397/641; Nucl. Data Tables **6** [1969] 1/127). – [70] O. Dragoun, H.C. Pauli, F. Schmutzler (Nucl. Data Tables **6** [1969] 235/351).

[71] I.M. Band, M.B. Trzhaskovskaya, M.A. Listengarten (At. Data Nucl. Data Tables **18** [1976] 433/57, **21** [1978] 1/48). – [72] V.F. Tursov (Nucl. Data Tables **10** [1971/72] 477/510). – [73] O. Dragoun, Z. Plajner, F. Schmutzler (Nucl. Data Tables **9** [1971] 119/33; At. Data Nucl. Data Tables **15** [1975] 49/56). – [74] W. Bambynek, B. Crasemann, R.W. Fink, H.U. Freund, H. Mark, C.D. Swift, R.E. Price, P.V. Rao (Rev. Mod. Phys. **44** [1972] 716/813). – [75] F.P. Larkins (At. Data Nucl. Data Tables **20** [1977] 311/87, **23** [1979] 587/8).

[76] J.A. Bearden, A.F. Burr (Rev. Mod. Phys. **39** [1967] 78/142). – [77] S.I. Salem, S.L. Panossian, R.A. Krause (At. Data Nucl. Data Tables **14** [1974] 91/109). – [78] M.P.M. Ramos, L.L. Riedinger, C.R. Bingham, M.W. Guidry, R.W. Lide, J.A. Vrba, J.L. Wood, R.W. Fink, E.H. Spejewski, R.L. Mlekodaj, H.K. Carter (Bull. Am. Phys. Soc. [2] **23** [1978] 946). – [79] W. Seelmann-Eggebert, G. Pfennig, H. Münzel, H. Klewe-Nebenius (Karlsruhe Chart of Nuclides, 5th Ed., KFK, Karlsruhe 1981). – [80] W.B. Ewbank, R.L. Haese, F.W. Hurley, M.R. McGinnis (Nucl. Data Sheets **24** [1978] 445/733).

[81] W.B. Ewbank, R.L. Haese, F.W. Hurley, M.R. McGinnis (Nucl. Data Sheets **16** [1975] Suppl.). – [82] H. Münzel (Gmelin Handbook "Uranium" Suppl. Vol. A2, 1980). – [83] H. Münzel (Gmelin Handbook "Francium" 1983, pp. 16/71). – [84] M.Ya. Kuznetsova, V.B. Brudanin, V.V. Kuznetsov, M. Milanov, Yu.V. Norseev, R.R. Usmanov, V.G. Chumin, Yu.V. Yushkevich (Izv. Akad. Nauk SSSR Ser. Fiz. **46** [1982] 2217/22; Bull. Acad. Sci. USSR. Phys. Ser. **46** No. 11 [1982] 153/8; JINR-P6-82-88 [1982]). – [85] M.Ya. Kuznetsova, V.B. Brudanin, V.G. Chumin (Izv. Akad. Nauk SSSR Ser. Fiz. **46** [1982] 2223/9; Bull. Acad. Sci. USSR Phys. Ser. **46** No. 11 [1982] 159/65; JINR-P6-82-123 [1982]).

[86] M.R. Schmorak (Nucl. Data Sheets **24** [1978] 117/73). – [87] T. Morek, W. Neubert, S. Chojnacki, K. Alexander, Z. Wilhelmi (JINR-P6-4494 [1969] 1/11; N.S.A. **23** [1969] No. 44884). – [88] P.B. Semmes, R.W. Fink (Bull. Am. Phys. Soc. [2] **27** [1982] 705, BC10). – [89] R.G. Helmer, C.W. Reich (Phys. Rev. [3] C **27** [1983] 2248/60). – [90] V.M. Vakhtel, N.A. Golokov, R.B. Ivanov, M.A. Mikhailova, A.F. Novogorodov, Yu.V. Norseev, V.G. Chumin, Yu.V. Yushkevich (Izv. Akad. Nauk SSSR Ser. Fiz. **45** [1981] 1861/4; Bull. Acad. Sci. USSR Phys. Ser. **45** No. 10 [1981] 46/9).

[91] V.M. Vakhtel, N.A. Golovkov, R.B. Ivanov, M.A. Mikhailova, V.G. Chumin (Izv. Akad. Nauk SSSR Ser. Fiz. **45** [1981] 1966/75; Bull. Acad. Sci. USSR Phys. Ser. **45** No. 10 [1981] 144/53). – [92] M.R. Schmorak (Nucl. Data Sheets **23** [1978] 287/352). – [93] W.E. Burcham (Proc. Phys. Soc. [London] A **67** [1954] 555/9, 733/4). – [94] U. Hagemann, W. Neubert, W. Schulze, F. Stary (Nucl. Phys. A **181** [1972] 145/52). – [95] P.K. Hopke, R.A. Naumann, E.H. Spejewski (Nucl. Phys. A **149** [1970] 63/4).

[96] T.P. Sjoreen, D.B. Fossan, U. Garg, A. Neskakis, A.R. Poletti, E.K. Warburton (Phys. Rev. [3] C **25** [1982] 889/98). – [97] T.P. Sjoreen (Diss. State Univ. N.Y. 1978, pp. 1/172; Diss. Abstr. Intern. B **39** [1978] 2376). – [98] A. Zelinsky, K. Zuber, J. Zuber, V.V. Kuznetsov, A. Kolaczkowskii, A. Latuszinski, Yu.V. Norseev, H.G. Ortlepp, I. Penev, A.W. Potempa (JINR-P6-8929 [1975] 1/20; N.S.A. **33** [1970] No. 2150; C.A. **84** [1976] No. 156694). – [99] A.W. Stoner (Diss. Univ. California 1956; UCRL-3471 [1956]). – [100] R.D. Griffioen, R.D. Macfarlane (Phys. Rev. [2] B **133** [1964] 1373/80).

[101] M.P. Webb (Nucl. Data Sheets **26** [1979] 145/205). – [102] E.W.A. Lingeman (Phys. Scr. **15** [1977] 205/12). – [103] R.W. Hoff, F. Asaro, I. Perlman (Bull. Am. Phys. Soc. [2] **4** [1959] 293/3). – [104] C.L. Duke, P.G. Hansen, O.B. Nielsen, G. Rudstam (Nucl. Phys. A **151** [1970] 609/33). – [105] V.B. Brudanin, N.A. Golovkov, B.S. Dzhelepov, R.B. Ivanov,

M.A. Mikhailova, A.V. Saulsky (Izv. Akad. Nauk SSSR Ser. Fiz. **46** [1982] 45/51; Bull. Acad. Sci. USSR Phys. Ser. **46** No. 1 [1982] 34/7).

[106] A.W. Stoner, E.K. Hyde (J. Inorg. Nucl. Chem. **4** [1957] 77/82). – [107] M.R. Schmorak (Nucl. Data Sheets **22** [1977] 487/544). – [108] B. Jonson, M. Alpsten, A. Appelqvist, G. Astner (Nucl. Phys. A **177** [1971] 81/102). – [109] N.A. Golovkov, Sh. Guetkh, B.S. Dzhelepov, Yu.V. Norseev, V.A. Khalkin, V.G. Chumin (Izv. Akad. Nauk SSSR Ser. Fiz. **33** [1969] 1622/30; Bull. Acad. Sci. USSR Phys. Ser. **33** [1970] 1489/96). – [110] D.H. Templeton, A. Ghiorso, I. Perlman (unpublished data from [29]).

[111] T. Kempisty, T. Morek, L.K. Peker, K. Petrozolin, S. Chojnacki (JINR-P6-5878 [1971] 1/10; N.S.A. **25** [1971] No. 50719). – [112] E.W.A. Lingeman, J.H.S. Andringa (IKO Progr. Rept. **1971/72** 60). – [113] E.K. Hyde, A. Ghiorso, G.T. Seaborg (Phys. Rev. [2] **77** [1950] 765/70). – [114] W.J. Treytl, E.K. Hyde, T. Yamazaki (Nucl. Phys. A **117** [1968] 481/508). – [115] J. Uhler, W. Forsling, B. Aström (Arkiv Fysik **24** [1963] 421/8).

[116] M.B. Lewis (Nucl. Data Sheets B **5** [1971] 243/83). – [117] N.A. Golovkov, R.B. Ivanov, Yu.V. Norseev, V.G. Chumin (JINR-R6-4615 [1969] 1/9; N.S.A. **24** [1970] No. 1821). – [118] J.P. Hummel (Diss. Univ. California 1956, pp. 1/161; UCRL-3456 [1956] 1/161; N.S.A. **12** [1958] No. 10033). – [119] J.W. Mihelich, A.W. Schardt, E. Segrè (Phys. Rev. [2] **95** [1954] 1508/16). – [120] M. Alpsten, A. Appelqvist, G. Alstner (Phys. Scr. **4** [1971] 137/50).

[121] L.J. Jardine, S.G. Prussin, J.M. Hollander (Nucl. Phys. A **233** [1974] 25/47). – [122] P.D. Croft, J.M. Alexander, K. Street Jr. (Phys. Rev. [2] **165** [1968] 1380/90). – [123] V.P. Afanasev, Ts. Vylov, N.A. Golovkov, I.I. Gromoya, A. Kolaczkowski, M.Ya. Kuznetsova, Yu.V. Norseev, M.I. Fominykh, V.I. Fominykh, V.G. Chumin (Izv. Akad. Nauk. SSSR Ser. Fiz. **37** [1973] 25/31; Bull. Acad. Sci. USSR Phys. Ser. **37** No. 1 [1973] 22/5). – [124] V.P. Afanasev, Ts. Vylov, N.A. Golovkov, I.I. Gromova, A. Kolaczkowski, M.Ya. Kuznetsova, Yu.V. Norseev, V.I. Fominykh, M.I. Fominykh, V.G. Chumin (Izv. Akad. Nauk SSSR Ser. Fiz. **37** [1973] 32/7; Bull. Acad. Sci. USSR Phys. Ser. **37** No. 1 [1973] 26/30). – [125] A. Charvet, R. Chery, Do Huu Phuoc, R. Duffait, A. Emsallem, G. Marguier (Compt. Rend. B **274** [1972] 290/3).

[126] I. Bergstrom, C.J. Herrlander, Th. Lindblad, V. Rahkonen, K.G. Renstelt, K. Westenberger (Z. Physik A **273** [1975] 291/304). – [127] T.P. Sjoreen, G. Schatz, S.K. Bhattacherjee, B.A. Brown, D.B. Fossan, P.M.S. Lesser (Phys. Rev. [3] C **14** [1976] 1023/32). – [128] T. Vylov, N.A. Golovkov, K.Ya. Gromov, I. Gromova, A. Kolachkovski, M.Ya. Kuznetsova, Yu.V. Norseev, V.G. Chumin (Izv. Akad. Nauk SSSR Ser. Fiz. **38** [1974] 701/9; Bull. Acad. Sci. USSR Phys. Ser. **38** No. 4 [1974] 31/7). – [129] B. Jonson, M. Alpsten, A. Appelqvist, B. Bengtsson, K.A. Johansson (Phys. Scr. **8** [1973] 142/8). – [130] T.H. Lindblad, C.G. Linden (Nucl. Instr. Methods **126** [1975] 397/406).

[131] M.J. Martin (Nucl. Data Sheets **22** [1977] 545/623). – [132] E.L. Kelly, E. Segrè (Phys. Rev. [2] **75** [1949] 999/1005). – [133] R.W. Hoff (Diss. Univ. California 1953, pp. 1/88; UCRL-2325 [1953] 1/88; N.S.A. **8** [1954] No. 408). – [134] L.J. Jardine, A.A. Shihab-Eldin (Nucl. Phys. A **244** [1975] 34/44). – [135] L.J. Jardine, S.G. Prussin, J.M. Hollander (Nucl. Phys. A **190** [1972] 261/83).

[136] R.W. Hoff, J.M. Hollander (Phys. Rev. [2] **109** [1958] 447/56). – [137] F. Schima, E.G. Funk Jr., J.W. Mihelich (Phys. Rev. [2] **132** [1963] 2650/6). – [138] S.G. Prussin, J.M. Hollander (Nucl. Phys. A **110** [1968] 176/92). – [139] K. Wikstrom, B. Fant, A. Filevich, K.G. Rensfelt, J. Sztarkier (Phys. Scr. **5** [1972] 126/8). – [140] B. Jonson, M. Alpsten, A. Appelqvist, B. Bengtsson, K.A. Johansson (Phys. Scr. **7** [1973] 147/59).

[141] D.F. Torgerson, R.D. Macfarlane (Phys. Rev. [3] C **2** [1970] 2309/18). – [142] B. Harmatz (Nucl. Data Sheets **34** [1981] 735/807). – [143] V. Rahkonen, I. Bergstrom, J. Blom-

qvist, O. Knuuttila, K.-G. Rensfelt, J. Sztarkier, K. Westenberg (Z. Physik A **284** [1978] 357/69). — [144] V. Rahkonen (Diss. Univ. Jyvaskyla, Finland, 1980, pp. 1/120; JYFL-10-1980 [1980] 1/120; INIS Atomindex **12** [1981] No. 594481). — [145] D.F. Torgerson, R.A. Gough, R.D. Macfarlane (Phys. Rev. [2] **174** [1968] 1494/9).

[146] Ts. Vylov, N.A. Golovkov, M. Gonusek, M.Ya. Kuznetsova, Yu.V. Norseev, H.-G. Ortlepp, V.G. Chumin (Izv. Akad. Nauk SSSR Ser. Fiz. **42** [1978] 765/72; Bull. Acad. Sci. USSR Phys. Ser. **42** No. 4 [1978] 64/9). — [147] S. Nagamiya, O. Hashimoto, Y. Yamazaki (Phys. Scr. **8** [1973] 95/8). — [148] V. Berg, C. Bourgeois, N. Perrin (Compt. Rend. B **276** [1973] 761/4). — [149] E.H. Appelman (Phys. Rev. [2] **121** [1961] 253/5). — [150] D.R. Corson, K.R. MacKenzie, E. Segrè (Phys. Rev. [2] **58** [1940] 672/8, **57** [1940] 459/9, 1087/7).

[151] P.D. Croft (Diss. Univ. California 1964, pp. 1/90; UCRL-11563 [1964] 1/90; N.S.A. **18** [1964] No. 34782). — [152] H.L. Garvin, T.M. Green, E. Lipworth, W.A. Nierenberg (Phys. Rev. Letters **1** [1958] 74/5). — [153] P.R. Gray (Phys. Rev. [2] **101** [1956] 1306/14). — [154] H.M. Neumann, I. Perlman (Phys. Rev. [2] **81** [1951] 958/62). — [155] G. Astner (Phys. Scr. **5** [1972] 31/40).

[156] G. Astner, V. Berg (Phys. Scr. **5** [1972] 55/7). — [157] K.H. Maier, J.R. Leigh, F. Pühlhofer, R.M. Diamond (Phys. Letters B **35** [1971] 401/4). — [158] I. Bergstrom, B. Fant, C.J. Herrlander, K. Wikström, J. Blomqvist (Phys. Scr. **1** [1970] 243/5). — [159] I. Bergstrom, B. Fant, C.J. Herrlander, P. Thieberger, K. Wikström (Phys. Letters B **32** [1970] 476/8). — [160] K.H. Maier, J.R. Leigh, F. Pühlhofer, R.M. Diamond (J. Phys. Colloq. [Paris] **32** [1971] C6-221/C6-222).

[161] L.J. Jardine (Phys. Rev. [3] C **11** [1975] 1385/91). — [162] W.J. Ramler, J. Wing, D.J. Henderson, J.R. Huizenga (Phys. Rev. [2] **114** [1959] 154/62). — [163] O. Hausser, T.K. Alexander, J.R. Beene, E.D. Earle, A.B. McDonald, F.C. Khanna, I.S. Towner (Hyperfine Interact. **2** [1976] 334/5). — [164] M.J. Martin (Nucl. Data Sheets **25** [1978] 397/432). — [165] M. Yanokura, H. Kudo, H. Nakahara, K. Miyano, S. Ohya, O. Nitoh (Nucl. Phys. A **299** [1978] 92/8).

[166] H. Ingwersen, W. Klinger, G. Schatz, W. Witthuhn (Phys. Rev. [3] C **11** [1975] 243/52). — [167] J.C. Ritter, W.G. Smith (Phys. Rev. [2] **128** [1962] 1778/82). — [168] M.M. Winn (Proc. Phys. Soc. [London] A **67** [1954] 949/50). — [169] W.B. Jones (Phys. Rev. [2] **130** [1963] 2042/3). — [170] P.L. Reeder (Phys. Rev. [3] C **1** [1970] 721/6).

[171] A.R. Barnett, J.S. Lilley (Phys. Rev. C **9** [1974] 2010/27). — [172] K. Valli, E.K. Hyde (Phys. Rev. [2] **176** [1968] 1377/89). — [173] M.J. Martin (Nucl. Data Sheets **27** [1979] 637/79). — [174] T.P. Sjoreen, U. Garg, D.B. Fossan, J.R. Beene, T.K. Alexander, E.D. Earle, O. Hausser, A.B. McDonald (Phys. Rev. [2] C **20** [1979] 960/8). — [175] K. Fransson, M. af Ugglas, P. Carle (Ann. Rept. Res. Inst. Phys. [Sweden] **1974/75** 147/7 from [29, 173]).

[176] J.D. Keys (Diss. McGill Univ., Montreal 1951). — [177] J. Borggreen, K. Valli, E.K. Hyde (Phys. Rev. [3] C **2** [1970] 1841/62). — [178] R.L. Hahn, M.F. Roche, K.S. Toth (Nucl. Phys. A **113** [1968] 206/14). — [179] W.W. Meinke, A. Ghiorso, G.T. Seaborg (Phys. Rev. [2] **75** [1949] 314/5). — [180] W.W. Meinke, A. Ghiorso, G.T. Seaborg (Phys. Rev. [2] **81** [1951] 782/98).

[181] J.D. McCoy (Soc. Sci. Fennica Commentationes Phys. Math. **30** No. 4 [1964] 1/37; N.S.A. **19** [1965] No. 21418; C.A. **62** [1965] 7320). — [182] K. Eskola (Phys. Rev. [3] C **5** [1972] 942/7). — [183] A. Ghiorso, W.W. Meinke, G.T. Seaborg (Phys. Rev. [2] **74** [1948] 695/6). — [184] B. Karlik, T. Bernert (Naturwissenschaften **32** [1944] 44). — [185] R.D. Griffion, R.D. Macfarlane (UCRL-10023 [1961] 50/1; N.S.A. **16** [1962] No. 20305).

[186] T. Nomura, K. Hiruta, M. Yoshie, O. Hashimoto (Phys. Rev. [3] C **9** [1974] 1168/73). – [187] B. Karlik, T. Bernert (Naturwissenschaften **31** [1943] 492). – [188] J.P. Brian, P. Chevallier (Compt. Rend. **260** [1965] 5251/4). – [189] J.P. Briand, M. Lefort (Phys. Letters **10** [1964] 90/2). – [190] J.P. Briand, P. Chevallier, A. Touati (Compt. Rend. B **271** [1970] 162/4).

[191] G. Bastin, C.F. Leang, R.J. Walen (J. Phys. Colloq. [Paris] **29** [1968] C1-181/C1-182). – [192] J.D. Bowman, E.K. Hyde, R.E. Eppley (LBL-1666 [1973] 4/6; N.S.A. **28** [1973] No. 22964, No. 22965). – [193] J.P. Briand, P. Chevallier, A. Touati (J. Phys. [Paris] **32** [1971] 101/5). – [194] H. Diamond, J.E. Gindler (J. Inorg. Nucl. Chem. **25** [1963] 143/64). – [195] F. Hagemann, L.I. Katzin, M.H. Studier, A. Ghiorso, G.T. Seaborg (Phys. Rev. [2] **72** [1947] 252).

[196] F. Hagemann, L.I. Katzin, M.H. Studier, G.T. Seaborg, A. Ghiorso (Phys. Rev. [2] **79** [1950] 435/43). – [197] A.C. English, T.E. Cranshaw, P. Demers, J.A. Harvey, E.P. Hincks, J.V. Jelley, A.N. May (Phys. Rev. [2] **72** [1947] 253/4). – [198] R.J. Walen (Compt. Rend. **255** [1962] 1604/5). – [199] K. Valli (Ann. Acad. Sci. Fennicae A VI No. 165 [1964] 1/42; N.S.A. **19** [1965] No. 16862). – [200] A.A. Vorobev, A.P. Komar, V.A. Korolev (Zh. Eksperim. Teor. Fiz. **39** [1960] 70/2; Soviet Phys.-JETP **12** [1961] 50/1).

[201] F.S. Stephens Jr. (Diss. Univ. California 1955, pp. 1/178; UCRL-2970 [1955] 1/178; N.S.A. **9** [1955] No. 4863). – [202] T.E. Cranshaw, J.A. Harvey (Can. J. Res. A **26** [1948] 243). – [203] C.F. Leang (Diss. Univ. Paris 1969, pp. 1/95; NP-17967 [1969] 1/95; N.S.A. **23** [1969] No. 46988; C.A. **72** [1970] No. 95633). – [204] B.S. Dzhelepov, R.B. Ivanov, M.A. Mikhailova, L.N. Moskvin, O.M. Nazarenko, V.F. Rodionov (Izv. Akad. Nauk SSSR Ser. Fiz. **31** [1967] 568/80; Bull. Acad. Sci. USSR Phys. Ser. **31** [1967] 563/74). – [205] G.D. Alkhazov, Yu.K. Zalite, M.L. Andersen, O.B. Nielsen (Pis'ma Zh. Eksperim. Teor. Fiz. **12** [1970] 7/9; JETP Letters **12** [1970] 4/5).

[206] D. Strominger, F.S. Stephens Jr., J.O. Rasmussen (Phys. Rev. [2] **103** [1956] 748/50). – [207] R.J. Walen (Compt. Rend. **227** [1948] 1090; J. Phys. Radium **10** [1949] 95). – [208] B. Karlik, T. Bernert (Naturwissenschaften **31** [1943] 298). – [209] E.K. Hyde, A. Ghiorso (Phys. Rev. [2] **90** [1953] 267/70). – [210] W.W. Daehnick, M.J. Spisak, R.M. Del Vecchio (Phys. Rev. [3] C **14** [1976] 2011/2).

[211] W.G. Davies, R.M. Devries, G.C. Ball, J.S. Forster, W. McLatchie, D. Shapira, J. Toke, R.E. Warner (Nucl. Phys. A **269** [1976] 477/92). – [212] C.E. Beneis Jr., P.F. Dittner, R.J. Silva, R.L. Hahn, J.R. Tarrant, L.D. Hunt, D.C. Hensley (Phys. Rev. [3] C **16** [1977] 1146/58). – [213] V.G. Chumin, M.P. Avotina, V. Varsharosh, I.I. Gromova, M.Ya. Kuznetsova, V.V. Kuznetsov, G.I. Lizurei, M. Milanov, Yu.V. Norseev, R. Usmanov, Yu.V. Yushkevich, A.F. Novgorodov, A. Abdumalikov (Izv. Akad. Nauk SSSR Ser. Fiz. **45** [1981] 1865/73; Bull. Acad. Sci. USSR Phys. Ser. **45** No. 10 [1981] 50/7; JINR-P6-12615 [1979]). – [214] T.P. Sjoreen, U. Garg, D.B. Fossan (Phys. Rev. [3] C **23** [1981] 272/80). – [215] T.P. Sjoreen, U. Garg, D.B. Fossan (Phys. Letters B **76** [1978] 397/9).

[216] V.G. Chumin, M.P. Avotina, Z. Gons, L.A. Kalmykova, M.Ya. Kuznetsova (Izv. Akad. Nauk SSSR Ser. Fiz. **45** [1981] 2102/6; Bull. Acad. Sci. USSR Phys. Ser. **45** No. 11 [1981] 71/4). – [217] L. Vasharosh, T. Vylov, N.A. Golovkov, I.I. Gromova, B.S. Dzhelepov, M.Ya. Kuznetsova, G.I. Lizurei, Yu.V. Norseev, I.I. Popova, V.P. Prikhodseva, R.R. Usmanov, V.G. Chumin, Ya.V. Yushkevich (Izv. Akad. Nauk SSSR Ser. Fiz. **43** [1979] 71/7; Bull. Acad. Sci. USSR Phys. Ser. **43** No. 1 [1979] 57/62). – [218] O. Dragoun, V. Brabec, A. Mastalka, A. Vovalik, M. Rysavy, M.Ya. Kuznetsova, Yu.V. Norseev (Nucl. Phys. A **391** [1982] 29/35). – [219] O. Dragoun, V. Brabec, A. Mastalka, A. Kovalik, M.Ya. Kuznetsova, Yu.V. Norseev (Czech. J. Phys. B **32** [1982] 711/2). – [220] V.M. Vakhtel, Ts. Vylov, N.A. Golovkov, B.S. Dzhelepov, A. Karakhodzhaev, V.V. Kuznetsov, M.Ya. Kuznetsova, M. Milanov, Yu.V. Nor-

seev, T.I. Popova, V.P. Prikhodtseva, V.G. Chumin, Yu.V. Yushkevich (Izv. Akad. Nauk SSSR Ser. Fiz. **45** [1981] 1841/7; Bull. Acad. Sci. USSR Phys. Ser. **45** No. 10 [1981] 29/33; JINR-P6-81-172 [1981]).

[221] V. Rahkonen, J. Hattula (Ann. Rept. Univ. Jyvaskyla [Finland] **1978/79** 3; JYFL-AR-1978 [1979] 3). — [222] B.S. Dzhelepov, M.Ya. Kuznetsova, T.I. Popova, V.P. Prikhodtseva, V.G. Chumin (Izv. Akad. Nauk SSSR Ser. Fiz. **47** [1983] 2/10; Bull. Acad. Sci. USSR Phys. Ser. **47** No. 1 [1983] 1/8; JINR-P6-81-173 [1981]). — [223] H.-E. Mahnke, W. Semmler, H. Grawe, H. Haas, R. Sielemann (Phys. Letters B **122** [1983] 27/30). — [224] C.L. Morris, H.A. Thiessen, W.J. Braithwaite, W.B. Cottingame, S.J. Greene, D.B. Holtkamp, I.B. Moore, C.F. Moore, G.R. Burleson, G.S. Blanpied, G.H. Daw, A.J. Viescas (Phys. Rev. Letters **45** [1980] 1233/4). — [225] J. Adam, A. Kuklik, A. Spalek, D. Venos, M.J. Kuznetsova (Czech. J. Phys. B **28** [1978] 857/64).

[226] L. Gueth, S. Gueth, E. Daroczy, B.S. Dzhelepov, Yu.V. Norseev, V.A. Khalkin (JINR-P6-4079 [1968] 1/13; N.S.A. **23** [1969] No. 8976). — [227] M.I. Babenkov, B.V. Bobykin, V.K. Petukhov (Izv. Akad. Nauk SSSR Ser. Fiz. **34** [1970] 1686/90; Bull. Acad. Sci. USSR Phys. Ser. **34** [1971] 1498/500). — [228] M.I. Babenkov, B.V. Bobykin, V.S. Zhdanov, V.K. Petukhov (J. Phys. B **15** [1982] 927/31). — [229] T. Lönnroth, V. Rahkonen (J. Phys. G **8** [1982] L153/L157). — [230] K. Abrahamsson, A. Filevich, K.G. Rensfelt, J. Sztarkier (Ann. Rept. Res. Inst. Phys. [Sweden] **1970** 5 from [142]).

[231] I. Bergstrom, B. Fant, A. Filevich, G. Linden, K.G. Rensfelt, J. Sztarkier, K. Wikström (Ann. Rept. Res. Inst. Phys. [Sweden] **1970** 96 from [142]). — [232] J.D. Bowman, R.E. Eppley, E.K. Hyde (Phys. Rev. [2] C **25** [1982] 941/51). — [233] T. Lönnroth, V. Rahkonen, B. Fant (Nucl. Phys. A **376** [1982] 29/44). — [234] G. Giorginis, H. Bohn, T. Faestermann, F.V. Feilitzsch, P. Kienle, C. Kozhuharov, T. Yamazaki (Nucl. Instr. Methods **188** [1981] 535/48). — [235] T.P. Sjoreen, U. Garg, D.B. Fossan (Phys. Rev. [3] C **21** [1980] 1838/46).

[236] B. Fant, T. Weckström, K. Fransson, C.J. Herrlander, K. Honkanen, A. Källberg, C.G. Linden, T. Lönnroth (Proc. 4th Intern. Conf. Nuclei Far From Stability, Helsingoer, Denmark, 1981, Vol. 2, pp. 644/8; CERN-81-09-Vol. 2 [1981] 644/8). — [237] H. Bohn, E. Endres, T. Faestermann, P. Kienle (Z. Physik A **302** [1981] 51/9). — [238] Y.A. Ellis (Nucl. Data Sheets **28** [1979] 619/38). — [239] G.T. Ewan, E. Hagberg, B. Jonson, S. Mattson, P. Tidemand-Petersson (Nucl. Phys. A **380** [1982] 423/37). — [240] K.S. Toth (Nucl. Data Sheets **21** [1977] 437/66).

[241] C. Maples (Nucl. Data Sheets **22** [1977] 207/21). — [242] G. Graeffe, P. Kauranen (J. Inorg. Nucl. Chem. **28** [1966] 933/8). — [243] Y.A. Ellis (Nucl. Data Sheets **17** [1976] 329). — [244] Y.A. Ellis (Nucl. Data Sheets **28** [1979] 639/53). — [245] N. Schulz, A. Chevallier, J. Chevallier, S. Khazrouni, C. Kraus, I. Fink (Phys. Rev. [3] C **28** [1983] 435/6).

[246] B.S. Dzhelepov, A.V. Zolotavin, R.B. Ivanov, M.A. Mikhailova, V.O. Sergeev (Izv. Akad. Nauk SSSR Ser. Fiz. **33** [1969] 1607/21; Bull. Acad. Sci. USSR Phys. Ser. **33** [1970] 1475/81). — [247] B.S. Dzhelepov, R.B. Ivanov, M.A. Mikhailova, V.O. Sergeev (Izv. Akad. Nauk SSSR Ser. Fiz. **36** [1972] 2080/8; Bull. Acad. Sci. USSR Phys. Ser. **36** [1973] 1832/9). — [248] J.K. Dickens, J.W. McConnell (Radiochem. Radioanal. Letters **47** [1981] 331/44). — [249] R.J. Walen (J. Phys. Radium **10** [1949] 95). — [250] K.S. Toth (Nucl. Data Sheets **21** [1977] 467/78).

[251] C. Maples (Nucl. Data Sheets **22** [1977] 223/42). — [252] A. Hachem (Diss. Univ. Nice, France, 1975). — [253] J.D. Stickler, K.J. Hofstetter (Phys. Rev. [3] C **9** [1974] 1064/71). — [254] R. Bimbot, M.F. Rivet (Phys. Rev. [3] C **8** [1973] 375/9). — [255] H. Münzel (in: Gmelin Handbook "Actinium" Suppl. Vol. 1, 1981, pp. 32, 35).

[256] B.N. Belayev, Yun-Yu Van, E.N. Sinotova, L. Nyemet, V.A. Kalkin (Radiokhimiya **2** [1960] 603/13; Radiochem. [USSR] **2** [1960/61] 243). – [257] X. Tarrago (J. Phys. Radium [8] **23** [1962] 1/6). – [258] M. Lindner, R.N. Osborne (Phys. Rev. [2] **103** [1956] 378/85). – [259] J. Visser, G.A. Brinkman, C.N.M. Bakker (Intern. J. Appl. Radiat. Isotop. **30** [1979] 745/8). – [260] W.A. Chalkin, E. Herrmann, J.W. Norseev, I. Dreyer (Chemiker-Ztg. **101** [1977] 470/81).

[261] J.P. Didelez, R.M. Lieder, H. Beuscher, D.R. Hänni, H. Machner, M. Müller-Veggian, C. Mayer-Böricke (Nucl. Phys. A **341** [1980] 421/39). – [262] D. Gardès, R. Bimbot, J. Maison, L. de Reilhac, M.F. Rivet (Phys. Rev. [3] C **18** [1978] 1298/316). – [263] Jun-Sheng Guo, Xi-Jun Sun, Xiao-Ji Xu, Jun Wang, Hong-Ye Liu (Kao Neng Wu Li Yu Heh Wu Li **2** [1978] 143/50; C.A. **89** [1978] No. 96140). – [264] W. Schier, J. Chervenak, A.C. DiRienzo, H. Enge, D. Grogan, J. Molitoris, M. Salomaa, A. Sperduto (Phys. Rev. [3] C **23** [1981] 261/71). – [265] R. Bimbot, D. Gardes, M.F. Rivet (Phys. Rev. [3] C **4** [1971] 2180/4).

[266] I.V. Kuznetsov, N.S. Maltseva, Ts.Yu. Oganesyan, A.M. Sukhov, V.A. Shchegolev (Yadern. Fiz. **8** [1968] 448/53; Soviet J. Nucl. Phys. **8** [1968] 260/3). – [267] C.F. Leang, G. Bastin-Scoffier (Compt. Rend. B **266** [1968] 629/32). – [268] H.M. Neumann (J. Inorg. Nucl. Chem. **4** [1957] 349/54). – [269] E.H. Appelman, E.N. Szoth, M.H. Studier (Inorg. Chem. **5** [1966] 766/9). – [270] L. Lindner, G.A. Brinkmann, T.H. Suer, A. Schimmel, J.T. Veenboer, F.H.S. Karten, J. Visser, C.J. Leurs (Radiopharm. Labelled Compd. Proc. Symp., Copenhagen 1973, Vol. 1, pp. 303/16; C.A. **81** [1974] No. 162521).

4 Production, Isolation, and Purification of Astatine Isotopes

H.W. Kirby
Mound Facility
Monsanto Research Corporation
Miamisburg, Ohio, USA

When astatine is needed for chemical or medical research, it is most conveniently produced by the bombardment of bismuth with accelerated helium ions. This reaction yields At isotopes with half-lives measured in hours, whereas the longest-lived At isotope in nature is 0.9-min ^{219}At. Not surprisingly, ^{219}At is also the only naturally occurring At isotope ever isolated by chemical means.

4.1 Preparation of ^{219}At from Natural Sources

^{219}At results from the 0.005% α-branching of 22-min ^{223}Fr, which is produced by the 1.38% α-branching of 22-a ^{227}Ac (Fig. 2-2, p. 12). ^{219}At could, in principle, be prepared directly from natural sources; however, ^{227}Ac, the logical source material, is extremely difficult to prepare in pure form, uncontaminated by rare earths. (The most concentrated sample ever recovered from nature consisted of 4 mg of La_2O_3 containing 1.2% Ac [1].)

Therefore, the preferred source is pure Ac produced by the nuclear reaction: $^{226}Ra(n, \gamma)\,^{227}Ra \rightarrow \,^{227}Ac + \beta^-$, and isolated by ion exchange or solvent extraction [2, 3]; see also [4] and the literature cited therein.

In a procedure described by Hyde and Ghiorso [5], ^{223}Fr (AcK), the parent of ^{219}At, is first isolated from the ^{227}Ac by Hyde's silicotungstic acid (HSiW) method [6, 7]; the Ac is dissolved in 15 mL of ice-cold concentrated HCl which has previously been saturated with HCl gas. A small amount of insoluble material may appear, which is centrifuged down. A few drops of 0.4 M HSiW are added to the clear supernate with vigorous stirring. A white crystalline precipitate of HSiW appears and is centrifuged down. After decantation, the remaining drops of solution are carefully removed with a pipet. The supernate, which contains the Ac, is set aside so that a fresh sample of AcK may be prepared later simply by adding more HSiW. The precipitate is dissolved in 1.0 mL of redistilled H_2O. This solution is drawn by suction through a short column of Dowex 50 cation resin (10 × 4 mm; 250 to 500 mesh; NH_4^+ form; 0.5 mL/min). The Fr is strongly adsorbed, but the nonionic HSiW passes through the column. The Fr is quickly desorbed by 0.5 mL of concentrated HCl. The purified carrier-free Fr is evaporated on a Pt filament. A Pt disk is mounted 1 to 2 mm above this filament and an electric current is passed through the filament to warm it and volatilize the At to the collector disk. Since Fr, Pb and Bi will also volatilize if the temperature is sufficiently high, the current is set at a level (previously determined by test runs) just high enough to volatilize the At without concomitant volatilization of the other nuclides. By this method, At can be accumulated on the collector disk during a 15-second milking period and analysis of the α-spectrum can begin within one-half minute after the end of the milking period.

4.2 Preparation of At Isotopes from Irradiated Probes

4.2.1 Procedures for Separation

4.2.1.1 Helium Ion Irradiated Bi

Irradiation of metallic Bi or Bi_2O_3 with accelerated α-particles produces At isotopes by the reaction: $^{209}Bi(\alpha, xn)\,^{213-x}At$, where $x = 2$, 3 or 4, depending on the energy of the bombarding α-particles [8 to 12] (Fig. 3-26, p. 64). Typically, metallic Bi is fused or vaporized in a layer 0.1 to 100 mg/cm^2 thick onto an Al or Au substrate [8, 10], [13 to 15]. Alternatively, a strip of Bi metal may be clamped to a Cu block [9, 16]. In either case, to minimize loss of At by volatilization from the molten Bi target, the back of the target is cooled by water and its face by an atmosphere of flowing or static He. When Bi_2O_3 is used, it is usually pressed into small holes drilled in the face of a thick Al or Cu target [17]. In this case, melting of the target and loss of At by volatilization is less likely. Nevertheless, the α-particle beam is defocussed to avoid local hot spots [14]. The α-energies may be controlled by inserting blank Al foils between the target and the particle beam [8, 10]. Irradiation of a thick Bi target with 43-MeV α-particles for one hour at an intensity of 1 μA yields 4.0 mCi of ^{210}At and 3.1 mCi of ^{211}At [18].

The threshold α-energy for the $^{209}Bi(\alpha, 2n)\,^{211}At$ reaction is ~20 MeV. The cross-section for the reaction reaches a peak at ~31 MeV. However, for several reasons, it is often desirable to hold the energy of the bombarding α-particles to <28.4 MeV, the threshold of the $^{209}Bi(\alpha, 3n)\,^{210}At$ reaction, so as to obtain ^{211}At in an isotopically and radiochemically pure form. This permits analysis by gross α-counting, since, by virtue of the short half-life of ^{211}Po, every disintegration of ^{211}At results in the prompt emission of an α-particle:

^{211}At 7.215 h	—— ε 58.06% ——→	^{211}Po 0.516 s
α 41.94% ↓		α ↓
^{207}Bi 32.2 a	—— ε 99+% ——→	^{207}Pb stable

Of greater importance is the fact that, although 8.3-h ^{210}At decays almost entirely by electron capture and β^+-emission, its decay product is the α-emitter, 138.4-d ^{210}Po. Because of its toxicity, ^{210}Po is an unacceptable contaminant if the At is intended for biological or medical use [19]; see also [44].

Only a few measurements have been reported with respect to the $^{209}Bi(\alpha,4n)^{209}At$ reaction as a function of α-energy. Barton et al. [9] obtained a good yield of ^{209}At with >60-MeV α-particles. Ramler et al. [10] reported a cross-section of 391 mb with 43.3 MeV α-particles. However, neither group was able to determine the threshold for the reaction principally because the yield of ^{211}At increases rapidly with decreasing beam energies and the tail of its 5.87 MeV α-group obscures the 5.65 MeV α-group of the diminishing ^{209}At. The threshold of the (α, 4n) reaction is yet to be determined. (The 34 MeV threshold value attributed by [14] and [20] to [21] could not be verified.) Nevertheless, this remains the best method for preparing large amounts of ^{209}At when isotopic purity is not required [22]. (Negligibly small amounts of radiochemically and isotopically pure ^{209}At can be obtained as a decay product of 2.7-min ^{213}Ra, which is produced by the bombardment of a Th target with 340 MeV protons [23].)

The irradiation of Bi with α-particles produces At isotopes with almost no contamination by genetically unrelated radioelements. Table 4/1, column A shows the radioactive composition of a typical Bi target 7 h after a 6-h irradiation with 28 to 30 MeV α-particles at a current of ~20 μA. The principal impurities, ^{66}Ga, ^{67}Ga and ^{65}Zn, result from (α, xn) and (α, pn) reactions on the Cu backing and sample holder. The other impurities are generated by (α, pxn) and (α, αxn) reactions on ^{209}Bi [16].

Either "wet" or "dry" methods can be used to isolate At from an irradiated Bi target. In the dry method, At is separated from molten Bi by distillation in air [9], vacuum [15]

Table 4/1
Radioactive Composition of an α-Bombarded Bi Target Before and After Wet and Dry Distillation [16].

	disintegrations in % relative to ^{211}At		
	A	B	C
nuclide	before processing[a)]	product of wet distillation[b)]	product of dry distillation[b)]
^{211}At	100	100	100
^{210}At	<20	<20	<20
^{211}Po	58	58	58
^{210}Po	<0.5	<0.01	<0.01
^{209}Po	<0.1	<0.01	<0.01
^{207}Po	<1	<0.01	<0.01
^{207}Bi	<0.5	<0.01	<0.01
^{206}Bi	<0.5	<0.01	<0.01
^{67}Ga	<20	<0.01	<0.01
^{66}Ga	<20	<0.01	<0.01
^{65}Zn	<0.1	<0.01	<0.01
^{69}Ge	<0.1	<0.05	<0.01

a) Measured 7 h after the end of irradiation. –
b) Measured immediately after distillation, i.e., 10 h after end of irradiation.

or a stream of inert gas [12, 24]. According to [9], the Bi target is dropped into a stainless steel crucible fitted on top with a water-cooled steel finger to which a collecting Pt disk is clamped. Before the condenser is put in place, volatile impurities may be removed by heating to 260 °C. At begins to volatilize at 271 °C, the melting point of Bi [13, 25]. At this temperature, the distilled At is relatively pure but the yield is low. For a quantitative yield, it is necessary to heat the target to 800 °C, at which temperature Po and Bi are also distilled. The At must then be purified by redistillation [14, 26]; see also [16, 27, 28] and Table 4/1, column C.

Several methods have been described that use extraction or carrier precipitation to isolate and purify the At from α-irradiated Bi targets [9, 14], [17 to 19]. In the "extractive distillation" method of Meyer and Rössler [16], the target is dissolved in 1 mL of hot concentrated H_2SO_4 and the hot (150 °C) solution is repeatedly extracted with $CHCl_3$. The organic solvent distills vigorously and carries the At into a cold trap. The $CHCl_3$ solution, containing 70 to 75% of the At, is slowly added to an aqueous solution of 0.2 M Na_2SO_3 at 80 °C. As the $CHCl_3$ distills slowly through the aqueous layer, the At is reduced to At^- and remains 90% in the aqueous fraction [29]. For a comparison of the results of wet and dry distillation, see Table 4/1, columns B and C.

4.2.1.2 Proton Irradiated Pb, Bi, Th, and U

Astatine isotopes are also produced by secondary nuclear reactions and spallation reactions resulting from the bombardment of Pb, Bi, Th or U targets with high energy protons [20, 30 to 39]. In this case, however, isotopes of a large number of elements other than At are formed, necessitating the use of more elaborate separation procedures than those required for α-bombarded Bi targets.

The separation scheme used at the Joint Institute for Nuclear Research (Dubna, USSR) for the isolation of At from Th targets irradiated with 660 MeV protons is shown in **Fig.** 4-1 [30]; see also [20, 31, 33, 35, 36, 39, 40].

According to Lefort and co-workers [41], proton-irradiated Th is dissolved in concentrated HCl containing a trace of HF. The solution is made 3 M in HCl and 5 mg of Te is added as H_2TeO_4. The solution (volume 3 mL) is treated with 2 mL of 10% $SnCl_2$ in HCl and heated at 100 °C for 10 min. This precipitates the Te, which carries down the At, Po, and some fission products, such as Au and As^{III}. The precipitate is centrifuged, washed, and redissolved in the minimum volume of concentrated HNO_3. The solution is made up to 3 mL with concentrated HCl and heated to 100 °C. An equal volume of concentrated HCl saturated with SO_2 and $N_2H_4 \cdot HCl$ is added, precipitating the Te with 99% of the At. Po remains in solution. The precipitate is redissolved in a drop of concentrated HNO_3 and

Fig. 4-1

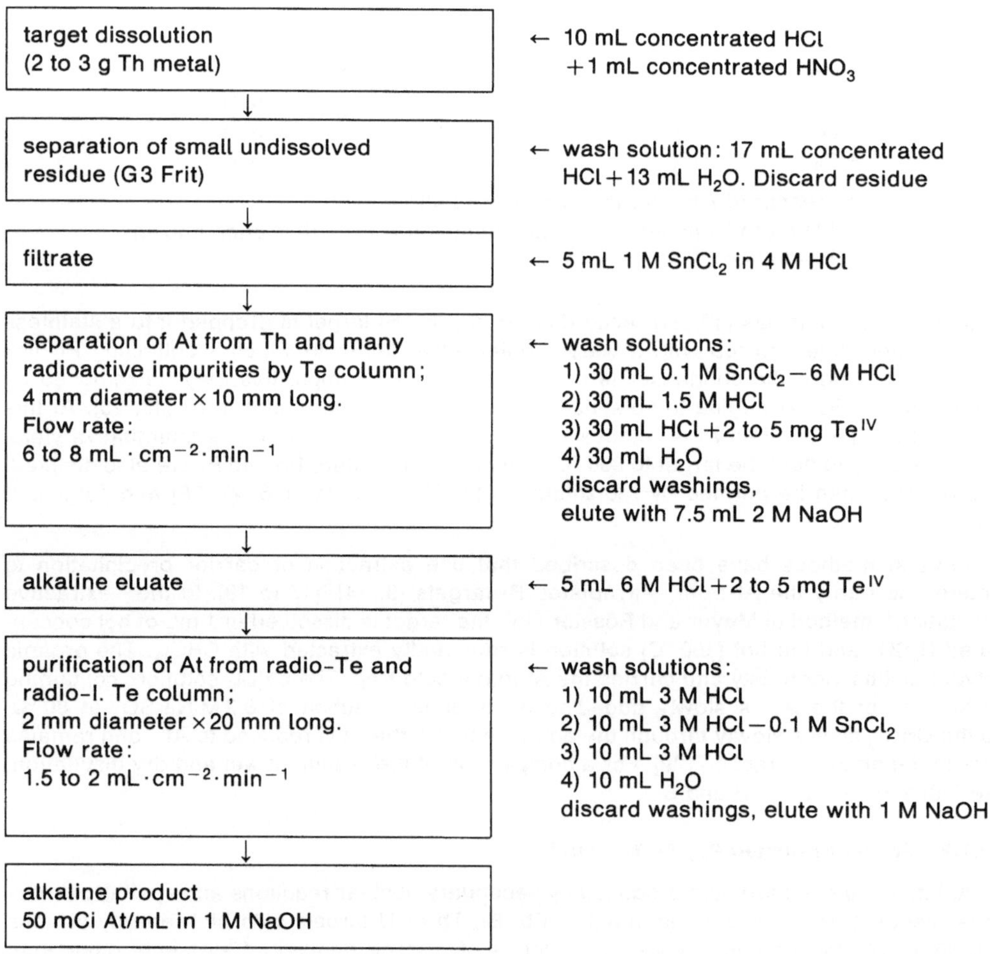

Scheme for Separation of Astatine from Thorium Targets Irradiated with 660-MeV Protons [30].

made alkaline with 6 M NaOH. Two mL of 5% Na stannite is added, precipitating the Te but not the At. After removal of the precipitate, NaF is added, the filtrate is acidified to 0.3 to 0.5 N with HNO_3, and At is deposited on Ag. The Ag is dissolved in HNO_3 and reprecipitated as AgCl, leaving At in the filtrate with an overall yield of 40 to 50%.

Alternatively, according to Merinis and Bouissières [42], the proton-irradiated Th target is introduced into a quartz tube which is heated by a furnace at one end and cooled at the other by flowing air or water. A plug of quartz wool acts as a filter between the two ends and a strip of Au, Pt or Ag on the cool side of the filter acts as a collector. A flow of O_2 is started, the furnace is heated to 1000 °C and the Th metal is quickly converted to ThO_2 powder. The At (mixed with I isotopes formed by Th fission) volatilizes, passes through the filter and collects on the first few mm of metal that it encounters.

In a gas-thermochromatographic method, developed by Vakhtel et al. [34], the quartz filter is replaced by one made of Ag foil ~50 μm thick with a surface area of 15 to 20 cm^2. This filter retains only a small fraction of the At but extracts I, Br, and Tc/Mo quantitatively from the gas stream. The method permits the isolation and purification in 10 min of 80% of the At from a 5-g U target irradiated with 660 MeV protons. Radioactive contamination is said to be <0.5%; cf. [43].

References:

[1] W.A. Lub (Compt. Rend. **204** [1937] 1417/8; Proc. Koninkl. Ned. Akad. Wetenschap. **40** [1937] 584/9; C.A. **1937** 4898, 8364). — [2] S. Peterson (in: G.T. Seaborg, J.J. Katz, W.M. Manning, The Transuranium Elements, McGraw-Hill, New York-Toronto-London 1949, pp. 1393/4). — [3] F. Hagemann (J. Am. Chem. Soc. **72** [1950] 768/71). — [4] H.W. Kirby (Gmelin Handbook "Actinium" Suppl. Vol. 1, 1981, pp. 83/95). — [5] E.K. Hyde, A. Ghiorso (Phys. Rev. [2] **90** [1953] 267/70).

[6] E.K. Hyde (J. Am. Chem. Soc. **74** [1952] 4181/4). — [7] H.W. Kirby (Gmelin Handbook "Francium" 1983, p. 75). — [8] E.L. Kelly, E. Segrè (Phys. Rev. [2] **75** [1949] 999/1005). — [9] G.W. Barton Jr., A. Ghiorso, I. Perlman (Phys. Rev. [2] **82** [1951] 13/9). — [10] W.J. Ramler, J. Wing, D.J. Henderson, J.R. Huizenga (Phys. Rev. [2] **114** [1959] 154/62).

[11] C. Aaij, W.R.J.M. Tschroots, L. Lindener, T.E.W. Feltkamp (Intern. J. Appl. Radiat. Isotop. **26** [1975] 25/30). — [12] V. Doberenz, D.G. Nhan, R. Dreyer, M. Milanov, Yu.V. Norseyev, V.A. Khalkin (Radiochem. Radioanal. Letters **52** [1982] 119/27). — [13] E.H. Appelman (UCRL-9025 [1960] 1/113; N.S.A. **14** [1960] No. 11501). — [14] E.H. Appelman (NAS-NS 3012 [1960] 1/29; N.S.A. **14** [1960] No. 20201). — [15] G.L. Johnson, R.F. Leininger, E. Segrè (J. Chem. Phys. **17** [1949] 1/10).

[16] G.-J. Meyer, K. Rössler (Radiochem. Radioanal. Letters **25** [1976] 377/90). — [17] A.H.W. Aten Jr., T. Doorgeest, U. Hollstein, H.P. Moeken (Analyst [London] **77** [1952] 774/7). — [18] H.M. Neumann (J. Inorg. Nucl. Chem. **4** [1957] 349/53). — [19] R.D. Neirinckx, J.A. Smit (Anal. Chim. Acta **63** [1973] 201/4). — [20] W.A. Chalkin [V.A. Khalkin], E. Herrmann, J.V. Norseev, I. Dreyer (Chemiker-Ztg. **101** [1977] 470/81).

[21] W. John Jr. (Phys. Rev. [2] **103** [1956] 704/13). — [22] E.K. Hyde, I. Perlman, G.T. Seaborg (The Nuclear Properties of the Heavy Elements, Prentice-Hall, Englewood Cliffs, N.J., 1964, Vol. 2, p. 1093). — [23] F.F. Momyer, E.K. Hyde (J. Inorg. Nucl. Chem. **1** [1955] 274/95). — [24] M.W. Parrott, W.M. Garrison, P.W. Durbin, M. Johnston, H.S. Powell, J.G. Hamilton (UCRL-3065 [1955] 1/8). — [25] E.H. Appelman (MTP [Med. Tech. Publ. Co.] Intern. Rev. Sci. Inorg. Chem. Ser. One **3** [1972] 181/98).

[26] E.H. Appelman (J. Am. Chem. Soc. **83** [1961] 805/7). – [27] E.H. Appelman, E.N. Sloth, M.H. Studier (Inorg. Chem. **5** [1966] 766/9). – [28] W.M. Garrison, J.D. Gile, R.D. Maxwell, J.G. Hamilton (Anal. Chem. **23** [1951] 204/5). – [29] K. Rössler, W. Tornau, G. Stöcklin (J. Radioanal. Chem. **21** [1974] 199/209). – [30] W.A. Chalkin [V.A. Khalkin], E. Herrmann (Isotopenpraxis **11** [1975] 333/40).

[31] B.N. Belyaev, V. Yun-Yui, E.N. Sinotova, L. Német, V.A. Khalkin (Radiokhimiya **2** [1960] 603/13; Radiochem. [USSR] **2** [1961] 92/102). – [32] V.D. Nefedov, Yu.V. Norseev, Kh. Savlevich, E.N. Sinotova, M.A. Toropova, V.A. Khalkin (Dokl. Akad. Nauk SSSR **144** [1962] 806/9; Proc. Acad. Sci. USSR Chem. Sect. **142/147** [1962] 507/10). – [33] Yu.V. Norseyev [Norseev], V.A. Khalkin (J. Inorg. Nucl. Chem. **30** [1968] 3239/43). – [34] V.M. Vakhtel, G.V. Vinel, Ts. Vylov, I.I. Gromova, A.F. Novgorodov, Yu.V. Norseev, V.A. Khalkin, V.G. Chumin (Radiokhimiya **18** [1976] 886/93; Soviet Radiochem. **18** [1976] 752/8; Isotopenpraxis **12** [1976] 441/6; C.A. **86** [1977] No. 48309, 62364). – [35] B.V. Kurchatov, V.N. Mekhedov, L.V. Chistiakov, M.Ya. Kuznetsova, N.I. Borisova, V.G. Solov'ev (Zh. Eksperim. Teor. Fiz. **35** [1958] 56/63; Soviet Phys.-JETP **8** [1959] 40/6).

[36] W. Fu-Chiung, K. Mêng-Hua, V.A. Khalkin (Radiokhimiya **4** [1962] 94/8; Soviet Radiochem. **4** [1962] 81/5). – [37] W. Yung-yü, V.A. Khalkin (Radiokhimiya **3** [1961] 662/6; Radiochem. [USSR] **3** [1962] 154/8). – [38] A.K. Lavrukhina, S.S. Rodin (Radiokhimiya **2** [1960] 82/93; Radiochem. [USSR] **2** [1961] 95/6). – [39] M. Bochvarova, D.K. Tyung, I. Dudova, Yu.V. Norseev, V.A. Khalkin (Radiokhimiya **14** [1972] 858/65; Soviet Radiochem. **14** [1972] 889/95). – [40] V.D. Nefedov, Yu.V. Norseev, M.A. Toropova, V.A. Khalkin (Usp. Khim. **37** [1968] 193/215; Russ. Chem. Rev. **37** [1968] 87/98; JINR-P-2895 [1966] 1/36; C.A. **66** [1967] No. 70899, **68** [1968] No. 100679).

[41] M. Lefort, G. Simonoff, X. Tarrago (Bull. Soc. Chim. France **1960** 1726/7; C.A. **1961** 7123). – [42] J. Merinis, G. Bouissières (Radiochim. Acta **12** [1969] 140/52). – [43] B. Bayar, I. Votsilka, N.G. Zaitseva, A.F. Novgorodov (Radiokhimiya **16** [1974] 329/36; Soviet Radiochem. **16** [1974] 333/8). – [44] S.M. Qaim, G. Stöcklin (Radiochim. Acta **34** [1983] 25/40).

4.2.2 Some Special Production and Work-Up Procedures of ^{211}At

K. Rössler
Institut für Chemie 1 (Nuklearchemie)
Kernforschungsanlage Jülich GmbH
Jülich, Federal Republic of Germany

Astatine can be produced by several types of nuclear reactions:
a) directed processes of α particles and ^{3}He with ^{209}Bi [1 to 5],
b) spallation reactions and secondary processes [6 to 13] and
c) indirectly via the decay of ^{211}Rn [4, 8, 9, 12 to 16].

The different methods and nuclear reactions are listed in Table 4/2, following a scheme in [17]; cf. also the somewhat modified review in [43].

The most commonly applied procedure is the activation of ^{209}Bi by α particles, either in the form of metallic foils [18, 19] or as Bi_2O_3 pressed pellets [20, 21]. The use of Bi_2O_3 reduces the problems of target heating, however, the α beam is less effectivly used for activation. The energy of the α particles must be kept at 28 MeV in order to avoid formation of ^{210}At. The cross section of some α- and ^{3}He-induced reactions in ^{209}Bi are plotted in Fig. 3-26, p. 64, according to [2, 5].

The cross sections for the $^{209}Bi(\alpha,2n)^{211}At$ and $^{209}Bi(\alpha,3n)^{210}At$ nuclear reactions were checked recently for the energy range between 21.1 and 29 MeV [4]. The values were in good agreement with those for ^{211}At by [2], but, in significant disagreement with those

Table 4/2
Nuclear Reactions for the Production of At Isotopes.

nuclear reaction	particle energy in MeV	cross section 10^{-27} cm^2 in mb	Ref.
direct nuclear reactions with light particles:			
^{209}Bi(α, xn)$^{213-x}$At	~30	~10^3	[1 to 4]
^{209}Bi(^{3}He, xn)$^{212-x}$At	~25	~2 to 200	[5]
spallation reactions and secondary processes:			
^{232}Th(p, sp)At	160	0.2 to 2	[6, 7]
^{232}Th(p, sp)At	660	~50	[8]
^{238}U(p, sp)At	660	~5	[9]
^{209}Bi[p(α, xn)]$^{213-x}$At	~480	~0.1	[10]
^{209}Bi[Ar(α, xn)]$^{213-x}$At	~250	~25	[7, 11 to 13]
^{209}Bi[Kr(α, xn)]$^{213-x}$At	~490	~15	[11 to 13]
indirect via decay of ^{211}Rn:			
^{209}Bi[Ar(Li, xn)]$^{216-x}$Rn	250 to 460	~5	[8, 9]
^{232}Th(p, sp)Rn	660	~40	[12 to 14]
^{209}Bi(^{6,7}Li, xn)$^{215,216-x}$Rn	35 to 40	10 to 100	[15]
^{209}Bi(^{7}Li, 5n)^{211}Rn	38 to 60	650	[16]
^{209}Bi(^{6}Li, 4n)^{211}Rn	60	<100	[42]

by [3]. As an interesting detail, the formation of ^{210}Po by the ^{209}Bi(α, t)^{210}Po was observed at ppm level in thick targets irradiated by 21 to 27.7 MeV α particles [4]. ^{210}Po was previously reported to be formed with cross sections $\leqq 3\times10^{-27}$ cm^2 (3 mb) for α energies down to 20 MeV via the (α, t) reaction [2]. The cross section of the (α, 2n) reaction at 28 MeV is relatively high ~7.5×10^{-25} cm^{2-} (750 mb) and theoretical thick target yields of 1.7×10^7 Bq · μA^{-1} · h^{-1} (0.6 mCi · μA^{-1} · h^{-1}) can be obtained [22]. The reaction with ^{3}He ions shows lower cross sections and ^{211}At may be severely contaminated with ^{210}At. Each group has its own array for the irradiation. The Jülich-Karlsruhe procedure is reported as a typical example in the following. 7 × 10 mm Bi foils of 200 μm thickness (bismuth of 99.9999% purity) were used. The irradiation was carried out "in beam" with a special target holder, shown in **Fig. 4-2**, p. 102, cf. [17, 23].

The crucial problem was the cooling of the target due to the low melting point of Bi (t = 271 °C) and the rather low thermal conductivity of 0.08 J · s^{-1} · cm^{-1} · K^{-1}. At the Karlsruhe KfK-cyclotron the target array shown below was equipped with a 4π-water cooling system. Beams of 20 μA 28 MeV α's were applied for 6 to 8 h without appreciably damaging the target. The activity at EOB amounted to 3.5×10^9 Bq (~0.1 Ci). The target could be handled without causing strong contamination, because the astatine produced at the surface had been removed by the cooling water. In target arrays for which 4π-water cooling is not possible, e.g., due to the limitations in α particle energy, control of the temperature conditions becomes necessary. In the Jülich CV28 compact cyclotron with α energies limited to maximum 28 MeV, 0.16 mm Bi layers electroplated on a 0.2 mm copper sheet were used. They were cooled at the rear side by a water current. The maximum beam intensity supported by the system was about 10 μA [24]. Still higher intensities could be attained by additional

Fig. 4-2

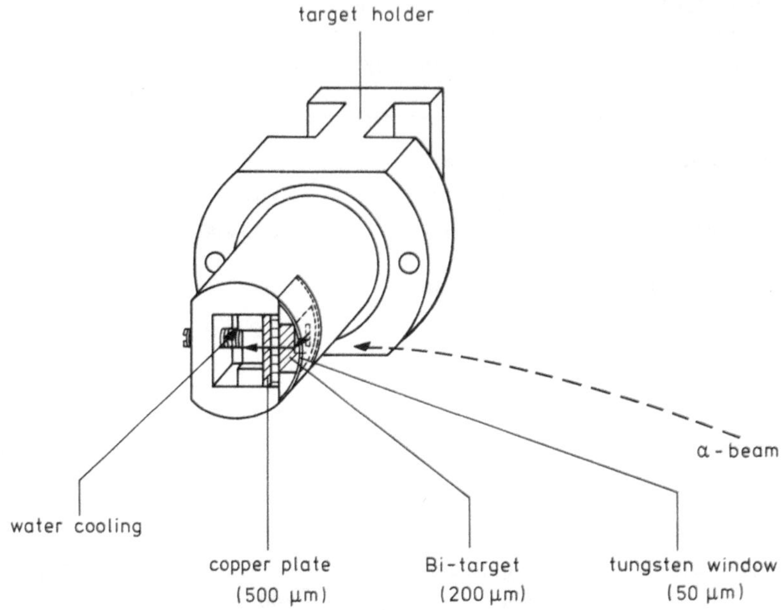

Schematic drawing of in-beam target holder [17, 23].

helium cooling of the front face of the target. A 0.08 mm Bi layer even supported 25 μA of a 28 MeV α beam, however, activation of the copper backing occurred which hampered the following separation procedures of target and backing. Details of the method can be found in [17, 24, 25].

Some more recent procedures are reported in [4, 26, 27]. A yield of 1.6×10^8 Bq·μA^{-1} (4.37 ± 0.06 mCi·μA^{-1}) was obtained for irradiation with <10 μA for <5 h by [4] and a yield of 1.5×10^7 Bq·$\mu A^{-1} \cdot h^{-1}$ (0.4 mCi·$\mu A^{-1} \cdot h^{-1}$) by [26].

Two major work-up procedures may be applied to the target: wet processing and dry distillation. In the first one, the target is dissolved in concentrated acid. At species, such as At^- or At^0, can be extracted or distilled out of the solution, adsorbed to metal surfaces such as platinum, tellurium etc., cf., e.g. [19, 25, 28 to 30]. **Fig.** 4-**3** shows the apparatus used in Jülich for the wet dissolving and distillation techniques [17, 24, 25, 30]. Astatine can be recovered in the form of At^0 and At^-.

Astatine was carried out off the dissolving vessel (1) with its concentrated sulfuric acid solution by an argon stream, most probably in the form of At^0(X). After passing through a washing flask in order to retain H_2SO_4 (2), the At^0 was trapped in three flasks cooled by an ice bath (3a to c). (3a) contained H_2O, (3b) 0.1 N Na_2SO_3 solution and (3c) some other reagent, e.g., 1 N $HNO_3 + K_2Cr_2O_7$. The astatine could be pipetted from flasks (2 to 3c) by means of syringes as At^0, At^- and, eventually, At^+ in aqueous solution. A fifth trap consisted of a $CaCl_2$/ice-cooled $CHCl_3$ reservoir (4). The astatine trapped in organic solution in the latter flask could be used directly for homogeneous organic labeling. Two liquid nitrogen-cooled charcoal traps prevented the escape of ^{211}At. During all operations,

Fig. 4-3

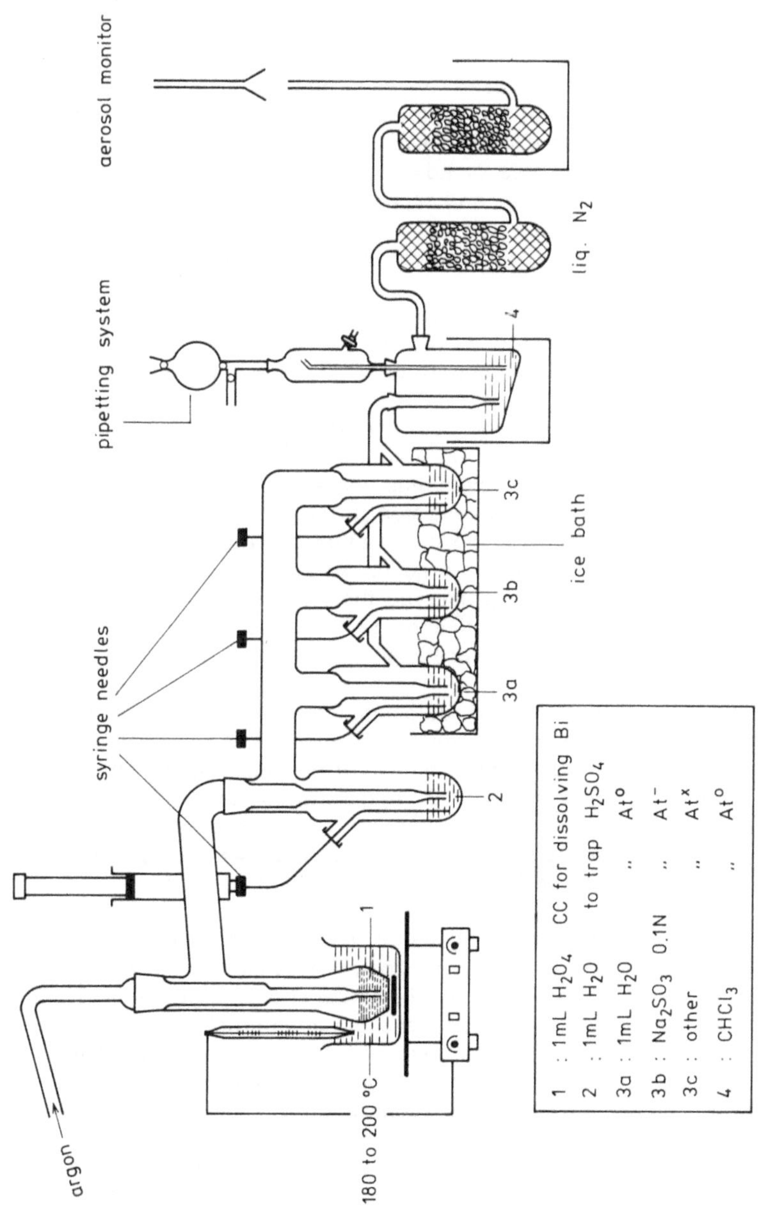

Apparatus for wet work-up of astatine [24, 25].

astatine was never found in the last trap, nor did the aerosol monitor detect any ^{211}At radioactivity. Concentrated sulfuric acid (96%) was prefered to nitric acid for dissolving the bismuth target because its high boiling point allowed to reach the temperature necessary to expell the astatine out of the mixture in vessel (1). The target did not dissolve readily in the cold acid and it was necessary to warm the vessel with a flame to initiate the process. After having bubbled argon for 10 min, 80 to 85% of the astatine initially present was extracted. Depending on the solutions used, between 30 and 70% were recovered in the flasks 3a to 4. The main advantage of this method is that a ready to use, highly chemically pure astatine is obtained in a rapid (~15 m) and clean way. Other methods of purification of astatine consist, e.g., in the use of adsorption columns filled with crystalline tellurium, where ^{211}At is retained specifically and separated from other spallation products [31].

Other wet chemical working-up procedures are reported in, e.g. [20, 32 to 34].

A typical example for a dry processing method is the distillation of astatine out of the metal, either at normal pressure or in vacuum [4, 12, 13, 17, 19, 24, 25, 33 to 36]. Fig. 10-14, p. 202, displays the dry distillation apparatus used in the Jülich experiments [17, 24, 25].

A fragment of the bismuth target was introduced into the quartz oven of the vacuum apparatus, cooled to dry ice temperature and evacuated at 0.1 Pa. Then the ampules were cooled by liquid nitrogen and the oven was heated for 30 min at 500 °C until the Bi began to sublime. The ^{211}At could be concentrated in the ampules or washed from the walls by $CHCl_3$. The Cl_2 supply served for the preparation of AtCl [25]. Reactants could be frozen onto the ^{211}At in the ampules and a reaction started by raising the temperature. Sealing the ampules before Cl_2 addition preserved the At^0 state. The distillation yield depended on the Bi sample size and the distillation temperature, reaching 60 to 80%. 15 to 35% were retained by the Bi, about 5% were adsorbed on the walls.

A novel one-step distillation and radiochemical recovery apparatus was developed [4], based on that reported above. It permitted elution of ^{211}At into small, controlled volumes and solvents required for subsequent radiopharmaceutical chemistry. Less than 2% ^{211}At remained in the target after a 50 min distillation at 650 to 680 °C. About 80% of the ^{211}At was collected on a silica gel column from a dry N_2/O_2 carrier gas. An overall radiochemical yield of $53 \pm 13\%$ was obtained for $^{211}At^-$. The methods were only reliable when the glassware and reagents were properly pretreated, dryed and cleaned [4].

Due to the broad spectrum of spallation reaction products, the processing of radon out of spallation targets is somewhat more complicated than that of the relatively radiochemically pure ^{211}At from the $^{209}Bi(\alpha,2n)$ reaction. Wet chemical procedures are reported, e.g., in [7, 31]. Thermochromatography and gas chromatography were developed as elegant methods for the separation of ^{211}Rn, also of ^{211}At, out of irradiated thorium targets [12, 13, 17, 37 to 41]. The Th target was heated in a quartz tube with a temperature gradient. The target was oxidized by streaming air at about 800 °C for 30 min. All volatile products were liberated. The separation due to the temperature gradient was increased by absorber materials. Astatine was absorbed at 100 °C on silver or platinum. The nobel gases were condensed in an active charcoal trap at dry ice temperature. The chromatographic separation was achieved on a 1 m copper column with 5 Å molecular sieves by raising the temperature to 400 °C. Almost 80% of the ^{211}Rn could be recovered [17].

A somewhat modified method was proposed for the preparation of a $^{211}Rn \rightarrow {}^{211}At$ generator, cf. [16], potentially for excitation labeling of radiopharmaceuticals. The $^{232}Th(p, sp)^{211}Rn$

nuclear reaction with 28.5 GeVp was applied. ^{211}Rn was collected 15 h post-irradiation by dissolution of the foil in concentrated acids and carrying it by a He stream through an Ag trap for purification into a charcoal trap at liquid nitrogen temperature [4].

References:

[1] E.L. Kelly, E. Segrè (Phys. Rev. [2] **75** [1949] 999/1005). — [2] W.J. Ramler, J. Wing, D.J. Henderson, J.R. Huizinga (Phys. Rev. [2] **114** [1959] 154/62). — [3] J.D. Stickler, K.J. Hofstetter (Phys. Rev. [3] C **9** [1974] 1064/71). — [4] R.M. Lambrecht, S. Mirzadeh (J. Labelled Compounds Radiopharm. **21** [1984] 1288/9; Intern. J. Appl. Radiat. Isotop. in press). — [5] E.T. Strom (UCRL-9372 [1960] 1/31; N.S.A. **15** [1961] No. 12080).

[6] B.V. Kurchatov, V.N. Mekhedov, L.V. Chistiakov, M.Ya. Kuznetsova, N.I. Borisova, V.G. Solovev (Zh. Experim. Teor. Fiz. **35** [1958] 56/63; Soviet Phys.-JETP **8** [1959] 40/6). — [7] B.N. Belyaev, Yung-Yu Wang, E.N. Sinotova, L. Német, V.A. Khalkin (Radiokhimiya **2** [1960] 603/13; Radiochem. [USSR] **2** [1960] 243). — [8] R. Bimbot, D. Gardes, M.F. Rivet (Phys. Rev. [3] C **4** [1971] 2180/4). — [9] R. Bimbot, M.F. Rivet (Phys. Rev. [3] C **8** [1973] 375/9). — [10] M. Lefort, G. Simonoff, X. Tarrago (Compt. Rend **248** [1959] 216/8).

[11] M. Lindner, R.N. Osborne (Phys. Rev. [2] **103** [1956] 378/85). — [12] V.M. Vakhtel, G.V. Vinel, Ts. Vylov, I.I. Gromova, A.F. Novgorodov, Yu.N. Norseev, V.A. Khalkin, V.G. Chumin (JINR-12-8996 [1975] 1/17; INIS Atomindex **8** [1977] No. 317841; C.A. **87** [1977] No. 158550). — [13] V.M. Vakhtel, G.V. Vinel, Ts. Vylov, I.I. Gromova, A.F. Novgorodov, Yu.V. Norseev, V.G. Chumin, V.A. Khalkin (Radiokhimiya **18** [1976] 886/92; Soviet Radiochem. **18** [1976] 752/8). — [14] V.A. Khalkin (Diss. Moscow State Univ. 1973). — [15] H. Freiesleben, H.C. Britt, J.R. Birkelund, J.R. Huizenga (COO-3496-44 [1974] 1/175, 97/101; N.S.A. **30** [1974] No. 22833, No. 22812).

[16] G.-J. Meyer, R.M. Lambrecht (Intern. J. Appl. Radiat. Isotop. **31** [1980] 351/5). — [17] G.J. Meyer (JUEL-1418 [1977] 1/107; C.A. **88** [1978] No. 71113). — [18] D.R. Corson, K.R. MacKenzie, E. Segrè (Phys. Rev. [2] **57** [1940] 1087). — [19] E.H. Appelman (UCRL-9025 [1960] 1/113; N.S.A. **14** [1960] No. 11501). — [20] G. Samson (Diss. Univ. Amsterdam 1971).

[21] A.H.W. Aten Jr., T. Doorgeest, U. Hollstein, H.P. Moeken (Analyst [London] **77** [1952] 774/7). — [22] G.L. Johnson, R.F. Leininger, E. Segrè (J. Chem. Phys. **17** [1949] 1/10). — [23] F. Schult, H. Bellemann (KFK-685 [1967]). — [24] A. Cavallero (Diss. Univ. Cathol. Louvain, Belg., 1981 [to be published as JUEL-Report in 1985]). — [25] G.-J. Meyer, K. Rössler (Radiochem. Radioanal. Letters **25** [1976] 377/90).

[26] G.-J. Beyer, R. Dreyer, H. Odrich, F. Rösch (Radiochem. Radional. Letters **47** [1981] 63/5). — [27] V. Doberenz, Dang Duc Nhan, R. Dreyer, M. Milanov, Yu.V. Norseev, V.A. Khalkin (Radiochem. Radioanal. Letters **52** [1982] 119/27). — [28] E.H. Appelman (NAS-NS-3012 [1960] 1/33; N.S.A. **14** [1960] No. 20201). — [29] V.A. Khalkin, E. Herrmann, Yu.V. Norseev, I. Dreyer (Chemiker-Ztg. **101** [1977] 470/81). — [30] K. Rössler, W. Tornau, G. Stöcklin (J. Radioanal. Chem. **21** [1974] 199/209).

[31] M. Bochvarova, Do Kim Tyung, I. Dudova, Yu.V. Norseev, V.A. Khalkin (Radiokhimiya **14** [1972] 858/65; Soviet Radiochem. **14** [1972] 889/95). — [32] A.H.W. Aten Jr. (Advan. Inorg. Chem. Radiochem. **6** [1964] 207/23). — [33] H.M. Neumann (J. Inorg. Nucl. Chem. **4** [1957] 349/53). — [34] R.D. Neirinckx, J.A. Smit (Anal. Chim. Acta **63** [1973] 201/4). — [35] E.H. Appelman, E.N. Sloth, M.H. Studier (Inorg. Chem. **5** [1966] 766/9).

[36] G.W. Barton, A. Ghiorso, J. Perlman (Phys. Rev. [2] **82** [1951] 13/9). — [37] A. Kolachkovskii, Yu.V. Norseev (JINR-P6-6923 [1973] 1/9; N.S.A. **27** [1973] No. 22317; C.A. **79** [1973]

No. 60427). — [38] B. Eichler, V.P. Domanov (JINR-P12-7928 [1974] 1/11; N.S.A. **32** [1975] No. 27691). — [39] B. Eichler (Radiochem. Radioanal. Letters **22** [1975] 147/55). — [40] L. Vasáros, Yu.V. Norseev, G.-J. Meyer, K. Berei, V.A. Khalkin (AED-CONF-77-411-001 [1977]; CONF-770964 [1977] 1/113; Abstr. 9th Intern. Hot Atom Chem. Symp., Blacksburg, Va., 1977; INIS Atomindex **9** [1978] No. 410596; Radiochim. Acta **26** [1979] 171/6).

[41] L. Vasáros, Yu.V. Norseev, G.-J. Meyer, K. Berei, V.A. Khalkin (Radiochim. Acta **26** [1979] 171/6). — [42] G.-J. Meyer (personal communication 1984). — [43] S.M. Qaim, G. Stöcklin (Radiochim. Acta **34** [1983] 25/40).

5 General Properties of Astatine

László Vasáros, Klara Berei

Central Research Institute for Physics
of the Hungarian Academy of Sciences
Budapest, Hungary

Only a very limited amount of direct evidence is available concerning the properties of astatine because it occurs naturally in trace amounts. In view of this, most of the data for astatine have been extrapolated or interpolated by various empirical or theoretical treatments from the values determined for the other halogens, or for the neighboring elements of astatine in the Periodic Table. Recent improvements in the theoretical computations have also made it possible to calculate some of the basic properties of astatine.

5.1 Electronic Structure

Similarly to the other halogens, the astatine atom lacks one electron to fill its valence shell. The electron configuration of the ground state astatine atom can be assigned as $[Xe]4f^{14}5d^{10}6s^{2}6p^{5}(^{2}P_{3/2})$ [1, 2].

The ground state electronic structure of astatine ions is also analogous to that for other halogens. The valence shell of the astatide ion has the noble gas configuration of $6s^{2}6p^{6}(^{1}S_{0})$ [3]. The ground state outer electron configuration of the At^{+} ion can be assigned as $6s^{2}6p^{4}(^{3}P_{2})$. For positive atomic ions, the electronic structure is given in Table 5/2, p. 109.

5.2 Electronegativity

The electronegativity (X) of astatine is given as 2.2 in Pauling's original table [4]. The subsequent improved and more precise electronegativity scales give similar results [1, 5, 6]. The X values based on a number of scales and calculated by different methods [1, 4 to 9] are summarized in Table 5/1. Electronegativities for higher oxidation states of astatine have been calculated on the basis of electrostatic forces, to be 1.74 and 2.048 for At^{5+} and At^{7+} ions, respectively [6].

Table 5/1
Electronegativity Values of Astatine Atom.

method	X values	Ref.
Pauling scale	2.2	[4]
Mulliken Scale	2.4	[1]
Gordy Scale	2.2	[5]
electrostatic forces	1.9	[6]
	2.42	[8]
extrapolation by Mendeleev's rule	2.25	[7]
$X=f(I)$	2.32	[9]

5.3 Atomic Volume

The atomic volume for astatine has been estimated by extrapolation from the plot of atomic volume vs. period of halogens as 27.5 cm^{3}/g-atom [10]. More recently the atomic information indices, derived from the known electronic configuration of halogens, have also been used for correlation with their atomic volume. The latter value for astatine 33.9 to 34.5 cm^3/g-atom was extrapolated from this dependence [11].

5.4 Atomic Refraction

The atomic refraction calculated based on the electronegativity of astatine was found to be 19.3 cm^3/g-atom [12].

5.5 Polarizabilities

The polarizability of astatide ion 8.3×10^{-24} cm^3 has been derived from its ionic radius [13]. Using theoretical methods of calculation, the total quadrupole polarizabilities have been estimated as 57.03699×10^{-40} cm^5 and 19.78638×10^{-40} cm^5 for the free and for the crystal astatide ions, respectively [14].

Theoretical calculations have also been performed to estimate the static dipole polarizability of neutral astatine atom in the ground state. The values obtained in this way are 4.36×10^{-24} cm^3 [15, 16], 7.88×10^{-24} cm^3 [17], and 5.76×10^{-24} cm^3 (38.9 $Bohr^3$)[1)] [18].

5.6 Ionization Energies

The first ionization energy (I_1) for the element with atomic number 85 was originally derived by Finkelnburg as 9.4 eV [19]. In subsequent papers the original value was modified to 9.5 ± 0.2 eV [20] and then to 9.2 ± 0.4 eV [21], the last value is usually quoted in the review literature [22]. This set of ionization energies was estimated by using regularities in the change of the screening constant ($\Delta\sigma$) from element to element in the Periodic Table. From the estimated value of $\Delta\sigma = 0.7$ for astatine and from the experimental I_1 quantities of elements adjacent to astatine (Po or Rn), the first ionization energy for astatine could be calculated [19 to 21]. Norseyev [Norseev] and Nefedov have estimated the value of I_1 as 9.5 eV from the linear dependence between the experimental parameter Z′ for halogens – derived from gas chromatographic retention data of halogenated aliphatic compounds – and their first ionization energy [23]. From the relationship between the vibrational frequency (ω_e) for halogen molecules and the first ionization energy of halogen atoms, the value of I_1 has been estimated to be 10.4 eV by Varshni [24] assuming Clark's predicted value of $\omega_e = 139.4$ cm^{-1} for astatine (see Section 5.11.4, p. 124). Ozhigov applied Mendeleev's rule of extrapolation to obtain 9.44 eV for the first and 16.71 eV for the second ionization energy of the astatine atom [7]. The electron binding energy values of the outermost P_3 ($6p_{3/2}$) subshell for the free At atom, i.e., its first ionization energy, obtained by interpolation (9.3 eV) and by theoretical calculation (8.6 eV), as given in Table 5/14, p. 122, are close to the values of I_1 mentioned above. In Parsons' "Handbook of Electrochemical Constants" a value of 9.6 eV is given [25]. The estimated first and higher ionization energies for the ground state astatine atom are summarized in Table 5/2.

The ionization energy for molecular astatine has been estimated by Kiser as 8.3 eV [26] in the same way as was done by Varshni [24]. The ionization energy for molecular halogens, as well as an estimated value of $\omega_e = 160$ cm^{-1} for At_2, were used in the calculation procedure [26]. Norseyev [Norseev] and Nefedov, using extrapolation based on the experimental parameter Z′ (see above), estimated the ionization energy for molecular astatine as 8.4 eV [23].

---

1) Originally quoted value is given in parentheses.

Table 5/2
Atomic Ionization Energies of Astatine.

I_n	ionization energy in eV	ground state configuration	Ref.
I_1	9.4	At^+ $6p^4$ ($^3P_2$)	[19]
	9.5		[20]
	9.2		[21]
	9.5		[23]
	10.4		[24]
	9.44		[7]
	9.6		[25]
I_2	20.1	At^{2+} $6p^3$ ($^4S_{3/2}$)	[21]
	17.3		[23]
	18.0		[25]
	16.71		[7]
I_3	29.3	At^{3+} $6p^2$ ($^3P_0$)	[21]
	30.0		[25]
I_4	41.0	At^{4+} $6p$ ($^2P_{1/2}$)	[25]
I_5	51.0	At^{5+} $6s^2$ ($^1S_0$)	[25]
I_6	78.0	At^{6+} $6s$ ($^2S_{1/2}$)	[25]
I_7	91.0	At^{7+} $5d^{10}$ ($^1S_0$)	[25]
I_8	138.0	At^{8+} $5d^9$ ($^2S_{1/2}$)	[25]

5.7 Oxidation States

From the general trend in the Periodic System, astatine is expected to possess a more electropositive character than the other halogens. Thus, the first investigators considered astatine to be a metal showing a closer resemblance to polonium than to iodine [27 to 29]. On the other hand, the volatility of astatine, its extractability with carbon tetrachloride [29], and its similarity in physiological behavior to iodine [30, 31] seemed to be consistent with its halogen character. Further systematic investigation of its aqueous solutions has shown that the valence states of astatine are similar to those of iodine. Johnson et al. described four oxidation states, viz. (−I), zero, and two positive valency states [32]. Appelman has confirmed and characterized (−I), zero, intermediate positive (I or III), and (V) valence states but found no evidence for the (VII) state [33, 34]. Later, perastate was obtained by using XeF_2 as the oxidizing agent [35] and also by other methods [36 to 39]. Methods of preparation for different oxidation states of astatine are described in Section 10.2.1, p. 211. In the following sections, the properties of these At species are briefly reviewed.

5.7.1 Astatine(−I)

The best defined valence state, astatine(−I), has been found to be stable in acidic and basic solutions containing sufficiently strong reducing agents [32 to 34, 40 to 45]. The astatide ion may be oxidized to the zero valence state in acidic solutions without reducing agents [40, 42]. Electromigration experiments [32, 46], coprecipitation with insoluble iodides [32 to 34, 42, 47, 48], paper electrophoresis [38, 40 to 42], paper chromatography [36, 49],

References on pp. 124/8

high pressure ion exchange chromatography [43], and free solution electrophoresis [39, 45, 50] have been used to characterize astatine(−I).

In contrast to the iodide ions, astatide has a strong tendency to adsorb on metallic silver [42, 47] and tellurium [29, 41, 42, 51] surfaces from acidic solutions containing reducing agents. Astatide, similarly to iodide, may be completely adsorbed on the reduced surfaces of metallic platinum from sulfuric acid solutions [52], and can also be characterized by the ability to replace halogens in simple halogenated aliphatic and aromatic compounds (see, e.g. [53]).

5.7.2 Astatine(0)

At^0 is the expected oxidation state of astatine when it is isolated by dry methods at high temperatures from bismuth after irradiation with α particles [32 to 34, 44, 54 to 58], and from thorium or uranium targets after bombardment with high energy protons [59 to 61]. At^0 is also assumed to be the oxidation state when astatine is redistilled at 500 °C from silver and platinum surfaces in closed glass ampules [33, 39, 50, 58].

When elemental astatine is dissolved in pure water [38, 44, 58, 62 to 64] or in nitric acid solutions [32, 33, 65], the retention of the zero oxidation state is expected. The zero valence state of astatine was also supposed when it is prepared from a cyclotron-irradiated bismuth target by conventional dissolution and extraction techniques [43, 44, 62 to 64].

Aqueous solutions of At^0 may be prepared by oxidation of astatide ion with dilute nitric acid, Fe^{3+}, I_2, VO_2^+, As^{3+} at pH<5, or $[Fe(CN)_6]^{3-}$ at pH<3 [32 to 34]. On the other hand, the zero oxidation state of astatine may also be formed from the higher oxidation states by reduction with Fe^{2+}, I^-, or VO^{2+} [32 to 34].

The zero oxidation state is characterized by its volatility, a tendency to be adsorbed by various metallic surfaces such as silver, gold, and platinum [32, 33, 38, 58, 60, 61] and less strongly by glass surfaces [33] both from vapors and solutions. At^0 can be extracted either from nitric acid solutions with isopropyl ether [54, 65, 66], CCl_4, C_6H_6 [32, 33, 54], n-C_6H_{14}, and c-C_6H_{12} [33, 65], or from sulfuric acid solutions with $CHCl_3$ [43, 44, 63 to 65]. At^0 is also characterized by reduced volatility from solutions as compared with the volatility of iodine, by varying degrees of coprecipitation with metal sulfides and hydroxides [32], and with metallic silver or tellurium in situ [33]. Astatine is partially carried by TlI and AgI as a result of adsorption of At^0 on the surface of precipitates from acidic solutions containing SO_2 [32, 33].

The exact chemical nature of At^0 in solutions, however, still remains uncertain. Merinis et al., during their experiments to produce interhalogen compounds of astatine (see Section 10.2.1.3.1, p. 224), described the astatine activity deposited in a gradient thermochromatographic tube at 16 °C, as At_2 [67]. More recently Otozai and Takahashi have claimed to identify the At_2 peak by gas liquid chromatography [68]. However, as several authors have pointed out, the existence of molecular astatine is excluded by its extremely low concentration under ordinary conditions of chemical experiments [33, 69 to 72]. Furthermore, the formation of At_2 does not seem to be realistic because any reaction which tends to split the At_2 molecule becomes thermodynamically favored at these low concentrations [65]. It is also unlikely that the zero valence state of astatine is present as a true $At^{\bullet}$ radical because of the assumed high reactivity of the latter [33, 71]. Since iodine is always present in excess, Aten supposed that in aqueous solutions At^0 exists as AtI [69]. A similar conclusion has been drawn by Visser and Diemer from their extraction experiments with At^0 [72]. Other authors have assumed that At^0 in aqueous solutions may react with organic impurities forming organic astatine compounds the exact nature of which depends on the medium [33, 34, 65, 71, 73].

Meyer et al. have investigated the reactivity of At^0 dissolved in neutral aqueous solutions with simple aromatic compounds. The hydrogen and chlorine substitution yields did not exceed 1% for benzene and chlorobenzene. A higher hydrogen replacement yield (49%) was observed with aniline [63, 64].

In the presence of elemental chlorine, bromine, and iodine, the zero oxidation state of astatine is represented by AtCl, AtBr, and AtI. These diatomic interhalogen compounds of astatine have been prepared both in solution and in the gaseous state [33, 44, 48, 63, 64, 69, 74 to 76], and have been characterized by their extractability with organic solvents [33, 48, 69, 74] and by their deposition temperature [67].

5.7.3 Astatine(I)

The time-of-flight spectrometry measurements of Appelman et al. have demonstrated the existence of At^+ ions in the gaseous phase (see Section 5.11.2, p. 121). This is so far the only direct method to identify any of the oxidation states of astatine [75]. The mass spectrum of At^+ is shown in Fig. 10-23, p. 225. Later, Golovkov et al. also detected At^+, formed in the plasma ion source of a mass separator, by its radioactivity [76].

The monovalent cationic form of astatine can be obtained and stabilized in nitric acid solutions containing $Cr_2O_7^{2-}$ as the oxidizing agent. The positive charge of astatine(I) species has been established by free solution electrophoresis [46, 50, 77], and its monovalent character by ion exchange chromatography [43, 77, 78].

The astatine(I) state has been assumed to be responsible for the formation of complex anions AtX_2^- (X = Cl, Br, I, SCN, CN, thiourea or its derivatives) [54, 79 to 84]. From the results of numerous experiments [46, 50, 77 to 82, 85 to 88], Khalkin et al. concluded that astatine(I) in acidic solutions containing $Cr_2O_7^{2-}$ can form a stable aqua complex $[(H_2O)_xAt]^+$ (x = 1 or 2) [86 to 88]. This assumption is well in line with the earlier findings of Neumann [54] and with the recent experimental results on the hydrogen substitution reactions of astatine with simple aromatic compounds. The existence of $[H_2OAt]^+$, as an electrophilic species, could also be proved from the positive results of aromatic H replacement [89].

5.7.4 Astatine(III)

The probable oxidation state of astatine(III) has been assumed for the anionic species formed by oxidation of astatine with elemental bromine [32, 33]. Dreyer et al. have proposed the existence of AtO^+ or $H_2AtO_2^+$, AtO_2^-, and $AtOX_2^-$ (X=Cl, Br, and I) species from the migration rates measured by free solution electrophoresis. In order to study the mobilities of these ions, At^{III} state was obtained by oxidation of At^0 with $S_2O_8^{2-}$ in $HClO_4$ solutions or with XeF_4 in neutral medium [36, 39, 50, 82]. In the course of investigating the properties of inorganic astatine species under oxidative conditions, Visser and Diemer suppose that etherates containing the At^{III} moiety are extracted with n-dibutyl ether from $S_2O_8^{2-}$ and H_2O_2 solutions. The expressed complex forming ability of astatine(III) is also discussed by these authors [72].

In the compounds $ArAtCl_2$ and Ar_2AtCl (Ar = C_6H_5 or p-$CH_3C_6H_4$) synthesized by Norseyev et al., the At^{III} state was obtained by oxidation of At^{-I} with Cl_2 [90, 91] (see Section 10.2.2.5, p. 254).

References on pp. 124/8

5.7.5 Astatine(V)

The AtO_3^- anion belongs to the well-characterized chemical form of astatine(V). This ionic species can be formed from the lower oxidation states under stronger oxidation conditions (i.e., in hot solutions containing the oxidizing agent) [32 to 34, 37 to 40, 50] or by acidifying the AtO_4^- containing solutions [39].

AtO_3^- ion was originally characterized by its tendency to coprecipitate with $AgIO_3$ [32 to 34], $Ba(IO_3)_2$, and $Pb(IO_3)_3$ [33, 34]. This interpretation became doubtful when the At^I state was also shown to coprecipitate with the insoluble iodates [80]. Identification by paper chromatography, paper electrophoresis [38, 40], and by free solution electrophoresis [39, 50] made the existence of the AtO_3^- anion certain.

The central astatine atom has an oxidation state of At^V in the organometallic compounds of $ArAtO_2$ ($Ar=C_6H_5$ or p-$CH_3C_6H_4$). These compounds have been prepared by Norseyev et al. by oxidation of $ArAtCl_2$ with hot NaOCl solution [90, 91].

5.7.6 Astatine(VII)

The perastatate ion, AtO_4^-, was first prepared by Khalkin et al. by oxidizing At^- with XeF_2 in a hot alkaline solution [35], in the same way that the formation of perbromate was carried out shortly before [92]. The AtO_4^- anion was identified and characterized by paper electrophoresis and by its coprecipitation with potassium and caesium metaperiodate [35]. Later, anodic oxidation [36] or oxidation with KIO_4 in neutral or alkaline solutions were also utilized to prepare AtO_4^- [37 to 39]. Paper chromatography [35, 36, 38], paper electrophoresis [38], and free solution electrophoresis [39, 50] have been applied for the identification of the perastatate ion.

AtO_4^-, similarly to IO_4^-, is stable only in neutral and alkaline solutions. The reduction of perastatate into astatate is completed in several minutes by heating in an acidic medium [38, 39].

5.8 Electrochemical Properties

5.8.1 Critical Deposition Potentials

The critical deposition potentials of astatine both at the cathode and anode from different aqueous solutions have been determined by Johnson et al. and are summarized in Tables 5/3 and 5/4. These values were obtained by extrapolation of deposition rates vs. potential curves to zero rate. The chemical forms of the deposited At species in the electrolytic experiments were not determined [32].

Table 5/3
Critical Deposition Potentials of Astatine at the Cathode [32].

solution	At concentration in pM	potential in V vs. NHE [a]
0.066 M HNO_3	0.28	−1.225
1.0 M HNO_3	0.05	−1.240
0.075 M H_2SO_4+0.1 M $Na_2Cr_2O_7$	0.60	−1.200
0.006 M HNO_3+3 mg Au	0.10	−1.220
0.066 M HNO_3	0.04	−1.220

[a] NHE = normal hydrogen electrode.

Table 5/4
Critical Deposition Potentials of Astatine at the Anode [32].

solution	At concentration in pM	potential in V vs. NHE[a)]
0.066 M HNO_3	0.24	−1.460
0.066 M HNO_3	0.53	−1.450
0.100 M HNO_3+0.1 M $K_2S_2O_8$	0.54	−1.445

[a)] NHE = normal hydrogen electrode.

5.8.2 Standard Electrode Potentials

On the basis of potentials of the redox couples employed by Johnson et al. for preparing the oxidation states of At^{-I}, At^{0}, At^{I}, and At^{V} [32], Latimer has constructed the first tentative standard electrode potential diagrams (in V) for astatine [93]:

acidic solution

$$HAtO_3 \xrightarrow{+1.4} HAtO \xrightarrow{+0.7} At_2 \xrightarrow{+0.2} At^-$$

basic solution

$$AtO_3^- \xrightarrow{+0.5} AtO^- \xrightarrow{0.0} At_2 \xrightarrow{+0.2} At^-$$

However, the uncertain potentials of the systems studied by Johnson et al. [32], as well as the problematic nature of the chemical form at At^0 (see Section 5.7.2, p. 110) make Latimer's potential diagram questionable. Later, in a more systematic study with suitably chosen redox couples Appelman determined the following reduction potential diagram for At^{-I}, At^{0}, At^{I}(?), At^{V}, At^{VII}(?) oxidation states in 0.1 M acid at 25 °C:

$$H_5AtO_6(?) \xrightarrow{>+1.6} AtO_3^- \xrightarrow{+1.5} HOAt(?) \xrightarrow{+1.0} At^0 \xrightarrow{+0.3} At^-$$

The exact valence of the positive oxidation state (or states) between At^0 and At^V could not be determined, and no evidence for At^{VII} state was found at that time [33, 34].

The standard electrode potential for At^0/At^{-I} couple has been estimated by Mendeleev's rule of extrapolation as 0.457 V [7]. More recently, ion exchange chromatography at a fixed redox potential has also been used to determine standard potentials for different redox couples of astatine. The potential values obtained by this technique were 0.335 V at 323 K and 0.85 V at 332 K for the At^0/At^{-I} and At^x/At^0 couples, respectively [94]. For the latter couple, At^x represents an intermediate oxidation state between At^0 and At^V.

5.9 Thermodynamic Data

The thermodynamic properties of astatine species have only been estimated by various theoretical and empirical calculations. Data on enthalpies, Gibbs free energies, entropies, and other thermodynamic properties of individual inorganic and organic astatine compounds are dealt with in Section 10.2, p. 210. In this section the thermodynamic data for different astatine species are summarized.

5.9.1 At^- Ion

Enthalpies, Gibbs free energies, entropies, and heat capacities at constant pressure for gaseous and hydrated At^- ions are summarized in Table 5/5. The heat of formation for gaseous At^- was first evaluated by Ladd and Lee from the ΔH_f^o values for gaseous alkali metal astatides by using the Born-Haber cycle [95]. Another ΔH_f^o value for gaseous At^- has been estimated from the lattice energy of alkali metal halides vs. ΔH_f^o for a gaseous halide ions plot [96]. The entropy of gaseous At^- was calculated by using the Sackur-Tetrode equation [96]. The thermodynamic data for gaseous At^- from the Handbook of Thermal Constants [97] are also given in Table 5/5.

Table 5/5
Thermodynamic Data for Gaseous and Hydrated At^- at 298.15 K.

At^-	ΔH_f^o	ΔG_f^o	S^o	C_p^o	Ref.
	in kJ/mol (kcal/mol)[a)]		in $J \cdot g\text{-atom}^{-1} \cdot K^{-1}$ ($cal \cdot g\text{-atom}^{-1} \cdot K^{-1}$)[a)]		
gaseous	−196.6 (−47.0)				[95]
	−190.8 (−45.6)	−255 (−53.8)	175.5 (41.95)		[96]
	−178.66 (−42.7)	−202.47 (−48.391)	175.4 (41.93)	20.79 (4.968)	[97]
hydrated	−8.4 (−2.0)	−7.95 (−1.9)	125.5 (30.0)		[96]

a) Originally quoted values are given in parentheses.

The heat of formation for hydrated At^- has been deduced from the calculated heat of hydration and heat of formation for gaseous At^- [96].

The entropy of astatide ion in a crystalline lattice has been given as 67.78 $J \cdot g\text{-atom}^{-1} \cdot K^{-1}$ (16.2 $cal \cdot g\text{-atom}^{-1} \cdot K^{-1}$)[1)] [98].

Krestov has calculated the change of thermodynamic functions on hydration for astatide ion in a series of papers [98 to 100]. The results are given below:

properties		values	Ref.
ΔH_{hydr}	in kJ/g-atom	182.42	[99]
	(kcal/g-atom)[a)]	(43.6)	
ΔS_{hydr}	in $J \cdot g\text{-atom}^{-1} \cdot K^{-1}$	−107.95	[98]
	($cal \cdot g\text{-atom}^{-1} \cdot K^{-1}$)[a)]	(−25.8)	
$(\Delta C_p)_{hydr}$	in $J \cdot g\text{-atom}^{-1} \cdot K^{-1}$	−66.94	[100]
	($cal \cdot g\text{-atom}^{-1} \cdot K^{-1}$)[a)]	(−16.0)	

a) Originally quoted values are given in parentheses.

An empirical method, based on the effective charge on the astatide ion, has been used to determine the heats of solvation (ΔH_{solv}) for At^- in different organic solvents [101]. The ΔH_{solv} values for aliphatic alcohols are given below together with the entropy changes on solvation (ΔS_{solv}) in these alcohols as evaluated by Krestov [102].

1) Originally quoted values are given in parentheses.

properties	CH_3OH	C_2H_5OH	n-C_3H_7OH	Ref.
ΔH_{solv} in kJ/g-atom	322.4	297.3	288.9	[101]
ΔS_{solv} in $J \cdot g\text{-atom}^{-1} \cdot K^{-1}$	−120.5	−135.14		[102]
($cal \cdot g\text{-atom}^{-1} \cdot K^{-1}$) [a]	(−28.8)	(−32.3)		

[a] Originally quoted values are given in parentheses.

5.9.2 Astatine Atom

5.9.2.1 Thermodynamic Properties

The fundamental thermodynamic properties for the gaseous astatine atom, taken from the literature [96, 97, 103], are summarized in Table 5/6 and show good agreement. In addition to the properties given in Table 5/6, the compilation of data by Stull and Sinke also gives the enthalpy, free energy function, and the logarithm of the equilibrium constant of formation for the temperature range 298.15 to 3000 K [103].

Table 5/6
Thermodynamic Properties of Gaseous Astatine Atom at 298.15 K.

properties	[96]	[97]	[103]
ΔH_f^o in kJ/g-atom	92	97.24	92.048
(kcal/g-atom) [a]	(22)	(23.24)	(22.000)
ΔG_f^o in kJ/g-atom	54	59.576	54.400
(kcal/g-atom) [a]	(13)	(14.239)	(13.002)
S^o in $J \cdot g\text{-atom}^{-1} \cdot K^{-1}$	187	186.98	186.94
($cal \cdot g\text{-atom}^{-1} \cdot K^{-1}$) [a]	(44.7)	(44.69)	(44.68)
C_p^o in $J \cdot g\text{-atom}^{-1} \cdot K^{-1}$		20.79	20.79
($cal \cdot g\text{-atom}^{-1} \cdot K^{-1}$) [a]		(4.968)	(4.97)

[a] Originally quoted values are given in parentheses.

The entropy for the solid state astatine atom has been estimated by Krestov as 60.67 $J \cdot g\text{-atom}^{-1} \cdot K^{-1}$ (14.5 $cal \cdot g\text{-atom}^{-1} \cdot K^{-1}$)[1] [96].

5.9.2.2 Electron Affinity

The classical method of determining electron affinity (EA) from the selected ΔH_f^o quantities for gaseous astatine atom and At^- [97] results in a value of −276.39 kJ/g-atom at 298.15 K. In similar calculations, the value $\Delta H_f^o(At) = 0.5\,D(At_2)$ – where $D(At_2)$ is the dissociation energy of molecular astatine (see Section 5.9.3.3, p. 119) – was used by Ladd and Lee, instead of the heat of formation for the astatine atom. The EA value given by these authors as −255.22 kJ/g-atom (−61 kcal/g-atom)[1] [95] is, therefore, not quite correct.

The atomic electron affinity of astatine has also been estimated by using different extrapolation methods [3, 7, 96], semiempirical [23] and empirical relationships [104], and theoretical calculations [105]. The EA value for astatine recommended by Hotop and Lineberger is

[1] Originally quoted values are given in parentheses.

given in [106]. The data are summarized in Table 5/7. With the exception of theoretical calculations, the atomic electron affinity values for astatine estimated by different methods are in reasonable agreement with those determined from thermodynamic data.

Table 5/7
Calculated Electron Affinity (EA) Values for Astatine.

method	EA in kJ/g-atom	Ref.
from thermodynamic data	−276.39	this work
	−255.2 (−61.0 kcal/g-atom)[a]	[95]
horizontal analysis	−270.2 (−2.80 eV)[a]	[3]
Mendeleev's rule of extrapolation	−267.8 (−64.0 kcal/g-atom)[a]	[7]
EA=f(Z′)	−270.2 (−2.8 eV)[a]	[23]
extrapolated from EA for halogens	−284.5 (−68.0 kcal/g-atom)[a]	[96]
EA=f(ω_e, μ)	−268.93 (−2.7872 eV)[a]	[104]
theoretical calculation	−318.4 (−3.3 eV)[a), b)]	[105]
recommended value	−270.2 (−2.8 eV)[a]	[106]

[a] Originally quoted values are given in parentheses.
[b] Approximated value derived from histogram bars.

The entropy change of atomic electron affinity (ΔS_{EA}) for astatine has been calculated by Krestov who deduced a value $\Delta S_{EA} = -41.99\ J \cdot g\text{-atom}^{-1} \cdot K^{-1}$ ($-10.036\ cal \cdot g\text{-atom}^{-1} \cdot K^{-1}$)[1] from the entropies of the gaseous astatine atom and At^-, the entropy of free electronic gas, and the entropy change on mixing [107].

5.9.3 Astatine Molecule

Although the existence of At_2 in typical chemical experiments so far remains uncertain (see Section 5.7.2, p. 110), many of its thermodynamic properties have been predicted. These data may be useful in further searches for the astatine molecule.

5.9.3.1 Phase Transition Properties

Melting and boiling temperatures (T_m and T_b), enthalpy and entropy changes on phase transitions (ΔH and ΔS) for the astatine molecule are listed in Table 5/8. Durand's prediction belongs to the first determinations of the physicochemical properties of astatine. His extrapolation was based upon the nearly linear dependence of T_m and T_b for noble gases and H_2 on the melting and boiling temperatures of halogen molecules in the corresponding rows of the Periodic Table. The value of 684 K for T_m, however, seems to be unrealistic being higher than that for T_b [108]. Ozhigov's T_m and T_b values have been extrapolated on the basis of Mendeleev's rule [7] and are in reasonable agreement with those given in different compilations [97, 103, 109, 110]. The T_m and T_b values estimated by Norseyev and Nefedov using the empirical parameter Z′ are somewhat lower compared with other corresponding data given in Table 5/8. However, the enthalpy of vaporization at the boiling

[1] Originally quoted value is given in parentheses.

temperature was found by these authors to be similar [23] to that given in the Handbook of Thermal Constants [97].

Table 5/8
Thermodynamic Data of Phase Transitions for At_2.

phase transition	T in K	ΔH in kJ/mol (kcal/mol) [a]	ΔS in $J \cdot mol^{-1} \cdot K^{-1}$ ($cal \cdot mol^{-1} \cdot K^{-1}$) [a]	Ref.
crystalline → liquid	684			[108]
	505.5			[7]
	473			[23]
	500	20.92 (5.0)	41.84 (10)	[97]
	575	23.85 (5.70)		[103]
	573			[109]
	575			[110]
crystalline → gaseous	0	86.91 (20.7)		[97]
	298.15	83.68 (20.0)	280.75 (67.1)	[97]
liquid → gaseous	572			[108]
	582			[7]
	543	50.21 12.0)		[23]
	590	54.39 (13.0)	92.05 (22.0)	[97]
	650	90.37 (21.6)		[103]
	608			[109]
	610			[110]

[a] Originally quoted values are given in parentheses.

Recently, Otozai and Takahashi determined the boiling temperature for At_2 from the GLC absolute retention volume and obtained 503 ± 3 K [68]. The T_b value determined by these authors is significantly lower than those given in Table 5/8 and does not seem to be reliable because of the uncertainty of the existence of At_2 (see Sections 5.7.2, p. 110, and 10.2.1.3.1, p. 224) under the given experimental conditions.

The heat of vaporization (ΔH_v) and heat of sublimation (ΔH_s) at the melting point for At_2 have been estimated by correlation of these quantities for halogens with their atomic numbers. The ΔH_v and ΔH_s values obtained in this way, as well as the literature value of the heat of melting (ΔH_m), were then adjusted using the relation $\Delta H_m = \Delta H_s - \Delta H_v$. The data estimated by Gerasimov et al. are given below [111].

$\Delta H_m = 17.57$ kJ/mol (2.1 kcal/g-atom)[1]

$\Delta H_s = 67.78$ kJ/mol (8.1 kcal/g-atom)[1]

$\Delta H_v = 50.21$ kJ/mol (6.0 kcal/g-atom)[1]

5.9.3.2 Thermodynamic Properties

The thermodynamic functions for molecular astatine have been estimated by Stull and Sinke by comparison with the corresponding thermodynamic properties of other halogens.

[1] Originally quoted values are given in parentheses.

 References on pp. 124/8

In their compilation, the data of the reference state and ideal gaseous state for At_2 over the entire temperature range from 298.15 to 3000 K are tabulated [103]. The selected and calculated values of the fundamental thermodynamic properties for crystalline solid and ideal gaseous At_2 at 298.15 K are given in the Handbook of Thermal Constants [97]. The enthalpy and Gibbs free energy of formation, entropy, and heat capacity at constant pressure for crystalline solid and ideal gaseous At_2 at 298.15 K from these two sources [97, 103] are given in Table 5/9. The heat of formation for At_2 has also been estimated by Kaganyuk based on the effective charge on astatine; he has obtained the value $\Delta H_f^\circ = 87.9$ kJ/mol [101].

Table 5/9
Thermodynamic Properties of At_2 at 298.15 K.

properties	crystalline solid		ideal gas	
	[97]	[103]	[97]	[103]
ΔH_f° in kJ/mol	0	0	83.68	90.37
(kcal/mol) [a]			(20.0)	(21.6)
ΔG_f° in kJ/mol	0	0	40.145	44.217
(kcal/mol) [a]			(9.595)	(10.568)
S° in $J \cdot mol^{-1} \cdot K^{-1}$	121.34	121.34	267.36	276.14
($cal \cdot mol^{-1} \cdot K^{-1}$) [a]	(29.0)	(29.0)	(63.9)	(66.0)
C_p° in $J \cdot mol^{-1} \cdot K^{-1}$	54.39	58.58	37.07	33.47
($cal \cdot mol^{-1} \cdot K^{-1}$) [a]	(13.0)	(14.0)	(8.86)	(8.0)

[a] Originally quoted values are given in parentheses.

Similar results were obtained by Kharitonov et al. using a statistical thermodynamic approach for determining the thermodynamic functions for gaseous molecular astatine. Estimated values of interatomic distance and vibrational frequency (see Section 5.11.4, p. 124) were used to calculate the enthalpy ($H_T^\circ - H_0^\circ$), internal energy ($U_T^\circ - U_0^\circ$), reduced isobar potential Φ^*, absolute entropy, and heat capacity at constant pressure for the temperature range from 298.15 to 1000 K [112]. These data are given in Table 5/10.

Table 5/10
Thermodynamic Functions for At_2 Calculated by Statistical Thermodynamic Approach [112].

T in K	H_T°-H_0° in J/mol	U_T°-U_0° in J/mol	$\Phi^* = -G_T^\circ$-H_0°/T in $J \cdot mol^{-1} \cdot K^{-1}$	S_T° in $J \cdot mol^{-1} \cdot K^{-1}$	C_p° in $J \cdot mol^{-1} \cdot K^{-1}$
298.15	10416	7937	241.2	276.1	37.1
400	14203	10878	251.6	287.8	37.2
500	17931	13774	259.5	295.4	37.3
600	21663	16675	266.1	302.2	37.3
700	25398	19578	271.7	307.9	37.4
800	29135	22483	276.5	312.9	37.4
900	32872	25389	280.8	317.1	37.4
1000	36611	28296	284.7	321.3	37.4

5.9.3.3 Dissociation Energy

From the decreasing tendency of dissociation energy (D) with increasing atomic number for molecular halogens (with the exception of F_2), the D value for At_2 is expected to be the lowest in this group of elements. Most of the dissociation energy data for At_2 have been estimated by using various empirical correlations connecting the D values for halogens with their other properties, such as ionization energy (I) [26], atomic number (Z) [113], period [113, 114], and experimental parameter (Z′) [23] (see Section 5.6, p. 108). Theoretical calculations of dissociation energy for At_2 have also been performed [115]. The D values from these sources are summarized in Table 5/11.

Table 5/11
Calculated Values of Dissociation Energy for At_2.

method of calculation	dissociation energy in kJ/mol		Ref.
lg D=f (lg I)	115.78	(1.2 eV)[a]	[26]
D=f (Z)	77.19	(0.8 eV)[a]	[113]
D=f (period)	77.19	(0.8 eV)[a]	[113]
	96.23	(23.0 kcal/mol)[a]	[114]
D=f (Z′)	112.13	(26.8 kcal/mol)[a]	[23]
theoretical	75.31	(18.0 kcal/mol)[a]	[115]
thermodynamic	108.78	(26 kcal/mol)[a]	[97]

[a] Originally quoted values are given in parentheses.

5.9.4 Positive Molecular Ion of Astatine, At_2^+

For this ionic species the enthalpy of formation has been deduced from the ionization energy of At_2 (see Section 5.6, p. 108) as 887 kJ/mol (212 kcal/mol)[1] [97].

The dissociation energy for At_2^+ derived from the thermodynamic cycle of dissociation and ionization, was 231.56 kJ/mol (2.4 eV)[1] [26]. A similar estimation with selected values for the dissociation energy of At_2 and the atomic and molecular ionization energies of astatine gave a value of 193.966 kJ/mol (46.359 kcal/mol)[1] [97].

5.9.5 At^+ and At^{2+} Ions

The heat of formation for gaseous At^+ and At^{2+} ions has been deduced from their ionization energies [96, 97], and is given below.

species	ΔH_f° in kJ/mol (kcal/mol)[a] [96]	[97]
At^+	1004 (240)	983 (235)
At^{2+}	2761 (660)	2084 (498)

[a] Originally quoted values are given in parentheses.

[1] Originally quoted values are given in parentheses.

References on pp. 124/8

5.10 Radii

5.10.1 Atomic Radius

The atomic radius (R) for astatine has been evaluated as 0.146 nm using Mendeleev's rule of extrapolation [7]. Estimation from the linear plot of atomic radii for halogens vs. their experimental parameter Z' (see Section 5.6, p. 108) leads to a value of 0.145 nm [23]. The values of 0.119 nm [17] and 0.127 nm [116] have been obtained by theoretical calculations.

5.10.2 Ionic Radii

5.10.2.1 At^- Ion

The radius (r) for the gaseous astatide ion has been estimated from the constancy of the ratio of radii for the isoelectronic ion pairs. The value of 0.197 nm calculated in this way by Krestov [96] seems to be low compared with those for other gaseous halogen ions given by Krasnov as 0.112 nm for F^-, 0.1683 nm for Cl^-, 0.1864 nm for Br^-, and 0.2119 nm for I^- [117].

Different methods have been applied to estimate the ionic crystal radius for At^- [7, 23, 96, 118 to 121]. The methods and the results of these calculations are summarized in Table 5/12.

Table 5/12
Calculated Values of Ionic Crystal Radius for At^-.

method	radius in nm	Ref.
from interionic distance	0.227	[117]
Mendeleev's rule	0.226	[7]
r=f (lg Z)	0.234	[118]
	0.261	[119]
r=f (r_{X^-}), X=halogen	0.230	[96]
r=f (Z')	0.229	[23]
r=f (lattice energy)	0.222	[120]

5.10.2.2 At^+ and At^{7+} Ions

Extrapolation according to Mendeleev's rule has been used to determine the radius of the gaseous At^+ ion. The value obtained in this way was 0.061 nm [7]. From the ratio $r_{At^+}/R_{At}=0.28$ determined by Korablev [122], the radius of At^+ can be estimated as 0.041 nm if one uses Ozhigov's value for the atomic radius of 0.146 nm [7].

The radius of At^{7+} has been determined by making use of the smooth regularity between radius and charge for isoelectronic ions. From the sequence Au^+, Hg^{2+}, Tl^{3+}, Pb^{4+}, and Bi^{5+} the extrapolated value for At^{7+} was 0.062 nm [123]. Norseyev and Nefedov estimated the radius of At^{7+} as 0.06 nm [23]. A value of 0.062 nm is given by Parsons [25].

5.11 Spectroscopic Properties

5.11.1 Atomic Absorption Spectrum

The ability to detect the atomic absorption spectrum is one of the very few ways of obtaining direct information on astatine. Using a highly sensitive spectroscopic detection method, McLaughlin was able to measure the absorption spectrum of astatine atoms. His method included the adaptation of capillary-absorption-cell spectroscopy, which allowed the detection of as little as 0.2 ng of gaseous astatine. The temperature of the quartz capillary absorption cell was kept at 600 °C to enhance the decomposition of molecules that might have contained astatine. In order to prove which of the recorded lines belongs to astatine, the intensities were monitored as a function of time. The decrease of line intensities for 244.401 nm and 216.225 nm were consistent with the radioactive decay of astatine. The two astatine lines were assigned by extrapolation from the lowest absorption lines for lighter halogens. The lines 224.401 and 216.225 nm were assigned to the transitions $^2P_{3/2} \rightarrow {}^4P_{5/2}$ and $^2P_{3/2} \rightarrow {}^4P_{3/2}$, respectively, between electron configurations $6p^5 \rightarrow 6p^4\,7s$ [124].

5.11.2 Mass Spectrometry

Using a highly sensitive (~40 atom/cm^3) time-of-flight spectrometer, Appelman et al. have directly detected the mass spectra of some astatine containing ions, such as At^+, HAt^+, CH_3At^+, and AtX^+ with X = Cl, Br, I (see Fig. 10-23, p. 225, and Fig. 10-26, p. 234), even though no evidence for At_2^+ and AtF^+ ions could be obtained [75].

More recently Golovkov et al., using a mass separator with plasma ion source, detected the following ionic species: AtH^+, $AtOH^+$, $AtOH_2^+$, AtX^+ (X = Cl, Br, I), and At_2^+ from their radioactivity. The formation of At_2^+ was explained as the result of a heterolytic reaction between At^+ and At^0 at the outlet of the plasma ion source. The existence of At_2^+ in this mass spectroscopic study was interpreted by means of its higher dissociation energy value compared with that of At_2 [76].

When introducing chlorine or bromine into the ion source in both cases, two lines of $AtCl^+$ and $AtBr^+$ could be observed with relative intensities corresponding to the relative abundance of the stable halogen isotopes in question [75, 76] (see Fig. 10-23, p. 225).

5.11.3 X-Ray Spectroscopic Properties

5.11.3.1 X-Ray Energies

In the electron capture branch of neutron-deficient radon isotopes, e.g., ^{208}Rn, ^{209}Rn, ^{210}Rn, ^{211}Rn, X-ray emission from the daughter astatine isotopes is expected. Indeed, X-rays with energies characteristic for astatine (K_{α_2} = 78.945 keV, K_{α_1} = 81.516 keV, and K_{β_1} = 92.30 keV) have been detected in the low energy region of γ spectra [125, 126]. Besides the experimental results, X-ray data for astatine are in a number of tables with computed values of X-radiation energies for the heavier elements (see, e.g. [127 to 129]). The X-ray energies of the principal K and L spectral lines for astatine, determined by interpolation – taken from the most frequently cited source, i.e., Bearden's X-ray wavelengths compilation [129], – are summarized in Table 5/13, p. 122. The relative intensities of these lines are also given in this table [127]. A complete tabulation of K and L X-ray energies and of their relative intensities for astatine are in [127].

5.11.3.2 X-Ray Atomic Energy Levels

Most of available atomic electron binding energy (E_b) values for ground state astatine have been estimated by interpolation in Z between the experimental binding energies of neighboring elements [127, 128, 130 to 133]. The X-ray atomic electron binding energy levels, given in Table 5/14, p. 122, are taken from the compilation of Sevier [132]. Electron binding energies for neutral astatine atoms have also been determined by theoretical calculations [134 to 136]. The theoretical data of Huang et al. [135] are also included in Table 5/14. The E_b values for the free astatine atom, deduced from X-ray data, are given in

References on pp. 124/8

the last column of Table 5/14 [133]. The theoretically calculated electron binding energy values are closer to those determined for the free astatine atom.

Table 5/13
The K and L X-Ray Energies for Astatine.

designation		X-ray energy in keV [129]	relative intensities in % [127]
K series			
α_2	KL_2	78.95	60.3
α_1	KL_3	81.52	100.0
β_3	KM_2	91.72	11.3
β_1	KM_3	92.30	22.0
β_2^{II}	KN_2	94.84	2.72
β_2^{I}	KN_3	94.99	5.41
L series			
β_3	L_1M_3	14.0670	33.55
β_1	L_2M_4	13.8760	131.50
γ_1	L_2N_4	16.2510	27.95
α_2	L_3M_4	11.3048	11.40
α_1	L_3M_5	11.4268	100.0 (5.21)[a]

[a] Relative intensity of the L_{α_1} line with respect to $K_{\alpha_1}=100\%$.

Table 5/14
Atomic Electron Binding Energies for Astatine.

level		$-E_b$ in eV		
		to Fermi level [132]	theor. calc. [135]	in free atom [133]
K	$(1s_{1/2})$	95724.0	95721.9	95729
L_1	$(2s_{1/2})$	17481.5	17496.0	17490
L_2	$(2p_{1/2})$	16777.3	16781.4	16782
L_3	$(2p_{3/2})$	14208.0	14213.4	14212
M_1	$(3s_{1/2})$	4311.7	4335.9	4320
M_2	$(3p_{1/2})$	3995.8	4009.2	4005
M_3	$(3p_{3/2})$	3410.5	3423.1	3420
M_4	$(3d_{3/2})$	2901.8	2909.6	2910
M_5	$(3d_{5/2})$	2780.7	2788.7	2788
N_1	$(4s_{1/2})$	1038.2	1054.9	1044
N_2	$(4p_{1/2})$	897.7	912.7	904
N_3	$(4p_{3/2})$	753.7	765.2	761

Table 5/14 [continued]

level	$-E_b$ in eV		
	to Fermi level [132]	theor. calc. [135]	in free atom [133]
N_4 $(4d_{3/2})$	527.6	538.1	535
N_5 $(4d_{5/2})$	500.1	510.6	508
N_6 $(4f_{5/2})$	207.0	212.6	217
N_7 $(4f_{7/2})$	200.8	206.3	211
O_1 $(5s_{1/2})$	193.4	206.6	196
O_2 $(5p_{1/2})$	145.6	154.3	153
O_3 $(5p_{3/2})$	113.6	124.4	123
O_4 $(5d_{3/2})$	40.9	47.9	48
O_5 $(5d_{5/2})$	37.4	43.9	44
P_1 $(6s_{1/2})$	15.0	24.5	19
P_2 $(6p_{1/2})$	5.7	11.9	11
P_3 $(6p_{3/2})$	2.8	8.6	9.3

5.11.3.3 K- and L-Shell Fluorescence Yields

The K-shell fluorescence yield (ω_K) has been estimated as 0.971 [137] and 0.969 [138]. The high value of ω_K indicates the high probability that a vacancy in the K-shell of astatine is filled via radiative transition. Theoretical calculations on the L-shell fluorescence yields have given the following results: $\omega_{L_1} = 0.129$ [139], $\omega_{L_2} = 0.422$, and $\omega_{L_3} = 0.380$ [140]. Similar values can be found in Krause's compilation: $\omega_{L_1} = 0.128$, $\omega_{L_2} = 0.415$, and $\omega_{L_3} = 0.399$ [138].

5.11.3.4 Natural Widths of Atomic K and L Levels and K_α X-Ray Lines

Semiempirical values of natural widths for K, L_1, L_2, and L_3 levels, have been estimated for atomic astatine from its corresponding theoretical radiative rate and estimated fluorescence yield. K_{α_1} and K_{α_2} X-ray line widths were calculated as the sum of the widths of the levels involved in the transition [141]. The semiempirical values of the natural widths for astatine atom are given below:

K	L_1	L_2	L_3	K_{α_1}	K_{α_2}
69.8	12.8	7.01	6.29	76.1	76.8

5.11.3.5 X-Ray Screening Constants

X-ray screening constants (σ_1 and σ_2), accounting for the screening of nuclear charge due to other electrons, can be calculated by Sommerfeld's classic relativistic energy formula [142]. In order to calculate the screening parameters in the X-ray spectra of astatine, the energy levels given by Bearden and Burr [130] were used. The value of σ_1 for the K level was estimated as 6.494 [143]; values of σ_2 for the $S_{1/2}$ terms of L_1, M_1, and N_1 levels were given as 2.5, 7.0, and 13.0, respectively [144]. The calculated values of σ_1 and σ_2 for various spin doublets are summarized in Table 5/15, p. 124 [145 to 148].

References on pp. 124/8

Table 5/15
X-Ray Screening Constants for Various Spin Doublets of Astatine.

screening constant	L_2L_3	M_2M_3	M_4M_5	N_2N_3	Ref.
σ_1	21.549	39.20	42.52	57.0	[145]
σ_2	3.580	8.64	13.60	16.7	[146, 147]
	3.550	8.00	13.60	17.0	[148]

5.11.4 Spectroscopic Constants of At_2

The ground state vibrational constant (ω_e) for the At_2 molecule was estimated using empirical relationships, combining the ω_e values with different atomic and molecular properties of astatine, such as atomic number (Z) [149], ionization energy and the principal quantum number of valence electrons (I, n) [150], electronegativity and the reduced mass (X, μ) [151, 152]. Theoretical calculations were also utilized [155, 153]. The ω_e values for the astatine molecule are given in Table 5/16. Lippincott's estimation of 99.0 cm^{-1} [115] seems to be too low compared with the other predicted ω_e values for At_2. From the sequence of ground state vibrational frequency values for other halogens [154], the estimated ω_e data for At_2 reported in [149 to 153] seem reasonable.

Table 5/16
Ground State Vibrational Constant Values for Astatine Molecule.

method	ω_e in cm^{-1}	Ref.
$\lg \omega_e = f(\lg 2Z)$	139.4	[149]
$\lg \omega_e = f(\lg n^2 I)$	172.9	[150]
$\omega_e = f(X/\mu^p)$	139.0	[151]
$\lg \omega_e = (\lg X/\mu^2)$	132.9	[152]
theoretical	141.0	[153]
	99.0	[154]

A theoretical calculation of the ground state anharmonic vibrational constant ($\omega_e x_e$) for the astatine molecule gave a value of 0.29 cm^{-1} [115].

From the ω_e values estimated by Clark [149] and by Tandon et al. [152], the force constant (k_e) for At_2 has been calculated as 1.213 N/cm (1.213 mdyn/Å)[1)] and 1.195 N/cm (1.195 mdyn/Å)[1)], respectively [112].

The interatomic distance (r_e) for At_2 has been calculated from the linear plot ln k_e vs. r_e for halogen molecules. Using the two estimated k_e values for At_2 (see above), the interatomic distance was 0.2967 and 0.2981 nm [112]. Theoretical calculations resulted in a value of 0.352 nm [115].

References:

[1] A.J. Downs, C.J. Adams (The Chemistry of Chlorine, Bromine, Iodine and Astatine, Pergamon, Oxford 1975, p. 1585). – [2] C.E. Moore (NBS-CIRC-467 [1958] 1/245; C.A. **1958**

1) Originally quoted values are given in parentheses.

11558). — [3] R.J. Zollweg (J. Chem. Phys. **50** [1969] 4251/61). — [4] L. Pauling (The Nature of the Chemical Bond, Cornell Univ. Press, New York 1960). — [5] W. Gordy, W.J.O. Thomas (J. Chem. Phys. **24** [1956] 439/44).

[6] Y. Zhang (Inorg. Chem. **21** [1982] 3886/9). — [7] E.P. Ozhigov (Zh. Obshch. Khim. **34** [1964] 3519/23; J. Gen. Chem. [USSR] **34** [1964] 3566/71). — [8] S.S. Batsanov (Zh. Strukt. Khim. **5** [1964] 293/301; J. Struct. Chem. [USSR] **5** [1964] 263/70). — [9] G.A. Korablev (Issled. Obl. Fiz. Khim. Perekhodnykh Elem. **1976** 133/7; C.A. **90** [1979] No. 92841). — [10] K. Bächmann, P. Hoffmann (Radiochim. Acta **15** [1971] 153/63).

[11] D. Bonchev, V. Kamenska (J. Phys. Chem. **85** [1981] 1177/86). — [12] I.L. Agafonov (Zh. Neorgan. Khim. **4** [1959] 1270/6; Russ. J. Inorg. Chem. **4** [1959] 573/5). — [13] T.S. Perova, V.S. Libov (Spektrokhim. Vnutri-Mezhmol. Vzaimodeistvii No. 2 [1978] 126/35; C.A. **91** [1979] No. 29774). — [14] K.D. Sen, P.T. Narasimhan (Phys. Rev. [3] B **15** [1977] 95/102). — [15] J. Thorhallsson, C. Fisk, S. Fraga (J. Chem. Phys. **49** [1968] 1987/8).

[16] R.R. Teachout, R.T. Pack (At. Data **3** [1971] 195/214). — [17] S. Fraga, J. Karwowski, K.M.S. Saxena (At. Data Nucl. Data Tables **12** [1973] 467/77). — [18] W.J. Stevens, M. Krauss (J. Phys. B **16** [1983] 2921/30). — [19] W. Finkelnburg (Z. Naturforsch. **2a** [1947] 16/20). — [20] W. Finkelnburg (Phys. Rev. [2] **77** [1950] 303).

[21] W. Finkelnburg, W. Humbach (Naturwissenschaften **42** [1955] 35/7). — [22] L.V. Gurvich, G.V. Karachevtsev, V.N. Kondrat'ev, Yu.A. Lebedev, V.A. Medvedev, V.K. Potapov, Yu.S. Khodeev (Energiya Razryva Khimicheskikh Svyazei: Potentsial Ionizatsii i Srodsvo k Elektronu, Nauka, Moscow 1974; C.A. **82** [1975] No. 47961). — [23] Yu.V. Norseev[1)], V.D. Nefedov (Issled. Khim. Tekhnol. Primen. Radioakt. Veshestv **1977** 3/8; C.A. **94** [1981] No. 3539). – [24] Y.P. Varshni (Z. Physik **135** [1953] 512/5). — [25] R. Parsons (Handbook of Electrochemical Constants, Butterworths, London 1959).

[26] R.W. Kiser (J. Chem. Phys. **33** [1960] 1265/6). — [27] D.R. Corson, K.R. MacKenzie, E. Segrè (Phys. Rev. [2] **57** [1940] 459). — [28] E. Segrè, K.R. MacKenzie, D.R. Corson (Phys. Rev. [2] **57** [1940] 1087). — [29] D.R. Corson, K.R. MacKenzie, E. Segrè (Phys. Rev. [2] **58** [1940] 672/8). — [30] J.G. Hamilton, M.H. Soley (Proc. Natl. Acad. Sci. U.S. **26** [1940] 483/9).

[31] J.G. Hamilton, M.H. Soley (J. Appl. Phys. **12** [1941] 314). — [32] G.L. Johnson, R.F. Leininger, E. Segrè (J. Chem. Phys. **17** [1949] 1/10). — [33] E.H. Appelman (UCRL-9025 [1960] 1/113; C.A. **1960** 16998). — [34] E.H. Appelman (J. Am. Chem. Soc. **83** [1961] 805/7). — [35] V.A. Khalkin[2)], Yu.V. Norseev, V.D. Nefedov, M.A. Toropova, V.I. Kuzin (Dokl. Akad. Nauk SSSR **195** [1970] 623/5; Dokl. Chem. Proc. Acad. Sci. USSR **190/195** [1970] 855/7).

[36] G.A. Nagy, P. Gróz, V.A. Halkin [Khalkin], Do-Kim Toung, J.V. Norszejev [Norseev] (KFKI-Kozl. **18** [1970] 173/6; C.A. **75** [1971] No. 14412). — [37] E. Kroó, I. Dudova, V.A. Halkin [Khalkin] (Magy. Kem. Folyoirat **82** [1976] 1/10; C.A. **84** [1976] No. 127325). — [38] I. Dreyer, R. Dreyer, Yu.V. Norseev, V.A. Chalkin [Khalkin] (Radiochem. Radioanal. Letters **33** [1978] 291/300). — [39] I. Dreyer, R. Dreyer, V.A. Chalkin [Khalkin] (Radiochem. Radioanal. Letters **35** [1978] 257/62). — [40] G.Á. Nagy, V.A. Halkin [Khalkin], J.V. Norszejev [Norseev] (Magy. Kem. Folyoirat **73** [1967] 191/4; C.A. **67** [1967] No. 50005).

[41] B.N. Belyaev, Van Yun-Yui, E.N. Sinotova, L. Német, V.A. Khalkin (Radiokhimiya **2** [1960] 603/13; Soviet Radiochem. [Jerusalem] **2** [1960] 92/102). — [42] M. Bochvarova,

1) Yu.V. Norseev, Yu.V. Norseyev, J.W. Norseev, J.V. Norszejev are the different transliterations of the same name.

2) V.A. Khalkin, W.A. Chalkin, V.A. Halkin are the different transliterations of the same name.

Do Kim Toung, I.V. Dudova, Yu.V. Norseev, V.A. Khalkin (Radiokhimiya **14** [1972] 858/65; Soviet Radiochem. **14** [1972] 889/95). — [43] K. Rössler, W. Tornau, G. Stöcklin (J. Radioanal. Chem. **21** [1974] 199/209). — [44] G.-J. Meyer, K. Rössler (Radiochem. Radioanal. Letters **25** [1976] 377/90). — [45] R. Dreyer, I. Dreyer, F. Rösch (Z. Chem. [Leipzig] **22** [1982] 54/6; C.A. **96** [1982] No. 150056).

[46] Do Kim Tuong, I.V. Dudova, V.A. Khalkin (Radiokhimiya **14** [1972] 766/7; Soviet Radiochem. **14** [1972] 790/2). — [47] A.H.W. Aten Jr., T. Doorgeest, U. Hollstein, H.P. Moeken (Analyst **77** [1952] 774/8). — [48] A.H.W. Aten Jr., J.G. van Raaphorst, G. Nooteboom, G. Blasse (J. Inorg. Nucl. Chem. **15** [1960] 198/9). — [49] I. Dreyer, R. Dreyer, Yu.V. Norseev, V.A. Chalkin [Khalkin] (Radiochem. Radioanal. Letters **33** [1978] 281/90). — [50] I. Dreyer, R. Dreyer, V.A. Chalkin [Khalkin] (Radiochem. Radioanal. Letters **36** [1978] 389/98).

[51] M. Lefort, G. Simonoff, X. Tarrago (Bull. Soc. Chim. France **1960** 1726/7). — [52] G.Á. Nagy, V.A. Halkin [Khalkin] (Magy. Kem. Folyoirat **81** [1975] 33/5; C.A. **82** [1975] No. 116686). — [53] K. Berei, L. Vasáros (in: S. Pataí, E. Rappoport, The Chemistry of Functional Groups, Suppl. D, Pt. 1, Wiley, New York 1983, pp. 405/40). — [54] H.M. Neumann (J. Inorg. Nucl. Chem. **4** [1957] 349/53). — [55] L. Lindner, G.A. Brinkman, T.H.G.A. Suèr, A. Schimmel, J.T. Veenboer, F.H.S. Karten, J. Visser, C.J. Leurs (Radiopharm. Labelled Compounds Proc. Symp., Copenhagen 1973, Vol. 1, pp. 303/16).

[56] C. Aaij, W.R.J.M. Tschroots, L. Lindner, T.E.W. Feltkamp (Intern. J. Appl. Radiat. Isotop. **26** [1975] 25/30). — [57] G.-J. Beyer, R. Dreyer, H. Odrich, F. Rösch (Radiochem. Radioanal. Letters **47** [1981] 63/5). — [58] V. Doberenz, D.D. Nhan, R. Dreyer, M. Milanov, Yu.V. Norseyev [Norseev], V.A. Khalkin (Radiochem. Radioanal. Letters **52** [1982] 119/27). — [59] B. Eichler (Radiochem. Radioanal. Letters **22** [1975] 147/55). — [60] V.M. Vakhtel, G.V. Vinel, Ts. Vylov, I.I. Gromova, A.F. Novgorodov, Yu.V. Norseev, V.A. Khalkin, V.G. Chumin (Radiokhimiya **18** [1976] 886/93; Soviet Radiochem. **18** [1976] 752/8).

[61] V.M. Vakhtel, G.V. Vinel, Ts. Vylov, I.I. Gromova, A.F. Novgorodov, Yu.V. Norseyev [Norseev], V.G. Chumin, V.A. Khalkin (Isotopenpraxis **12** [1976] 441/6). — [62] G.-J. Meyer (JUEL-1076-NC [1974] 1/61; C.A. **82** [1975] No. 78910). — [63] G.-J. Meyer, K. Rössler, G. Stöcklin (AED-CONF-404-029 [1975] 1/9; C.A. **85** [1976] No. 139398). — [64] G.-J. Meyer (JUEL-1418 [1977] 1/107; C.A. **88** [1978] No. 71113). — [65] E.H. Appelman (NAS-NS-3012 [1960] 1/29; C.A. **1960** 19191).

[66] Yung-Yu Wang, V.A. Khalkin (Radiokhimiya **3** [1961] 662/6; Radiochem. [USSR] **3** [1961] 258). — [67] J. Merinis, Y. Legoux, G. Bouissières (Radiochem. Radioanal. Letters **11** [1972] 59/64). — [68] K. Otozai, N. Takahashi (Radiochim. Acta **31** [1982] 201/3). — [69] A.H.W. Aten Jr. (in: O.R. Frisch, F.A. Paneth, F. Laves, P. Rosbaud, Beiträge zur Physik und Chemie des 20. Jahrhunderts, Vieweg, Braunschweig 1959, pp. 121/3). — [70] A.H.W. Aten Jr. (Advan. Inorg. Chem. Radiochem. **6** [1964] 207/23).

[71] E.H. Appelman (MTP [Med. Tech. Publ. Co.] Intern. Rev. Sci. Inorg. Chem. Ser. One **3** [1972] 181/98). — [72] G.W.M. Visser, E.L. Diemer (Radiochim. Acta **33** [1983] 145/51). — [73] A.G. Maddock (in: Mellor's Comprehensive Treatise on Inorganic and Theoretical Chemistry, Suppl. II, Pt. I, Longmans, London 1956, pp. 1064/79). — [74] E.H. Appelman (J. Phys. Chem. **65** [1961] 325/31). — [75] E.H. Appelman, E.N. Sloth, M.H. Studier (Inorg. Chem. **5** [1966] 766/9).

[76] N.A. Golovkov, I.I. Gromova, M. Janicki, Yu.V. Norseyev [Norseev], L. Vasáros (Radiochem. Radioanal. Letters **44** [1980] 67/78). — [77] Fu-Chun Wang, Yu.V. Norseev, V.A. Khalkin, Tao-Nan Ch'ao (Radiokhimiya **5** [1963] 351/5; Soviet Radiochem. **5** [1963] 318/22). — [78] Do Kim Tuong, I.V. Dudova, V.A. Khalkin (Radiokhimiya **15** [1973] 548/53; Soviet Radio-

chem. **15** [1973] 552/6). – [79] Yu.V. Norseyev [Norseev], V.A. Khalkin (J. Inorg. Nucl. Chem. **30** [1968] 3239/43). – [80] Fu-Chun An, N.G. Krylov, Yu.V. Norseev, Tao-Nan Ch'ao, V.A. Khalkin (Soosazhdenie Adsorbtsiya Radioakt. Elem. **1965** 80/8; C.A. **63** [1965] 7695).

[81] Yu.V. Norseev, Tao-Nan Ch'ao, V.A. Khalkin (Radiokhimiya **8** [1966] 497/504; Soviet Radiochem. **8** [1966] 461/6). – [82] I. Dreyer, R. Dreyer, V.A. Chalkin [Khalkin], M. Milanov (Radiochem. Radioanal. Letters **40** [1979] 145/53). – [83] R. Dreyer, I. Dreyer, M. Pfeiffer, F. Rösch (Radiochem. Radioanal. Letters **55** [1982] 207/13). – [84] R. Dreyer, I. Dreyer, F. Rösch, S. Fischer (Z. Chem. [Leipzig] **23** [1983] 346/7; C.A. **100** [1984] No. 13379). – [85] Yu.V. Norseev, V.A. Khalkin, Tao-Nan Ch'ao (Izv. Sibirsk. Otd. Akad. Nauk SSSR Ser. Khim. Nauk **1965** No. 3, pp. 21/7; C.A. **64** [1966] 16676).

[86] W.A. Chalkin [V.A. Khalkin], E. Herrmann (Isotopenpraxis **11** [1975] 333/40). – [87] W.A. Chalkin [V.A. Khalkin], E. Herrmann, J.W. Norseev, I. Dreyer (Chemiker-Ztg. **101** [1977] 470/81; C.A. **88** [1978] No. 28663). – [88] M. Milanov, V. Doberenz, V.A. Khalkin, A. Marinov (J. Radioanal. Nucl. Chem. Articles **83** [1984] 291/9). – [89] L. Vasáros, Yu.V. Norseyev [Norseev], D.D. Nhan, V.A. Khalkin (Radiochem. Radioanal. Letters **54** [1982] 239/47). – [90] V.D. Nefedov, Yu.V. Norseev, Kh. Savlevich, E.N. Sinotova, M.A. Toropova, V.A. Khalkin (Dokl. Akad. Nauk SSSR **144** [1962] 806/9; Dokl. Chem. Proc. Acad. Sci. USSR **142/147** [1962] 507/10).

[91] V.D. Nefedov, Yu.V. Norseev, M.A. Toropova, V.A. Khalkin (Usp. Khim. **37** [1968] 193/215; Russ. Chem. Rev. **37** No. 4 [1968] 87/98). – [92] E.H. Appelman (J. Am. Chem. Soc. **90** [1968] 1900/1). – [93] W.M. Latimer (The Oxidation States of the Elements and their Potentials in Aqueous Solutions, Prentice-Hall, New York 1953). – [94] A. Cavallero (Diss. Univ. Cath. Louvain, Belg., 1981, pp. 1/100; JUEL-Report to be published in 1985). – [95] M.F.C. Ladd, W.H. Lee (J. Inorg. Nucl. Chem. **18** [1961] 163/5).

[96] G.A. Krestov (Radiokhimiya **4** [1962] 690/6; Soviet Radiochem. **4** [1962] 612/7). – [97] V.P. Glushko, V.A. Medvedev, G.A. Bergman, L.V. Gurvich, V.S. Yungman, A.F. Vorobev, V.P. Kolesov, L.A. Reznitskii, B.V. Mikhailov, G.L. Galchenko, M.Kh. Karapetyants (Termicheskie Konstanty Veshchestv, Pt. I, Nauka, Moscow 1966; C.A. **65** [1966] 3097). – [98] G.A. Krestov (Izv. Vysshikh Uchebn. Zavedenii Khim. Khim. Tekhnol. **6** [1963] 754/61; C.A. **60** [1964] 8712). – [99] G.A. Krestov (Teor. Eksperim. Khim. **1** [1965] 479/84; C.A. **64** [1966] 15078). – [100] G.A. Krestov (Izv. Vysshikh Uchebn. Zavedenii Khim. Khim. Tekhnol. **6** [1963] 228/32; C.A. **59** [1963] 12247).

[101] D.S. Kaganyuk (Radiokhimiya **25** [1983] 814/6; Soviet Radiochem. **25** [1983] 778/80). – [102] G.A. Krestov (Izv. Vysshikh Uchebn. Zavedenii Khim. Khim. Tekhnol. **11** [1968] 762/5; C.A. **70** [1969] No. 7001). – [103] D.R. Stull, G.C. Sinke (Advan. Chem. Ser. No. 18 [1956]). – [104] V. Gopal (Current Sci. [India] **36** [1967] 456/7). – [105] L.A. Cole, J.P. Perdew (Phys. Rev. [3] A **25** [1982] 1265/71).

[106] H. Hotop, W.C. Lineberger (J. Phys. Chem. Ref. Data **4** [1975] 539/76). – [107] G.A. Krestov (Izv. Vysshikh Uchebn. Zavedenii Khim. Khim. Tekhnol. **10** [1967] 391/7; C.A. **68** [1968] No. 7040). – [108] J.F. Durand (Bull. Soc. Chim. France [5] **3** [1936] 1382/8). – [109] R.E. Honig, D.A. Kramer (RCA Rev. **30** [1969] 285/305; C.A. **71** [1969] No. 64413). – [110] R.C. Weast, M.J. Astle (CRC Handbook of Chemistry and Physics, 62nd Ed., CRC Press, Boca Raton, Fla., 1981/82, p. B-7).

[111] Ya.I. Gerasimov, V.M. Glazov, V.B. Lazarev, V.V. Zharov, A.S. Pashinkin (Dokl. Akad. Nauk SSSR **235** [1977] 846/9; Dokl. Phys. Chem. Proc. Acad. Sci. USSR **232/237** [1977] 753/5). – [112] Yu.Ya. Kharitonov, O.V. Bazileva, T.V. Gerzha (Izv. Vysshikh Uchebn. Zavedenii Khim. Khim. Tekhnol. **22** [1979] 939/43; C.A. **91** [1979] No. 199848). – [113] J. Drowart, R.E. Honig (J. Phys. Chem. **61** [1967] 980/5). – [114] S.G. Kim (Zh. Fiz. Khim. **53** [1979]

736/7; Russ. J. Phys. Chem. **53** [1979] 414/3). – [115] E.R. Lippincott, M.O. Dayhoff (Spectrochim. Acta **16** [1960] 807/34).

[116] E. Clementi, D.L. Raimondi, W.P. Reinhardt (J. Chem. Phys. **47** [1967] 1300/7). – [117] K.S. Krasnov (Zh. Neorgan. Khim. **3** [1958] 1993/8; Russ. J. Phys. Chem. **3** No. 9 [1958] 1/9). – [118] W.H. Zachariasen (in: G.T. Seaborg, J.J. Katz, The Actinide Elements, McGraw-Hill, New York 1954, pp. 769/96). – [119] L. Genov (Zh. Obshch. Khim. **29** [1959] 689/91; J. Gen. Chem. [USSR] **29** [1959] 683/5). – [120] A.N. Zhitomirskii (Zh. Strukt. Khim. **11** [1970] 1082/6; J. Struct. Chem. [USSR] **11** [1970] 1008/11).

[121] H.D.B. Jenkins, K.P. Thakur (J. Chem. Educ. **56** [1979] 576/7). – [122] G.A. Korablev (Zh. Strukt. Khim. **6** [1965] 323/4; J. Struct. Chem. [USSR] **6** [1965] 306/7). – [123] L.H. Ahrens (Geochim. Cosmochim. Acta **2** [1952] 155/69; C.A. **1952** 8475). – [124] R. McLaughlin (J. Opt. Soc. Am. **54** [1964] 965/7). – [125] Ts. Vylov, N.A. Golovkov, K.Ya. Gromov, I.I. Gromova, A. Kolaczkowski, M.Ya. Kuznetsova, Yu.V. Norseev, V.G. Chumin (Izv. Akad. Nauk SSSR Ser. Fiz. **38** [1974] 701/9; Bull. Acad. Sci. USSR Phys. Ser. **38** No. 4 [1974] 31/7).

[126] Ts. Vylov, N.A. Golovkov, M. Gonusek, M.Ya. Kuznetsova, Yu.V. Norseev, H.G. Ortlepp, V.G. Chumin (Izv. Akad. Nauk SSSR Ser. Fiz. **42** [1978] 765/72; Bull. Acad. Sci. USSR Phys. Ser. **42** No. 4 [1978] 64/9). – [127] E. Storm, H. Israel (Nucl. Data Tables A **7** [1970] 565/681). – [128] S. Hagström, C. Nordling, K. Siegbahn (in: K. Siegbahn, Alpha-, Beta- and Gamma-Ray Spectroscopy, Vol. 1, North-Holland, Amsterdam 1968, pp. 845/56). – [129] J.A. Bearden (Rev. Mod. Phys. **39** [1967] 78/124). – [130] J.A. Bearden, A.F. Burr (Rev. Mod. Phys. **39** [1967] 125/42).

[131] F.T. Porter, M.S. Freedman (J. Phys. Chem. Ref. Data **7** [1978] 1267/84). – [132] K.D. Sevier (At. Data Nucl. Data Tables **24** [1979] 323/71). – [133] W. Lotz (J. Opt. Soc. Am. **60** [1970] 206/10). – [134] J.P. Desclaux (At. Data Nucl. Data Tables **12** [1973] 311/406). – [135] Keh-Ning Huang, M. Aoyagi, M.H. Chen, B. Crasemann, H. Mark (At. Data Nucl. Data Tables **18** [1976] 243/91).

[136] M.H. Chen, B. Crasemann, M. Aoyagi, Keh-Ning Huang, H. Mark (At. Data Nucl. Data Tables **26** [1981] 561/74). – [137] W. Bambynek, B. Crasemann, R.W. Fink, H.-U. Freund, H. Mark, C.D. Swift, R.E. Price, P.V. Rao (Rev. Mod. Phys. **44** [1972] 716/813). – [138] M.O. Krause (J. Phys. Chem. Ref. Data **8** [1979] 307/27). – [139] B. Crasemann, M.H. Chen, V.O. Kostroun (Phys. Rev. [3] A **4** [1971] 2161/64). – [140] M.H. Chen, B. Crasemann, V.O. Kostroun (Phys. Rev. [3] A **4** [1971] 1/7).

[141] M.O. Krause, J.H. Oliver (J. Phys. Chem. Ref. Data **8** [1979] 329/38). – [142] A. Sommerfeld (Atombau und Spektrallinien, Vol. 1, Vieweg, Braunschweig 1944). – [143] U.D. Misra, R.N. Singh, B.G. Gokhale (J. Phys. B **12** [1979] 1775/80). – [144] B.G. Gokhale, U.D. Misra (J. Phys. B **11** [1978] 2077/86). – [145] B.G. Gokhale, U.D. Misra (J. Phys. B **10** [1977] 3599/606).

[146] B.D. Padalia (J. Phys. B **2** [1969] 811/3). – [147] B.D. Padalia, S.G. Phadnis (J. Phys. B **2** [1969] 1094/6). – [148] A.F. Burr, J.K. Carson (J. Phys. B **7** [1974] 451/9). – [149] C.H.D. Clark (Trans. Faraday Soc. **33** [1937] 1398/401). – [150] K. Majumdar, Y.P. Varshni (Indian J. Phys. **28** [1954] 103/8).

[151] Z. Hussain (Can. J. Phys. **43** [1965] 1690/2). – [152] S.P. Tandon, M.P. Bhutra, P.C. Mehta (Z. Physik. Chem. [Leipzig] **245** [1970] 230/5). – [153] H. Dunken, H. Müller (Z. Chem. [Leipzig] **3** [1963] 73). – [154] G. Herzberg (Molecular Spectra and Molecular Structure, I: Spectra of Diatomic Molecules, Van Nostrand, Toronto 1953).

6 Analytical Chemistry of Astatine

H.W. Kirby
Mound Facility
Monsanto Research Corporation
Miamisburg, Ohio, USA

6.1 Introduction

Astatine (At), element 85, has no stable isotopes. Its longest-lived isotope, ^{210}At, has a half-life of only 8.3 h and a specific activity of 1.8 Ci/μg. Consequently, studies of At chemistry have necessarily been limited to trace levels on the order of 10^{-12} to 10^{-15} M. General reviews of At chemistry are given by [1 to 16].

Astatine is a halogen and one might expect, from its position in Group VII of the Periodic Table, that its chemical properties could be deduced or extrapolated from those of iodine. However, in the last rows of the Periodic System, the similarities between adjacent members are often greater than those with their lower homologues [17, 19]. Thus, At has marked metallic properties and, in some respects, seems to resemble Po (element 84) more than it does I. Furthermore, since the trace level chemistry of I sometimes differs significantly from its own macroscopic chemistry, analogies between At and I are likely to be questionable at best [18]. The anomalous behavior of At and I at trace levels has been attributed, at least in part, to reactions with trace impurities [19 to 21].

Five oxidation states are known. Appelman [22] has proposed the following scheme for the electrode potentials of At species in 0.1 M acid:

$$At^- \xrightarrow{-0.3\,V} At^0 \xrightarrow{-1.0\,V} At^{+x} \xrightarrow{-1.5\,V} AtO_3^- \xrightarrow{<-1.6\,V} H_5AtO_6(?)$$

Elemental astatine, At^0, is the form obtained when At is distilled from Bi at elevated temperatures. According to Johnson et al. [19], it is the oxidation state most easily prepared and most stable in aqueous solution. At^0 is characterized by its volatility from solution, its tendency to be adsorbed on metal and glass surfaces and its extractability by organic solvents, such as CCl_4 and C_6H_6. It is also characterized by extremely irreproducible behavior [41]. However, in the presence of molecular I_2, the dominant species are interhalogen compounds (e.g., AtI or AtI_2^-) and the results are more reproducible [23 to 25].

At^0 is reduced to the astatide ion, At^-, by metallic Zn, $SnCl_2$ or SO_2 in acid solution [17, 19], by $HSnO_2^-$ in alkaline solution [26], by ferrocyanide at pH>2 and by AsO_2^- at pH>5 [22]. At^- is oxidized to At^0 by dilute HNO_3, VO_2^+, Fe^{3+}, I_2, arsenate at pH<5 or ferricyanide at pH<3 [2, 21, 22]. At^- coprecipitates quantitatively with AgI, TlI and other insoluble iodides [19]. In acid solution, At^-, unlike I^-, is quantitatively adsorbed on Ag and Tl surfaces [26, 27].

The astatate ion, AtO_3^-, is formed by treating lower oxidation states with strong oxidizing agents, such as alkaline hypochlorite, periodate, persulfate or Ce^{IV} in acid solution. Heating is sometimes necessary for complete oxidation. AtO_3^- is quantitatively coprecipitated with $AgIO_3$, $Ba(IO_3)_2$ and $Pb(IO_3)_2$ [4, 19, 22].

An oxidation state (or states) of unknown valence, intermediate between 0 and +5, has been designated At^{+x} [2, 22]. It is presumed to be either At^+ or At^{3+} or an indeterminate mixture of the two. At^{+x} is formed by the oxidation of aqueous At^0 with relatively mild oxidants, such as bromine and dichromate, or, in the presence of light, Fe^{3+} and VO_2^+. At^{+x} is neither extracted by CCl_4 nor carried by $AgIO_3$ [19, 22].

Finally, the perastatate ion, AtO_4^-, is prepared by oxidation of At^- in alkaline solution at 100 °C with XeF_2. It is characterized by its behavior during paper electrophoresis (**Fig.** 6-**1**) and by cocrystallization with KIO_4 and $CsIO_4$ [28].

The aqueous reactions of astatine are summarized in **Fig.** 10-3, p. 187 [6, p. 1588].

Fig. 6-1

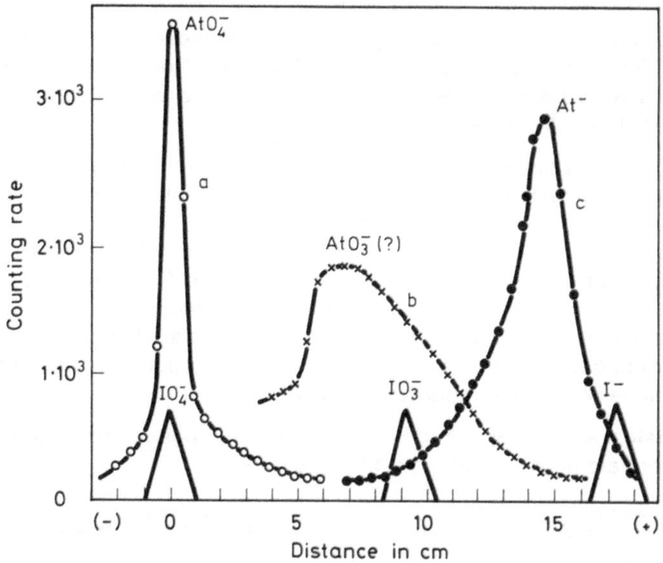

Electrophoresis of various forms of astatine [28] obtained (a) in the oxidation of astatide by xenon difluoride (b) in the reduction of the oxidized form with sulfur dioxide (c) in the reduction of the oxidized form with rongalite or in the case of alkalinization of a preparation of $[At(x)Cl_2]^-$ Whatman No. 1 paper, 0.1 M Na_2SO_4, 50 V/cm, 30 min.

6.2 Detection and Quantification

All At isotopes are short-lived (Table 6/1). The longest-lived isotope is 8.3-h ^{210}At; however, 7.2-h ^{211}At is the isotope most often used in chemical and biological research because of its favorable nuclear properties and the relative ease with which it can be prepared in radiochemically and isotopically pure form (cf. Section 4.2.1.1, p. 95).

^{211}At can be assayed by gross α-counting in a gas-flow ionization chamber or proportional counter [2], or in a ZnS or liquid scintillation counter [5, 29, 34]. If the isotopic purity of the sample is in doubt, α-pulse height analysis is possible [11, 26, 31].

However, in all of these methods (except liquid scintillation counting), sample preparation techniques are critical. The sample must be dried before it is presented to the detector and care must be taken to avoid loss of material during evaporation. According to Appelman [3], if an At solution in ~2 M HCl is evaporated to dryness on a Pt or Ag foil under an infrared lamp, a reproducible and adherent At deposit is obtained, but erratic losses of At occur under most other conditions. On the other hand, Wang and co-workers [30] evaporated 0.5 M HCl solutions of At to dryness on Pt and Ag surfaces at ~95 °C on a boiling

Table 6/1
Nuclear Properties of Astatine Isotopes [40].

A	$T_{1/2}$	decay modes (% of decay)	principal α-energies in MeV (% of α)	principal γ-energies in keV (relative intensity)
196	0.3 s	α	7.055	
197	0.4 s	α	6.959	
198	4.9 s	α	6.748	
198m	1.5 s	α	6.849	
199	7.2 s	α	6.639	
200g	42 s	α (53) $\varepsilon+\beta^+$ (47)	6.466 (60) 6.415 (40)	
200m	4.3 s	α	6.536	
201	1.50 min	α (71) $\varepsilon+\beta^+$ (29)	6.342	571.0
202	3.0 min	α (15) $\varepsilon+\beta^+$ (85)	6.227 (60) 6.133 (40)	441.3 (33) 569.7 (94) 675.3 (100)
202	2.6 min	α	6.227?	
203	7.3 min	α (31) $\varepsilon+\beta^+$ (69)	6.086	245.9 (48) 639.3 (97) 641.4 (53) 1002.0 (86)
204	9.3 min	α (4.4) $\varepsilon+\beta^+$ (95.6)	5.947	426.0 (66) 516.3 (95) 684.5 (100)
205	26.2 min	α (10) ε (87), β^+ (3)	5.903	669.4 (30) 719.3 (100)
206	31.4 min	α (0.96) ε (82), β^+ (17)	5.703	395.5 (42) 476.9 (82) 700.3 (100)
207	1.80 h	α (~10) $\varepsilon+\beta^+$ (~90)	5.759	300.7 (29) 588.4 (44) 814.5 (100)
208	1.63 h	α (0.55) $\varepsilon+\beta^+$ (99.45)	5.641 (96.8) 5.626 (2.1) 5.586 (0.9)	177.0 (50) 660.0 (92) 685.0 (99)
209	5.42 h	α (4.1) ε (95.9)	5.647	545.0 (62) 782.0 (58)

References on pp. 138/9

Table 6/1 [continued]

A	$T_{1/2}$	decay modes (% of decay)	principal α-energies in MeV (% of α)	principal γ-energies in keV (relative intensity)
210	8.3 h	α (0.175) $\varepsilon+\beta^+$ (99+)	5.524 (31) 5.442 (30) 5.361 (26)	245.3 (79) 1180.4 (100) 1482.9 (48)
211	7.214 h	α (41.94) ε (58.06)	5.866 (~100)	669.6
212	0.315 s	α	7.679 (83.5) 7.618 (15.1)	
212 m	0.122 s	α	7.897 (33.1) 7.837 (65.3)	
213	0.11 μs	α	9.080	
214	~2 μs	α	8.819	
215	0.10 ms	α	8.026	
216	0.30 ms	α	7.800	
216 m	unknown	α	7.960	
217	32.3 ms	α (99+) β^- (0.012)	7.067 (99.0)	218 259 334 592
218	1.5 to 2.0 s	α (99.9) β^- (0.1)	6.757 (3.6) 6.694 (90) 6.654 (6.4)	
219	0.9 min	α (~97) β^- (~3)	6.28	

water bath with no loss of material. Belyaev et al. [31] prepared sources for α-spectrometry by evaporation of a nitrate solution on a polished Pt disk and reported that losses of At caused by volatility were <2 or 3%, see also [32].

When the At is mixed with relatively large amounts of inert material (as in coprecipitation experiments), counting losses due to self-absorption may become significant. This problem can be circumvented by preparing "infinitely thick" samples; i.e., samples so thick that none of the α-particles emitted from the bottom of the sample are counted. However, errors of up to 10% may result from the use of this technique [5] and, of course, only relative counting rates can be measured, see also [33].

Absolute α-counting is possible by liquid scintillation counting [34]. Surprisingly, this method is seldom used although it is easy to apply and relatively insensitive to sample preparation technique. (It should be noted, however, that liquid scintillation counting is incapable of distinguishing between α-particles and hard β^--particles.) Autoradiography, using the At α-radiation, has also been used in biological research [29].

The simplest and most convenient detection method is by X- and γ-ray scintillation counting with an NaI(Tl) well crystal. This nondestructive method requires no special preparation technique to avoid loss of material and is subject to relatively few manipulative errors. If the output of the photomultiplier tube is fed to a single- or multichannel pulse height analyzer and the window or region of interest set to count only the ^{211}Po X-rays at ~80 keV, the inherently high background of the NaI crystal can be reduced to tolerable levels.

Lefort et al. [26] used both α- and X-ray spectrometry to check the purity of the At recovered from a proton-irradiated Th target, since spallation and fission reactions produce stable and radioactive isotopes of nearly all the elements from P to Pa. (Especially high effective cross sections are observed for isotopes of Mo, Te, I and Cs.) A typical α-spectrum of ^{211}At and its daughter, ^{211}Po, is given in **Fig. 6-2**; as little as 1 part in 1000 of other α-emitters from Bi to U would have been detected. The X-ray/γ-ray spectra of ^{210}At and ^{211}At are shown in **Fig. 6-3**, p. 134, see also [31].

Fig. 6-2

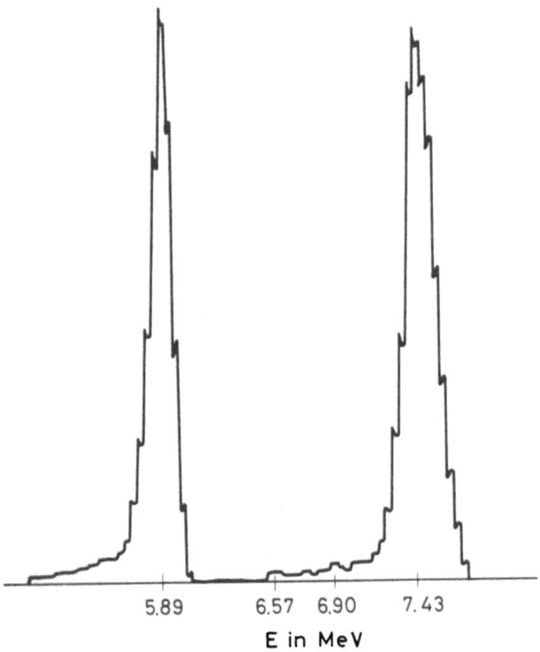

Alpha spectrum of ^{211}At (5.89 MeV) and ^{211}Po (7.43 MeV) [26].

6.3 Separation from Other Elements

6.3.1 By Distillation

The volatility of At was one of its earliest observed properties [17, 35]. Its application to the separation of At from irradiated Bi targets has been discussed in Section 4.2.1, p. 95.

The following method was used by Barton et al. [36] for the separation of short-lived At isotopes:

Fig. 6-3

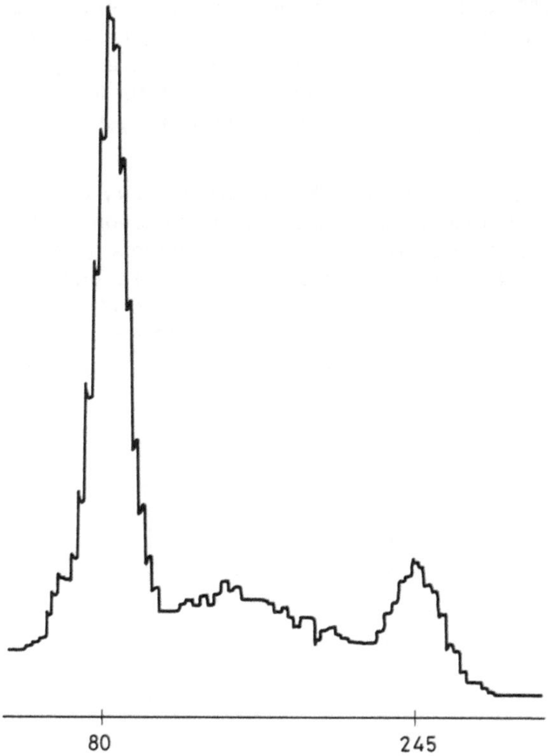

X-ray/gamma-ray spectrum of At isotopes separated from an irradiated Th target [26].

The Bi target is dropped into a stainless steel crucible fitted with a water-cooled steel finger to which a collecting Pt disk is clamped. When the Bi is kept slightly above its melting point (as measured by a thermocouple fitted into a well in the crucible), within a few seconds At distills onto the collecting plate. Po does not distill in appreciable quantities until considerably higher temperatures are reached.

According to Appelman [2, 3], temperatures as high as 800 °C may be required for nearly quantitative removal of the At. At such temperatures, from 40 to 80% of the At can be removed from the target, but it is likely to be contaminated with Po as well as with some Bi. Deposits on Pt or Ag are strongly adherent; those on Al much less so. To purify the At, Appelman recommends placing the Pt plate containing the At and some Bi or Bi_2O_3 in a tube on the end of an all-glass vacuum system incorporating a U tube. The plate is heated to 120 °C to drive off volatile impurities. The U tube is cooled with liquid N_2 and the temperature of the plate is raised to 500 °C for ~10 min. The vacuum is broken and the At is recovered from the U tube by dissolution with H_2O or any other aqueous solution, see also [19, 37, 38, 43, 45, 46].

In the case of solutions, the degree of volatility appears to be dependent on the nature of the solution (Table 6/2) as well as on the kind of surface (Table 6/3) from which the At is being evaporated. From the data in Table 6/2, it appears that At is distilled most

Table 6/2
Distillation of Astatine Solutions [19].

initial solution (10 mL)	percent of At left in still
16 M HNO_3	86
5 M HNO_3 + 3 mg I^-	98 [a]
12 M HCl	70
1:1 H_2SO_4	91
60% $HClO_4$	101
0.5 M H_2SO_4 + 0.05 M Fe^{2+}	7.4
CCl_4	64
benzene	62

[a] First 10% of distillate contained all of the I_2 but only 2% of the At.

Table 6/3
Effect of Substrate on Astatine Volatility [30].

material	percent of At remaining after drying 0.1 mL of 0.5 M HCl At solution at ~90 °C
Pt	100
Ag	100
stainless steel (1Kh18N9T)	85±5
Sn	82±5
Cu	81±4
polyvinyl chloride	67±3
Pb	58±15
Ta	12±4
polyethylene	11±2
glass	<5

effectively from 0.5 M H_2SO_4 containing some Fe^{2+}. According to Belyaev and co-workers [31], the presence of 5 to 10 mg of Te in the sulfate solution sharply inhibits the distillation of At. Before attempting to distill the At, therefore, it is necessary to separate it from the Te by some other methods, such as extraction with diisopropyl ether [36].

Kurchatov et al. [27] used the following procedure to separate At isotopes from Bi targets irradiated with 180 to 480 MeV protons:

> A bombarded specimen was dissolved in a minimum amount of concentrated HNO_3 and diluted to 30 to 40 mL by 10% HCl. After the addition of 0.25 mL of a 2% solution

References on pp. 138/9

of Te in HNO_3, At was coprecipitated with Te by 1 N $SnCl_2$. The precipitate was washed and the Te was dissolved in 2 mL of concentrated H_2SO_4 to which two drops of concentrated HNO_3 had been added. The solution was diluted to 80 to 100 mL with H_2O and 2 g of $FeSO_4$ added. The At was distilled, with 10 mL of the solution remaining in the distilling flask. The distillate was received in a vessel containing a solution of NaOH. Five mg of Te was added and precipitated by addition of a stannite for purification from the residual Po and β^-- and γ-emitters.

In the codetermination of ^{127}I, ^{131}I and ^{211}At in biological materials, Durbin et al. [42] digested the tissue with a mixture of chromic acid and 18 N H_2SO_4, then distilled both the I and At quantitatively from the digestion mixture after reduction with oxalic acid and $FeSO_4$. Recovery of ^{211}At is said to be $90 \pm 2.8\%$.

6.3.2 By Solvent Extraction

At^0 and At interhalogen compounds are readily extracted from acid solutions into nonpolar organic solvents but, according to [19, 23], the partition coefficients tend to be somewhat erratic. On the other hand, Wang and Khalkin [41] reported no erratic behavior in the extraction by diisopropyl ether (DIPE), of At which has previously been oxidized by Cl_2 at 25 °C or by 12 M HNO_3 or Ce^{IV} at 100 °C.

In a method developed by Barton et al. [36] for the removal of At from an irradiated Bi target, the target was dissolved so as to yield a solution in concentrated HCl. $FeSO_4$ was added to insure that the At was reduced to the zero state. This solution was contacted with DIPE and the DIPE was washed with dilute H_2SO_4 or HCl. Based on the absence of Po and Bi α-activity, the At was pure. Samples of the DIPE solution could then be evaporated on Pt or stainless steel disks for radiation measurement.

A solution containing pure At isotopes begins to grow Po, Bi and Pb isotopes. According to [36], these may be removed quantitatively by adding one-tenth volume of 20% tributyl phosphate in isobutyl ether to the DIPE solution and back-extracting into a 2 M HNO_3–4 M HCl aqueous solution. At remains quantitatively in the organic phase.

DIPE is also the basis of a method developed by Belyaev et al. [31] to isolate At from targets irradiated with 660 MeV protons:

One g of metallic Bi or Th is dissolved in 5 mL of concentrated HNO_3 by heating in a flask with a reflux condenser. To this solution is added 40 mL of 8 M HCl saturated with Cl_2. The extraction is carried out with 60 mL of DIPE. The organic layer is washed twice with 15 mL 8 M HCl. The At is reextracted from the ether by 40 mL of a 0.1 M solution of sodium stannite in 2 M NaOH. Te carrier is added and the At is coprecipitated with the Te several times from both acid and alkaline media. The At is finally separated from the Te by extraction with DIPE. The DIPE is washed twice with 1.5 to 2 M HCl after which the At is back-extracted with H_2O.

If the irradiated target is Pb instead of Bi or Th, the Pb (1 g) is dissolved in 5 mL of 6 M HNO_3 and 20 mL of 4 M HCl is added to the resulting solution. The solution is cooled and the precipitated $PbCl_2$ is filtered off. The solution is then acidified by the addition of 20 mL of concentrated HCl saturated with Cl_2. The subsequent operations are the same as for Bi targets.

Neumann [25] uses the following procedure to isolate At from irradiated Bi:

The target is dissolved in HNO_3 and the solution is concentrated by boiling, but not taken to dryness, as a little HNO_3 does not interfere. Concentrated HCl is added to bring the HCl concentration to 8 M, and the solution is cooled and extracted with isopropyl ether. (Isopropyl ether that has previously been equilibrated with 8 M HCl is somewhat better than pure ether for this purpose, since its use minimizes volume changes.) The yields in a single extraction are better than 90%, the presence of Bi^{III} salts increasing the extractability. Washing the ether phase with 8 M HCl removes ~10% of the activity. Less than 0.01% of the Bi is retained in this procedure.

Appelman [3] recommends that the HCl used to wash the At-containing ether should be ~1 M in HNO_3. If no nitrate is present, up to half the At may be back-extracted.

The "extractive distillation" method of Meyer and Rössler is described in Section 4.2.1.1, p. 100; see also Table 4/1, p. 97.

6.3.3 By Coprecipitation

As described above, coprecipitation is often used as a preliminary step when At must be separated from large amounts of fission products as well as from the target itself; cf. [27, 31]. Garrison et al. [39] used a single precipitation with Te to isolate At for its determination in biological materials:

A sample of At-containing tissue (<10 g wet weight) is placed in a 100-mL beaker and digested in a minimum volume of 9 N $HClO_4$ containing 20% by volume of 16 N HNO_3. After the organic material has been oxidized, the clear solution is evaporated to 10 to 15 mL of concentrated $HClO_4$. The solution is cooled and diluted to 3 N. Five mg of Te as tellurous acid and 1 mL of 12 N HCl are added. A stream of SO_2 is passed through the solution, precipitating metallic Te, which carries the At quantitatively. The Te is centrifuged, washed three times with distilled water and transferred to a porcelain counting dish. After drying at 70 °C, the dish is counted for α-activity.

Lefort and co-workers [26] used a two-stage coprecipitation procedure to isolate At from a proton-irradiated Th target as well as from Po, fission products and other elements formed by spallation (Pa, Ac, Ra):

The metallic Th target is attacked by concentrated HCl in the presence of a trace of HF. The solution is diluted to a concentration of 3 M in HCl and ~5 mg of Te as $Te(OH)_6$ is added. To this solution, 2 mL of 10% $SnCl_2$ in HCl is added. After 10 min on a water bath, the Te is completely precipitated. It carries down the At, the Po and some of the fission products, such as Au and As^{III}. After centrifugation, decantation and washing, the precipitate is taken up in the minimum amount of concentrated HNO_3 and diluted to 3 mL with concentrated HCl.

To eliminate Po and a large fraction of the fission products, the At is again coprecipitated with Te after reduction by SO_2, a less powerful reducing agent: The solution is heated to 100 °C on a water bath. An equal volume of concentrated HCl saturated with SO_2 and $N_2H_4 \cdot HCl$ is added. The Te precipitates, carrying down 99% of the At, while leaving the Po in solution. Finally, to separate the At from the Te, the precipitate is redissolved in a drop of concentrated HNO_3, the solution is made alkaline with NaOH and 2 mL of 5% Na stannite is added. The Te precipitates, leaving the At in solution. The precipitate is removed and NaF is added to the filtrate to prevent the colloidal precipitation of Sn on reacidification of the solution, since At is strongly adsorbed on such precipitates. The solution is now made 0.3 to 0.5 N in HNO_3 and the At is deposited on Ag foil by spontaneous electrodeposition, see also [39].

According to the authors, the purity of the At deposit is usually very high, as attested by the α- and X-ray spectra (cf. **Fig.** 6-**2**, p. 133, and **Fig.** 6-**3**, p. 134).

6.3.4 By Chromatography

Wang and co-workers [30] used a 50×2 mm column of the cation exchanger, Dowex 50X8, to separate At from Te. The column was washed with 5 to 10 mL of 8 M HCl containing Cl_2. The free column volume was 0.1 mL.

The starting sample, which consisted of 5 to 15 mg of elemental Te with coprecipitated At, was dissolved in 0.2 to 0.3 mL of 8 M HCl containing Cl_2. (The purpose of the Cl_2 was to oxidize the Te.) The solution was passed through the resin bed at a rate of 1 to 2 drops per min. The Te was completely eluted with 0.3 mL of 8 M HCl (Cl_2). More than 99% of the Po (resulting both from inadequate preliminary purification and from the decay of At during the experiment) was eluted with the Te. After elution of the Te, the At was desorbed from the resin with Cl_2 water at a rate of 3 to 4 drops per min. About 7% of the At was not adsorbed from the starting solution, but eluted with the wash solution; 78±8% of the At was eluted in the first 10 drops (~0.3 mL) and ~7% of the At remained on the resin after the column had been washed with 2 to 4 mL of Cl_2 water.

According to Bochvarova et al. [44], Te may be regarded as an ion exchanger that selectively adsorbs the At^- ion on the surface of crystals in acid medium. In alkaline solution, adsorption is suppressed by OH^- ions; consequently, solutions of alkalies are effective eluents for astatides sorbed on Te. The authors found that a system of two columns, filled with Te powder (grain size 100 to 200 mesh) permits the isolation, within 1.5 to 2 h, of At from 2 to 3 g of Th irradiated with 660 MeV protons. A high degree of separation from I and Po isotopes was also obtained; random radioactive contaminations were <5%; cf. Fig. 4-1, p. 98.

References:

[1] E. Anders (Ann. Rev. Nucl. Sci. **9** [1959] 203/20). — [2] E.H. Appelman (UCRL-9025 [1960] 1/113; N.S.A. **14** [1960] No. 11501). — [3] E.H. Appelman (NAS-NS-3012 [1960] 1/33; N.S.A. **14** [1960] No. 20201). — [4] E.H. Appelman (MTP [Med. Tech. Publ. Co.] Intern. Rev. Sci. Inorg. Chem. Ser. One **3** [1972] 181/98; C.A. **76** [1972] No. 120925). — [5] A.H.W. Aten Jr. (Advan. Inorg. Chem. Radiochem. **6** [1964] 207/23).

[6] A.J. Downs, C.J. Adams (in: J.C. Bailar, H.J. Emeléus, R. Nyholm, A.F. Trotman-Dickenson, Comprehensive Inorganic Chemistry, Vol. 2, Pergamon, Oxford 1973, pp. 1573/94). — [7] R. Dreyer, F. Rösch, G.J. Beyer (Wiss. Fortschr. **32** [1982] 251/5; C.A. **97** [1982] No. 151718). — [8] C.S. Garner (in: A.C. Wahl, N.A. Bonner, Radioactivity Applied to Chemistry, Wiley, New York 1951, pp. 179/243). — [9] M. Haïssinsky (in: P. Pascal, Nouveau Traité de Chimie Minérale, Vol. 16, Masson, Paris 1960, pp. 659/66). — [10] E.K. Hyde (J. Phys. Chem. **58** [1954] 21/6).

[11] E.K. Hyde (J. Chem. Educ. **36** [1959] 15/21). — [12] W.A. Chalkin [V.A. Khalkin], E. Herrmann (Isotopenpraxis **11** [1975] 333/40). — [13] W.A. Chalkin [V.A. Khalkin], E. Herrmann, J.V. Norseev, I. Dreyer (Chemiker-Ztg. **101** [1977] 470/81). — [14] A.K. Lavrukhina, A.A. Pozdnyakov (Analiticheskaya Khimiya Tekhnetsiya, Prometiya, Astatina i Frantsiya, Nauka, Moscow 1966, pp. 1/307; Analytical Chemistry of Technetium, Promethium, Astatine and Francium, Ann Arbor Sci., Ann Arbor 1970, pp. 227/60). — [15] J. Sedlet (Treatise Anal. Chem. II **6** [1964] 487/501).

[16] V.D. Nefedov, Yu.V. Norseev, M.A. Toropova, V.A. Khalkin (Usp. Khim. **37** [1968] 193/215; Russ. Chem. Rev. **37** [1968] 87/98; JINR-P-2895 [1966] 1/36; C.A. **66** [1967] No. 70899, **68** [1968] No. 100679; N.S.A. **21** [1967] No. 6). — [17] D.R. Corson, K.R. MacKenzie, E. Segrè (Phys. Rev. [2] **58** [1940] 672/8). — [18] A.H.W. Aten Jr., T. Doorgeest, U. Hollstein,

H.P. Moeken (Analyst **77** [1952] 774/7). – [19] G.L. Johnson, R.F. Leininger, E. Segrè (J. Chem. Phys. **17** [1949] 1/10). – [20] M. Kahn, A.C. Wahl (J. Chem. Phys. **21** [1953] 1185/9).

[21] M.L. Good, R.R. Edwards (J. Inorg. Nucl. Chem. **2** [1956] 196/200). – [22] E.H. Appelman (J. Am. Chem. Soc. **83** [1961] 805/7). – [23] E.H. Appelman (J. Phys. Chem. **65** [1961] 325/31). – [24] E.H. Appelman, E.N. Sloth, M.H. Studier (Inorg. Chem. **5** [1966] 766/9). – [25] H.M. Neumann (J. Inorg. Nucl. Chem. **4** [1957] 349/53).

[26] M. Lefort, G. Simonoff, X. Tarrago (Compt. Rend. **248** [1959] 216/8, **250** [1960] 106/8; Bull. Soc. Chim. France **1960** 1726/7; C.A. **1959** 12047, **1960** 12813, **1961** 7123). – [27] B.V. Kurchatov, V.N. Mekhedov, L.V. Chistiakov, M.Ya. Kuznetsova, N.I. Borisova, V.G. Solov'ev (Zh. Eksperim. Teor. Fiz. **35** [1958] 56/63; Soviet Phys.-JETP **8** [1959] 40/6). – [28] V.A. Khalkin, Yu.V. Norseev, V.D. Nefedov, M.A. Toropova, V.I. Kuzin (Dokl. Akad. Nauk SSSR **195** [1970] 623/5; Dokl. Chem. Proc. Acad. Sci. USSR **190/195** [1970] 855/7). – [29] J.G. Hamilton, C.W. Asling, W.M. Garrison, K.G. Scott (Univ. Calif. Publ. Pharmacol. **2** [1953] 283/344). – [30] Wang Fu-Chiung, Kang Mêng-Hua, V.A. Khalkin (Radiokhimiya **4** [1962] 94/8; Soviet Radiochem. **4** [1962] 81/5).

[31] B.N. Belyaev, Van Yun-Yui, E.N. Sinotova, L. Német, V.A. Khalkin (Radiokhimiya **2** [1960] 603/13; Soviet Radiochem. **2** [1960] 92/102). – [32] Yu.V. Norseev, Chao Tao-Nan', V.A. Khalkin (Radiokhimiya **8** [1966] 497/504; Soviet Radiochem. **8** [1966] 461/6). – [33] A.H.W. Aten Jr., J.G. van Raaphorst, G. Nooteboom, G. Blasse (J. Inorg. Nucl. Chem. **15** [1960] 198/9). – [34] J.K. Basson (Anal. Chem. **28** [1956] 1472/4). – [35] E. Segrè, K.R. MacKenzie, D.R. Corson (Phys. Rev. [2] **57** [1940] 1087).

[36] G.W. Barton Jr., A. Ghiorso, I. Perlman (Phys. Rev. [2] **82** [1951] 13/9). – [37] M.W. Parrott, W.M. Garrison, P.W. Durbin, M. Johnston, H.S. Powell, J.G. Hamilton (UCRL-3065 [1955] 1/8; N.S.A. **9** [1955] No. 7063). – [38] G.-J. Meyer, K. Rössler (Radiochem. Radioanal. Letters **25** [1976] 377/90). – [39] W.M. Garrison, J.D. Gile, R.D. Maxwell, J.G. Hamilton (Anal. Chem. **23** [1951] 204/5). – [40] C.M. Lederer, V.S. Shirley (Table of Isotopes, 7th Ed., Wiley, New York 1978).

[41] Wang Yung-yü, V.A. Khalkin (Radiokhimiya **3** [1961] 662/6; Soviet Radiochem. [Jerusalem] **3** [1961] 154/8; Radiochem. [USSR] **3** [1961] 258). – [42] P.W. Durbin, J.G. Hamilton, M.W. Parrott (UCRL-2792 [1954] 1/7; N.S.A. **9** [1955] No. 1477). – [43] V. Doberenz, D.G. Nhan, R. Dreyer, M. Milanov, Yu.V. Norseyev [Norseev], V.A. Khalkin (Radiochem. Radioanal. Letters **52** [1982] 119/27). – [44] M. Bochvarova, D.K. Tyung, I. Dudova, Yu.V. Norseev, V.A. Khalkin (Radiokhimiya **14** [1972] 858/65; Soviet Radiochem. **14** [1972] 889/95). – [45] J. Merinis, G. Bouissières (Radiochim. Acta **12** [1969] 140/52).

[46] V.M. Vakhtel, G.V. Vinel, Ts. Vylov, I.I. Gromova, A.F. Novgorodov, Yu.V. Norseev, V.A. Khalkin, V.G. Chumin (Radiokhimiya **18** [1976] 886/93; Soviet Radiochem. **18** [1976] 752/8; Isotopenpraxis **12** [1976] 441/6; C.A. **86** [1977] No. 62364).

7 Handling of Astatine

K. Rössler
Institut für Chemie 1 (Nuklearchemie)
Kernforschungsanlage Jülich GmbH
Jülich, Federal Republic of Germany

7.1 Introduction

Astatine, the fifth halogen and homologue of iodine, is one of the best studied radioelements. Since its discovery in 1940 by Corson, MacKenzie and Segrè [1 to 5] after more than 16 years of intensive search and discussion, cf. e.g. [6 to 15, 31], about 800 publications have been devoted totally or partially to the physical and chemical properties of that element. The majority belongs to the domain of nuclear physics. The 24 isotopes of astatine ranging from ^{196}At to ^{219}At are interesting for their production cross sections by various nuclear reactions, for their very differentiated decay properties and the multiplicity of nuclear excitation levels, cf. e.g. [16 to 25]. Astatine isotopes are often observed as intermediate or final products of heavy ion reactions and spallation processes. They exhibit many special features with respect to their neighbor elements, cf. e.g. [26, 27]. Irrespective of the manifold of nuclear physical experiments an intensive review devoted entirely to astatine does not exist hitherto.

About 310 publications (until February 1985) are concerned with the chemical properties, including nuclear medical applications. This number is somewhat astonishing when considering the fact that astatine and especially its mostly studied isotope ^{211}At ($T_{1/2}$=7.22 h) can only be produced via cyclotrons which are able to accelerate α particles to energies ≧28 MeV [16, 17]. Most of the work in the past has indeed been carried out in larger nuclear establishments, such as Berkeley, Argonne, and Brookhaven (USA), Dubna (USSR), IKO (now NIKEF) Amsterdam (The Netherlands), Jülich (Germany) and others, or in close collaboration with them. However, the interesting behavior of that element (sometimes like a halogen another time like a metal, even more ambivalent than its lighter homologue iodine) and the potential application of its isotope ^{211}At for α therapy of tumors and immuno-suppression have prompted this relatively large number of papers. A plot of the annual number of publications on astatine chemistry including nuclear medical applications as a function of the year of appearance is shown in **Fig. 7-1**. It exhibits a slight maximum in 1954 (many papers of the Berkeley medical group), but also a steady increase in recent times. In contrast to nuclear physics, there are many reviews on the chemical properties (about 45). The most important ones can be found in [28 to 44]. About 70 papers are devoted to nuclear medical applications; 70 to problems of organic and biochemistry, mostly studies of labeling and the stability of carbon-astatine bonds; another 70 to problems of inorganic chemistry, especially the stability of the different valence states ranging from −1 to +7. More than 50 publications are concerned with physicochemical applications including extrapolations of thermodynamic data. Even atomic beam work has been carried out. Most of this work has been performed by single persons or small groups only. Astatine chemistry has never been the goal of a massive coordinated effort. Each group has published between five and fifteen papers, then work had been stopped and people marched out of the field. The only exception is the Dubna group, which in close collaboration to nuclear physicists continues astatine research for almost two decades. Active groups in February 1985 are besides Dubna: Brookhaven National Laboratory in cooperation with Harvard Medical School (USA); Zentralinstitut für Kernforschung, Dresden-Rossendorf and Technische Universität, Dresden (DDR); Central Institute of Physics of the Hungarian Academy of Sciences, Budapest (Hungary) in cooperation with Dubna; University of Cambridge, School of Clinical Medicine,

Fig. 7-1

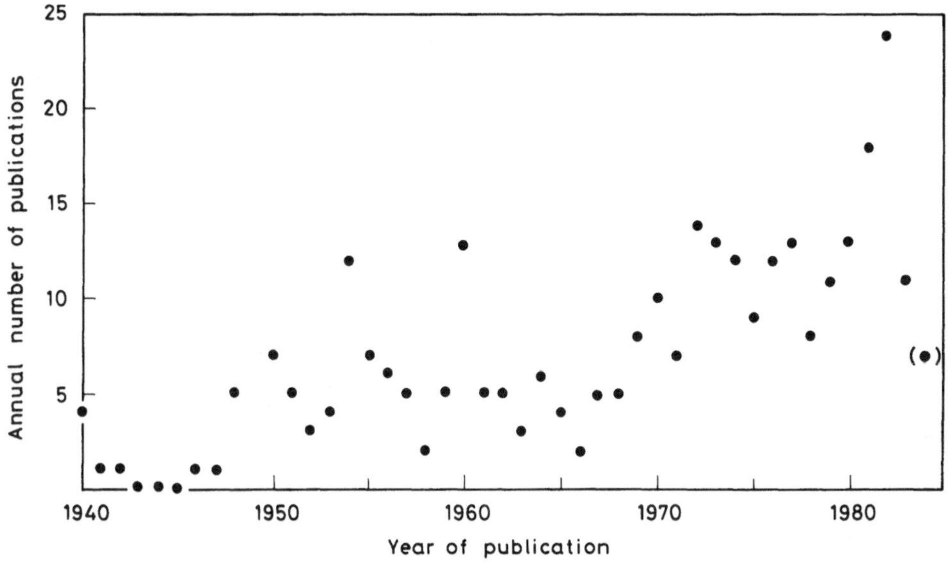

Annual number of publications on astatine chemistry including nuclear medical applications as a function of the year of appearence.

University of Birmingham and AERE Harwell (U.K.). Even a Chinese group has started some astatine work in 1983/84 [46]. Altogether about 150 scientists are listed as co-authors of astatine chemical and medical publications.

Chemical work on astatine is almost entirely restricted to the relatively long-lived isotopes ^{209}At ($T_{1/2}$=5.4 h), ^{210}At ($T_{1/2}$=8.3 h) and ^{211}At ($T_{1/2}$=7.2 h) which can easily be produced by α irradiation of bismuth. The last is by far the most important one. The isotopes ^{209}At and ^{210}At form unwanted radiochemical impurities especially in nuclear medical studies. The other shorter-lived isotopes have barely been studied, mainly since they are not as easily accessible. One exception is the isotope ^{217}At ($T_{1/2}$=32.3 ms) which is applied in studies of atom-molecule reactions (atomic beam experiments) of HAt with, e.g., Cl and Br. Here, the short half-life is advantageous to achieve a better detectability [47 to 56].

In the following, information on the safe and successful handling of astatine is given. This is based on the literature, the general chemical and physical properties of that radioelement and the author's own experience in about 10 years of work.

References:

[1] D.R. Corson, K.R. MacKenzie (Phys. Rev. [2] **57** [1940] 250). – [2] D.R. Corson, K.R. MacKenzie, E. Segrè (Phys. Rev. [2] **57** [1940] 459). – [3] D.R. Corson, K.R. MacKenzie, E. Segrè (Phys. Rev. [2] **57** [1940] 1087). – [4] D.R. Corson, K.R. MacKenzie, E. Segrè (Phys. Rev. [2] **58** [1940] 672). – [5] D.R. Corson, K.R. MacKenzie, E. Segrè (Nature **159** [1947] 24).

[6] P.D. Foote (Nature **114** [1924] 789). – [7] F.H. Loring, J.G.F. Druce (Chem. News **131** [1925] 305). – [8] G. Hevesy (Kgl. Danske Videnskab. Selskab Mat. Fys. Medd. **7** No. 11

[1926] 1/11). – [9] G. von Hevesy, R. Hobbie (Z. Anorg. Allgem. Chem. **208** [1932] 107/12). – [10] F. Allison, E.J. Murphy, E.R. Bishop, A.L. Sommer (Phys. Rev. [2] **37** [1931] 1178/80).

[11] F. Allison (J. Chem. Educ. **10** [1933] 71/8). – [12] E. Buch Andersen (Kgl. Danske Videnskab. Selskab Mat. Fys. Medd. **16** No. 5 [1938] 1/22). – [13] H. Hulubei, Y. Cauchois (Compt. Rend. **209** [1939] 39/42). – [14] W. Minder (Helv. Phys. Acta **13** [1940] 144/52). – [15] B. Karlik, T. Bernert (Naturwissenschaften **30** [1942] 685/6).

[16] W.J. Ramler, J. Wing, D.J. Henderson, J.R. Huizenga (Phys. Rev. [2] **114** [1959] 154/62). – [17] E.T. Strom (UCRL-9372 [1960]; N.S.A. **15** [1961] No. 12080). – [18] M. Lefort, G. Simonoff, X. Tarrago (Compt. Rend. **248** [1959] 216/8). – [19] M. Lefort, G. Simonoff, X. Tarrago (Nucl. Phys. **19** [1960] 173/83). – [20] R. Bimbot, M.F. Rivet (Phys. Rev. [3] C **8** [1973] 375/9).

[21] L.J. Jardine, S.G. Prussin (Nucl. Phys. A **190** [1972] 261/83). – [22] L.J. Jardine, S.G. Prussin (Nucl. Phys. A **233** [1974] 25/47). – [23] L.J. Jardine (Phys. Rev. [3] C **11** [1975] 1385/91). – [24] V. Paar (Phys. Rev. [3] C **11** [1975] 1432/42). – [25] I. Bergström, C.J. Herrlander, T. Lindblad, V. Rakhonen, K.G. Rensfelt, K. Westerberg (Z. Physik A **273** [1975] 291/304).

[26] M. Epherre, G. Audi, C. Thibault, R. Klapisch, G. Huber, T. Tauchard, H. Wollnik (Nucl. Phys. A **340** [1980] 1/12). – [27] F.A. Janouch, R.J. Liotta (Nucl. Phys. A **334** [1980] 427/44). – [28] M. Haïssinsky (Bull. Soc. Chim. France **1949** 668/78). – [29] G.L. Johnson, R.F. Leininger, E. Segrè (J. Chem. Phys. **17** [1949] 1/10). – [30] E.K. Hyde (J. Phys. Chem. **58** [1954] 21/6).

[31] A.G. Maddock (Mellor's Comprehensive Treatise on Inorganic and Theoretical Chemistry, Vol. 2, Suppl. 1, Longmans Green, London 1956, pp. 1064/79; Usp. Khim. **29** [1960] 1388/406). – [32] K.W. Bagnall (Chemistry of the Rare Radioelements Polonium-Actinium, Butterworths, London 1957). – [33] E. Anders (Ann. Rev. Nucl. Sci. **9** [1959] 203/20). – [34] E.K. Hyde (J. Chem. Educ. **36** [1959] 15/21). – [35] M. Haïssinsky (in: P. Pascal, Le Nouveau Traité de Chimie Minérale, Vol. 16, Masson, Paris 1960, pp. 659/66).

[36] A.H.W. Aten Jr. (Advan. Inorg. Chem. Radiochem. **6** [1964] 207/23). – [37] J. Sedlett (Treatise Anal. Chem. II **6** [1964] 487/609). – [38] V.D. Nefedov, Yu.V. Norseev, M.A. Toropova, V.A. Khalkin (Usp. Khim. **37** [1968] 193/215; Russ. Chem. Rev. **37** [1968] 87/98). – [39] A.K. Lavrukhina, A.A. Pozdnyakov (Analiticheskaya Khimiya Tekhnetsiya, Prometiya, Astatina i Frantsiya, Nauka, Moscow 1966, pp. 228/59; C.A. **67** [1967] No. 29087; Analytical Chemistry for the Elements Technetium, Promethium, Astatine, and Francium, Ann Arbor-Humphrey, Ann Arbor, Mich., 1970, pp. 227/60; N.S.A. **25** [1971] No. 15470). – [40] E.H. Appelman (MTP [Med. Tech. Publ. Co.] Intern. Rev. Sci. Inorg. Chem. Ser. One **3** [1972] 181/98; C.A. **76** [1972] No. 120925).

[41] A.J. Downs, C.J. Adams (The Chemistry of Chlorine, Bromine, Iodine, and Astatine, Pergamon, Elmsford, N.Y., 1975, pp. 1/488; Comprehensive Inorg. Chem. **26** [1973] 1573/94). – [42] W.A. Chalkin [Khalkin], E. Herrmann (Isotopenpraxis **11** [1975] 333/40). – [43] W.A. Chalkin [Khalkin], E. Herrman, I.V. Norseev, I. Dreyer (Chemiker-Ztg. **101** [1977] 470/81). – [44] K. Berei, L. Vasáros (in: S. Patai, Z. Rappoport, The Chemistry of Functional Groups, Suppl. D, Wiley, New York 1983, pp. 405/40). – [45] K. Berei, L. Vasáros (KFKI-1984-29 [1984]; Chem. Halides Pseudo-Halides Azides **1** [1983] 369/403; C.A. **100** [1984] No. 208645).

[46] L. Cheng, D. Qunxia, Z. Jimin, Z. Yanyan, X. Lun, L. Boli, J. Yutai, L. Zhenghao, T. Zhigang (He Huaxue Yu Fangshe Huaxue **6** [1984] 244). – [47] J.R. Grover, E. Lebowitz, E. Baker (J. Inorg. Nucl. Chem. **31** [1969] 3705/20). – [48] J.R. Grover, F.M. Kiely, E. Lebowitz,

E. Baker (Rev. Sci. Instr. **42** [1971] 293/302). — [49] J.R. Grover, H.V. Lilenfeld (Nucl. Instr. Methods **105** [1972] 189/96). — [50] J.R. Grover, H.V. Lilenfeld (Rev. Sci. Instr. **43** [1972] 690/2).

[51] J.R. Grover, C.R. Iden (J. Chem. Phys. **61** [1974] 2157/9). — [52] J.R. Grover, C.R. Iden, F.E. Schubert, J.T. Muckerman (Proc. 9th Intern. Conf. Phys. Electron. At. Collisions, Seattle, Wash., 1975, Abstr. No. 331). — [53] J.R. Grover, C.R. Iden, H.V. Lilenfeld (J. Chem. Phys. **64** [1976] 4657/71). — [54] J.R. Grover, C.R. Iden, H.V. Lilenfeld, F.M. Kiely, E. Lebowitz (Rev. Sci. Instr. **47** [1976] 1098/108). — [55] J.R. Grover, D.E. Malloy, J.B.A. Mitchell (Proc. 7th Intern. Symp. Mol. Beams, Riva del Garda, Italy, 1979, p. 90).

[56] J.R. Grover, D.E. Malloy, J.B.A. Mitchell (J. Chem. Phys. **76** [1982] 362/77).

7.2 Special Chemical Behavior of Astatine

Any essay on astatine chemistry has to consider the fundamental facts that it is a radioelement with relatively short-lived isotopes and that it occupies a position in the periodic system where the properties of a halogen and those of a metal may be mixed. Despite the presence of the short-lived isotopes ^{215}At ($T_{1/2}=0.1$ ms), ^{218}At ($T_{1/2}=2$ s) and ^{219}At ($T_{1/2}=$ 0.9 min) in side branches of natural decay chains [1], astatine can be considered as an artificial radioelement. The radioactivity of ^{211}At produced in typical experiments via the classical ^{209}Bi$(\alpha,2n)^{211}$At nuclear reaction [2] amounts to about 2×10^9 Bq (50 mCi). This corresponds to 8×10^{13} atoms or about 10^{-10} mol. Typical radioactivities of mL samples in the experiments are by a factor of 10^2 to 10^3 lower. The 10^{11} to 10^{12} atoms correspond to 10^{-10} to 10^{-9} M solutions (0.0002 to 0.002 ppb). The maximum radioactivity which could be produced by optimizing the conditions in cyclotrons with very intense beams would amount to 1 Ci ^{211}At corresponding to about 10^{15} atoms. Astatine cannot be stored and accumulated due to its short half-life. This excludes the production of weighable amounts of this element. Likewise, it seems rather improbable that compounds containing two astatine atoms can be synthesized, such as, e.g., At_2. This also includes the fact that disproportionation reactions, typical for all other halogens, do not take place. The reports on macroscopic measurements, such as optical absorption spectra [3] and chromatographic separation of At_2 [4] (see Fig. 10-5, p. 189) must be taken with some caution. On the other hand, sensitive mass spectrometry seems to be able to detect the trace amounts [5]. In general, all information is deduced from the distribution of astatine in the interacting system within which it has been placed, followed by measurement of the radioactivity. The extrapolation of the results to deduce macroscopic behavior, e.g., in order to compare with other halogens, is somewhat uncertain. A testimony to this comes from the frequent observations implying that the chemistry of iodine at very low concentrations — but still orders of magnitude greater than that of astatine — differs notably from its macro-chemistry, cf., e.g. [6, 7]. Different from other radioelements such as Tc or Pm, there is no suitable carrier for astatine. The most nearest one, iodine, is not very satisfactory [8, 9]. The reactions with impurities can play a detrimental role at this concentration level. The chemistry and analysis may furthermore be complicated by the general properties of carrier-free substances such as adsorption to walls and filters, coprecipitation and formation of radiocolloids. Recent experiments on potential changes in chemical behavior at very low ^{211}At concentrations did not yield any differences in electrodeposition on silver foils or in coprecipitation of $^{211}At^-$ with AgI when varying the number of astatine atoms between 10^4 to 10^7 [10].

The second important factor is the chemical ambiguity of astatine between halogen and metal character. Indications for a tendency to behave like a heavy metal are, e.g., the formation of a positively charged astatinium cation At^+ or $[H_2OAt]^+$, cf. [11 to 15], the formation of pseudohalide compounds [16], complexes of astatine cations [12, 13, 18], complex anions of trivalent astatine [17, 18] as well as complexes with a variety of organic solvents [19, 20]. The complex forming character of astatine in organic chemistry is demonstrated best by the high yield of astatination of aromatic substances via the decomposition of diazonium salts, as compared to the corresponding reactions with radioiodine [20 to 23]. On the other hand, there are indications from radiopolarography that astatine is not amalgamated by mercury but rather adsorbed to the Hg droplets [24], cf. also [25].

Among other special chemical properties of astatine are the self-irradiation effects at higher concentrations, which will be treated in Chapter 9, p. 179.

Another characteristic feature of astatine chemistry is the easy oxidation of the frequently used astatide ions by impurities. This implies that At^- can only be stabilized in slightly reducing medium, e.g., 0.1 M Na_2SO_3 solutions [27 to 29]. In his pioneering experiments on the oxidation states of astatine in aqueous solutions, Appelman reports the potential scheme at pH 1 as follows [28]:

$$At^- \xrightarrow{-0.3} At^0 \xrightarrow{-1.0} HOAt(?) \xrightarrow{-1.5} AtO_3^- \xrightarrow{<-1.6} H_5AtO_6(?)$$

Astatide is oxidized to At^0 at 25 °C by the iodide-iodine couple (emf = −0.60 V), by a 1:1 ferrocyanide/ferricyanide mixture (1:1) at pH < 3 (emf < −0.4 V) and by the As^{III}-As^V couple at pH < 4 (emf < 0.3 V). The ferrocyanide-ferricyanide couple reduces At^0 to At^- at low acidity and ionic strength if the concentration of ferrocyanide is several times that of ferricyanide (emf > −0.4 V). The As^{III}-As^V couple does likewise at pH > 4 (emf > 0.3 V) [28]. Thus, the redox potential of the At^--At^0 couple should be written at least: 0.3 ± 0.1 V. Compared to the corresponding values for the $X^-/{}^1/_2X_2$ couples of the other halogens in 0.1 M H^+:

Fig. 7-2

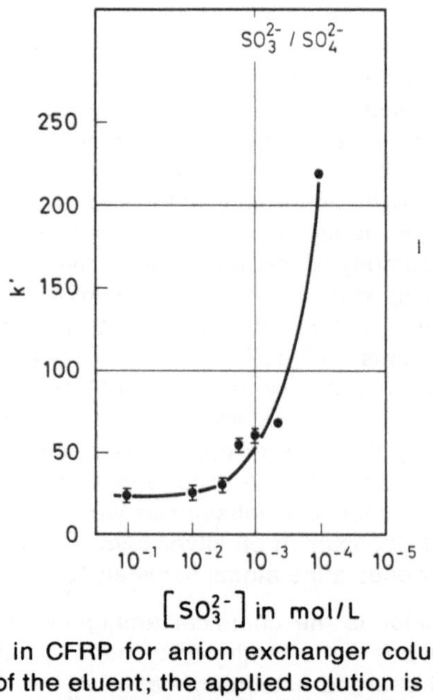

Capacity factor k' of $^{211}At^-$ in CFRP for anion exchanger columns at 323 K as dependent on the SO_3^{2-} concentration of the eluent; the applied solution is 1 M $NaNO_3 + 10^{-3}$ M SO_4^{2-} + 10^{-5} to 10^{-1} M SO_3^{2-} [24, 25, 30].

Fig. 7-3

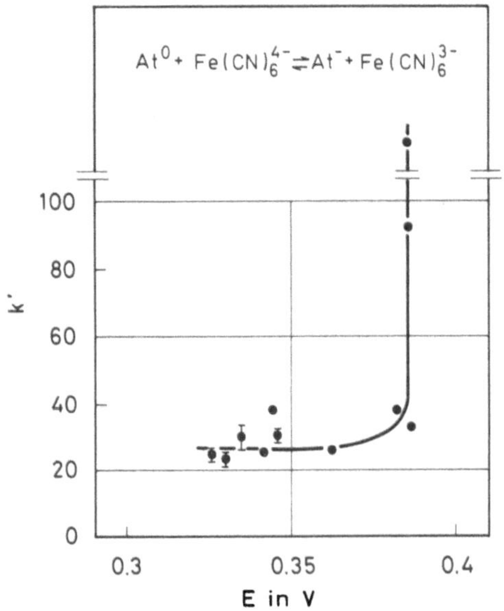

Capacity factor k′ of $^{211}At^-$ in CFRP for anion exchanger columns at 323 K as dependent on the ferrocyanide/ferricyanide redox couple in the eluent 1 M $NaNO_3 + 10^{-2}$ M hexacyanoferrate II/III [24, 25, 30].

I (0.62), Br (1.09) and Cl (1.40) at 25 °C, it is rather low and, thus, facilitates the oxidation. It remains somewhat astonishing that till 1980 this important, fundamental value never has been checked again. In 1980/81 two new attempts [24, 25] were made to redetermine the redox potentials with methods other than coprecipitation or solvent extraction used by Appelman [28].

High pressure liquid chromatography of At^- and At^+ was performed on anion and cation exchanger resins, respectively, with eluents containing the redox couples: SO_3^{2-}/SO_4^{2-} and $Fe(CN)_6^{4-}/Fe(CN)_6^{3-}$ for At^- and Cr^{III}/Cr^{VI} for At^+, all in acidic solutions. This new method, chromatography at fixed redox potential (CFRP), allowed the determination of the redox potentials for the reactions: $At^0 + e^- \rightarrow At^-$ to 0.335 ± 0.05 V (323 K) and for $At^+ + e^- \rightarrow At^0$ to 0.85 ± 0.05 V (332 K), cf. **Fig.** 7-**2**, **Fig.** 7-**3**, and **Fig.** 7-**4**, p. 146 [24, 25, 30]. Even if the redox couples are not totally reversible, neither were those of Appelman. There is a good agreement with the older values, cf. Table 7/1, p. 147. A second method, radiopolarography, was modified for the special properties of astatine [24]. The mercury drop of the working electrode was collected shortly before falling down and measured for its astatine radioactivity. This "hanging drop" radiopolarography yielded a redox potential of 0.225 ± 0.025 V for the reaction $At^0 + e^- \rightarrow At^-$ [24], cf. **Fig.** 7-**5**, p. 147. Table 7/1 compares the redox potentials obtained by the three different methods: coprecipitation or solvent extraction, CFRP, and hanging drop radiopolarography; included are some values for the iodide/iodine couple for comparison. Despite the disadvantages of the three methods, the agreement is reasonable. Redox potentials may be somewhat lower than reported by Appelman, which may result in easier oxidation of At^- to At^0 and somewhat more difficult oxidation to the astatinium cation.

Fig. 7-4

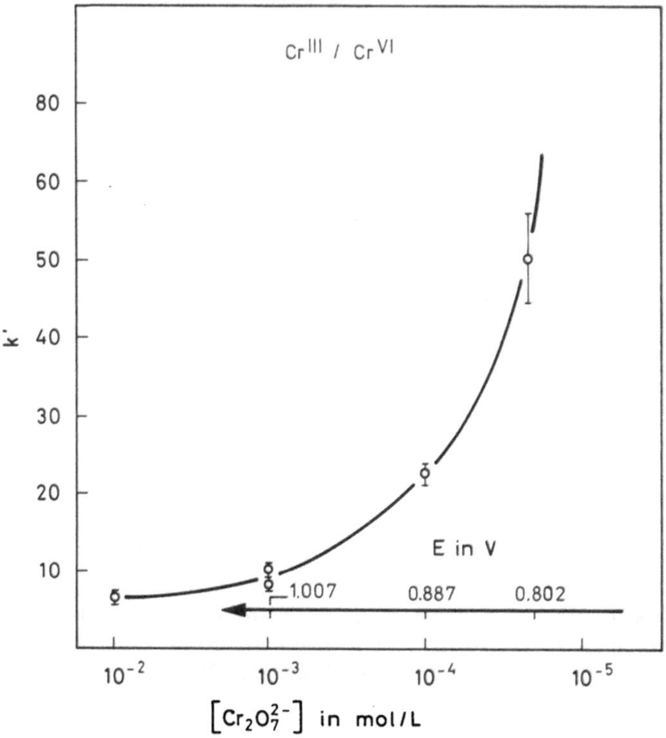

Capacity factor k′ for At^+ in CFRP for cation exchanger columns at 332 K as dependent on the concentration of $Cr_2O_7^{2-}$ ions in the eluent; Cr^{III}/Cr^{VI} redox couple; the applied solution is 1 M $HNO_3 + 10^{-3}$ Cr^{3+} + variable M $Cr_2O_7^{2-}$ [24, 25, 30].

Table 7/1
Comparison of Redox Potentials of Some Iodine and Astatine Species.

$^1/_2 I_2 + e^- \rightleftharpoons I^-$	$^{211}At^0 + e^- \rightleftharpoons {}^{211}At^-$	$^{211}At^+ + e^- \rightleftharpoons {}^{211}At^0$
0.620 V (298 K) E_0 [31]	0.3±0.1 V (298 K) (coprecipitation, solvent extraction) [28]	1.00±0.05 V (298 K) (coprecipitation, solvent extraction) [28]
0.64±0.02 V (323 K) no carrier added ^{131}I (CFRP) [24, 25, 30]	0.335±0.05 V (298 K) (CFRP) [24, 25, 30] 0.225±0.025 V (298 K) (hanging drop radiopolarography) [24, 30]	0.85±0.05 V (332 K) (CFRP) [24, 25, 30]

Fig. 7-2, p. 144, shows that an SO_3^{2-} concentration of 10^{-2} M is sufficient to retain astatine in the At^- state in neutral or weekly acidic solutions. Likewise, a $Cr_2O_7^{2-}$ concentration of 10^{-3} to 10^{-2} M is sufficient to oxidize At^0 to its monovalent positive state in 1 M HNO_3 solution, cf. Fig. 7-4.

Fig. 7-5

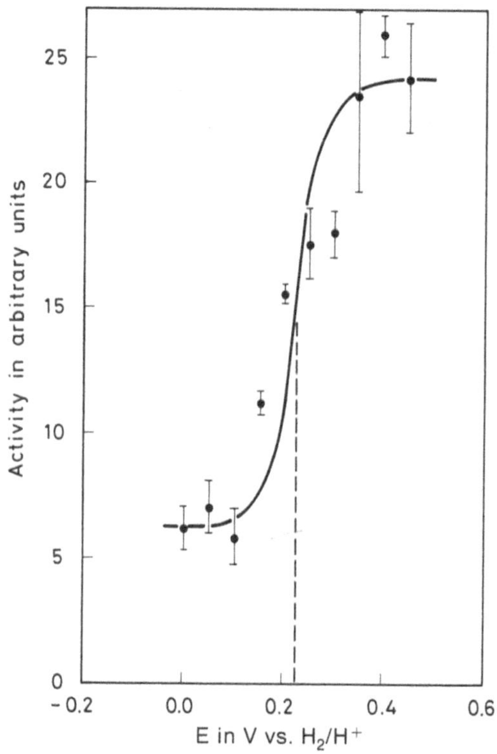

Half-wave potential of hanging drop radiopolarography of the system $At^{0}+e^{-} \rightleftharpoons At^{-}$ [24, 30].

Due to the special chemical behavior of carrier-free astatine, it is advisable to use chromatographic methods for the analysis of the chemical states, such as, e.g., radiogaschromatography, high pressure liquid chromatography, thin layer chromatography and electrophoretic techniques. These methods allow a sufficient separation of microscopic as well as of carrier-free species. One of the first applications of HPLC to astatine chemistry [11] revealed an interesting relationship between inorganic and organic compounds of astatine and its halogen homologues. It is, however, necessary to work in a standard fashion. For astatine, the eluents must be extremely pure. Water has to be triply distilled and filtered through membranes. The same is true for any other solvent or solution. The use of millipore filters can be recommended. The columns have to be washed out with the eluent much longer than for normal macroscopic runs. Heating of the columns in the temperature range from 320 to 345 K improves the mobility of astatine species and, thus, the chromatographic separation. An interesting effect was observed for the analysis of astatinated aromatic compounds. Radiogaschromatography using a heated injection block (100 to 150 °C) and heated columns (100 to 120 °C) led to relatively reproducible results, even if it was not always sure that all the radioactivity injected had left the column [20 to 23, 26]. Repeating some of these experiments and applying radio-HPLC yielded in several cases, annoying ghost peaks of astatine complexes with the substrate and the solvent which superimposed the proper reaction products. HPLC works under milder conditions than GC. Thus, metastable aggregates and complexes are more likely to survive, in HPLC which will be destroyed

References on p. 148

in GC. It is necessary to check for interference of "wild" astatine compounds with the distribution of the "wanted" products.

References:

[1] E.K. Hyde, A. Ghiorso (Phys. Rev. [2] **90** [1953] 267/70). – [2] D.R. Corson, K.R. MacKenzie (Phys. Rev. [2] **57** [1940] 250). – [3] R. McLaughlin (J. Opt. Soc. Am. **54** [1964] 965/7). – [4] K. Otozai, N. Takahachi (Radiochim. Acta **31** [1982] 201/3). – [5] E.H. Appelman, E.N. Sloth, M.H. Studier (Inorg. Chem. **5** [1966] 766/9).

[6] M. Kahn, A.C. Wahl (J. Chem. Phys. **21** [1953] 1185/9). – [7] H.M. Eiland, M. Kahn (J. Phys. Chem. **65** [1961] 1317/20). – [8] G.L. Johnson, R.F. Leininger, E. Segrè (J. Chem. Phys. **17** [1949] 1/10). – [9] A.H.W. Aten Jr., T. Doorgeest, U. Hollstein, H.P. Moeken (Analyst [London] **77** [1952] 774/7). – [10] B. Rumler, N. Trautmann, B. Herrmann (IKMz-84-1 [1984] 24).

[11] K. Rössler, W. Tornau, G. Stöcklin (J. Radioanal. Chem. **21** [1974] 199/209). – [12] W.A. Chalkin [Khalkin], E. Herrmann (Isotopenpraxis **11** [1975] 330/40). – [13] W.A. Chalkin [Khalkin], E. Hermann, J.V. Norseev, I. Dreyer (Chemiker-Ztg. **101** [1977] 470/81; C.A. **88** [1978] No. 28663). – [14] L. Vasáros, J.V. Norseev, D.D. Nhan, V.A. Khalkin (Radiochem. Radioanal. Letters **54** [1982] 239/48). – [15] M. Milanov, V. Doberenz, V.A. Khalkin, A. Marinov (J. Radioanal. Nucl. Chem. **83** [1984] 291/9).

[16] R. Dreyer, I. Dreyer, M. Pfeiffer, F. Rösch (Radiochem. Radioanal. Letters **55** [1982] 207/4). – [17] I. Dreyer, R. Dreyer, V.A. Chalkin [Khalkin], M. Milanov (Radiochem. Radioanal. Letters **40** [1979] 145/54). – [18] K. Berei, L. Vasáros (KFKI-1984-29 [1984]; Chem. Halides Pseudo-Halides **1** [1983] 309/403; C.A. **100** [1984] No. 208645). – [19] G.W.M. Visser, E.L. Diemer (Radiochim. Acta **33** [1983] 145/51). – [20] G.-J. Meyer, K. Rössler, G. Stöcklin (Radiochem. Radioanal. Letters **21** [1975] 247/59).

[21] G.-J. Meyer, K. Rössler, G. Stöcklin (J. Labelled Compounds Radiopharm. **12** [1976] 449/58). – [22] G.-J. Meyer (Diss. Univ. Köln, FRG, 1977; JUEL-1418 [1977]). – [23] G.-J. Meyer, K. Rössler, G. Stöcklin (J. Am. Chem. Soc. **101** [1979] 3121/3). – [24] A. Cavallero (Diss. Univ. Cathol. Louvain, Belgium, 1981, pp. 1/102; JUEL-Report to be published in 1985). – [25] A. Cavallero, K. Rössler (Lect. Meeting Ges. Deut. Chemiker Fachgr. Nuklearchem. Nucl. Radiat. Radiochem., Jülich, FRG, 1980, Abstr., p. 40; INIS-MF-6397 [1980] 40; INIS Atomindex **12** [1981] No. 598948).

[26] G.-J. Meyer, K. Rössler, G. Stöcklin (Radiochim. Acta **24** [1977] 81/5). – [27] E.H. Appelman (Diss. Univ. California 1960; UCRL-9025 [1960] 1/113; N.S.A. **14** [1960] No. 11501). – [28] E.H. Appelman (J. Am. Chem. Soc. **83** [1961] 805/7). – [29] E.H. Appelman (NAS-NS-3012 [1960] 1/33; N.S.A. **14** [1960] No. 20201). – [30] A. Cavallero, K. Rössler (Radiochim. Acta [1985] in press).

[31] G. Desideri, L. Lepri, D. Heimler (in: A.J. Bard, Encyclopedia of Electrochemistry of the Elements, Vol. 1, Chapter 1/3, Dekker, New York 1973).

7.3 Physical Properties, Radioactivity Assay and Working Time Schedule

The decay schemes of ^{210}At and ^{211}At are reported in [1 to 3]. Both isotopes are measured in general via the X-rays of ^{210}Po and ^{211}Po, respectively, which are the same. A Ge(Li) detector recording of these lines is shown in **Fig. 7-6** [4]. In order to discriminate between the two radioisotopes it is necessary to count the less frequent γ rays: 670 and 687 keV for ^{211}At with a relative abundance of 0.004 and 0.26, and 245 and 1180 keV for ^{210}At with a relative abundance of 79 and 100, respectively [5]. They also can be assayed via the

Fig. 7-6

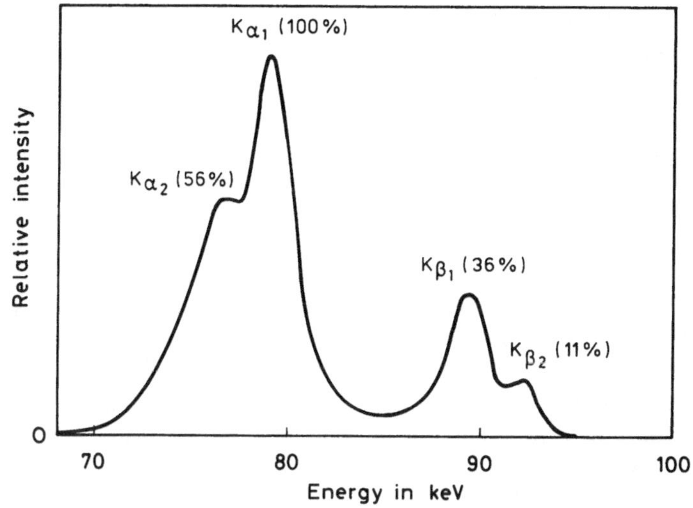

X-ray spectrum of Po, taken from [4].

Table 7/2
Major γ Lines of Radioisotopes Produced by α Irradiation of Bi Metal.

nuclide	half-life	major γ lines in keV (relative abundance, cf. [5])
^{211}At	7.2 h	670 (0.004), 687 (0.26) (also X-rays of Po: 77, 79, 90, 92 keV)
^{210}At	8.3 h	117 (0.7), 245 (79), 461 (weak), 528 (1.1), 818 (1.7), 853 (1.4), 957 (1.8), 1180 (100), 1436 (29), 1483 (47.7), 1599 (14) (also X-rays of Po: 77, 79, 90, 92 keV)
^{211}Po	0.5 s	570 (0.5), 900 (0.5)
^{210}Po	138.4 d	803 (0.001)
^{207}Bi	33.4 a	570 (98), 1064 (77), 1777 (7) (also X-rays of Pb: 73, 75, 85, 87 keV)
^{206}Bi	6.2 d	184 (19), 344 (24), 398 (10), 497 (15), 516 (40), 537 (29), 803 (100), 881 (68), 895 (15), 1095 (13), 1719 (34) (also X-rays of Pb: 73, 75, 85, 87 keV)
^{67}Ga	78 h	185 (20), 209 (2), 300 (15), 394 (4)
^{66}Ga	9.4 h	511 (114), 834 (6), 1039 (37)
^{65}Zn	244 d	511 (3), 1116 (50)
^{69}Ge	39 h	511 (68), 574 (12), 870 (10), 1107 (26)

characteristic γ lines of their decay products. Table 7/2, p. 149, shows a list of radioisotopes and their typical γ and X rays, which may be present in a typical α-activated Bi target. The main impurities ^{66}Ga, ^{67}Ga and ^{65}Zn can recoil out of the frequently used copper backing of the cyclotron target or the target holder as a consequence of (α, pxn) processes. A contribution of the reaction $^{209}Bi(d,n)^{210}Po$, due to pollution of the α beam by residual deuterons [6], may also contribute to the final α activity observed after total decay of ^{211}At.

In old samples, the ^{207}Bi ($T_{1/2}$=33.4 a) interfers because its emits the X-rays of lead that cannot be discriminated from those of polonium. The sample should then be assayed again a few days later after total decay of astatine and the residual activity can be subtracted.

^{210}At and ^{211}At can be measured in NaI scintillator or Ge(Li) semiconductor detectors via the ^{210}Po or ^{211}Po X-rays. α counting has been applied in only a few cases because samples with equal self-absorption are difficult to prepare and liquid scintillation counting is somewhat too tedious compared to the simple X-ray assay. α counting may, however, be suited for measurement of very low ^{211}At activities, cf., e.g. [7].

The short half-life of ^{211}At is advantageous with respect to contamination, incorporation, etc. It imposes, however, problems for the chemist in view of the working time schedule. The typical total radioactivity of 2×10^9 Bq (50 mCi) allows measurement over 6 and even more half-lives, corresponding to 2 to 3 d. Our own experience shows that it is better to continue as long as possible from the start. There is some danger that the samples are chemically changed during to long standing. The next morning, a good experiment often cannot be repeated again.

References:

[1] L.J. Jardine (Phys. Rev. [3] C **11** [1975] 1385/91). – [2] S.G. Prussin, J.M. Hollander (Nucl. Phys. A **110** [1968] 176/92). – [3] L.J. Jardine, A.A. Shihab-Eldin (Nucl. Phys. A **244** [1975] 34/44). – [4] A. Cavallero (Diss. Univ. Cathol. Louvain, Belgium, 1981; JUEL-Report to be published in 1985). – [5] G. Erdtmann, W. Soyka (JUEL-1003-AC [1973]).

[6] R. Bimbot, Y. Le Beyec (Rev. Phys. Appl. **4** [1969] 393/6). – [7] B. Rumler, N. Trautmann, B. Herrmann (IKMz-84-1 [1984] 24).

7.4 Radiation Risks, Health Physics and Protection

^{211}At must a priori be regarded as a harmful radioisotope. Each decay leads to the emission of an α particle either directly or via electron capture to ^{211}Po [1]. The high linear energy transfer of the α particles in water containing tissue results in a high biological efficiency [2]. Astatine may be dangerous upon incorporation or upon adsorption to the epidermis, and even when only brought near to it (eyes!). The mean energy of 6.8 MeV is dissipated to a limited volume of the tissue with a radius of about 60 μm. The maximum dose received after total decay of 3.7×10^4 Bq (1 μCi) ^{211}At in form of a point source amounts to 150 rad per g [3]. Compared to the radiation effects of the α particles, those caused by the X-rays of ^{211}Po, as well as by the γ's emitted during β^+ and ε decay of the relatively long-lived ^{207}Bi ($T_{1/2}$=33.4 a) are negligible. The recoil of the products from the α decay, ^{207}Bi and ^{207}Pb, with energies of 100 to 120 keV leads to very localized effects only. It is treated in detail in the chapter on irradiation and self-irradiation of astatine compounds, see Chapter 9, p. 179.

The situation is somewhat different for ^{210}At. Its main decay line (99.83%) leads to the long-lived ^{210}Po ($T_{1/2}$=138.4 d) [5, 6]. This radionuclide is relatively radiotoxic due to the emission of 100% 5.3 MeV α particles. Especially in nuclear medical studies, it constitutes an unwanted isotopic impurity. Its formation should be avoided by careful selection of radioactivation conditions.

The first nuclear medical experiments with astatine by Hamilton and Soley revealed the uptake of $^{211}At^-$ by the thyroid gland of guinea pigs [7 to 10]. Uptake curves were reported to be similar to those of radioiodine. The kidneys acted as the chief channel of elimination. Potential application for α therapy of diseased thyroid glands was discussed. A series of experiments followed which described the partial destruction of the thyroid glands of rats and monkeys [2, 11 to 21]. The injected ^{211}At radioactivities (in form of $^{211}At^-$) ranged from 3×10^5 to 1×10^7 Bq (about 10 to 250 μCi) corresponding to about 1.2×10^3 to 4×10^4 Bq (0.03 to 1 μCi) per gram body weight of rats. The administration of ^{211}At at a dose level $\geqq 1.2\times10^4$ Bq (0.3 μCi) per gram body weight caused changes and destruction of the thyroidal glands and of lymphatic tissues of rats, the formation of myxedema in monkeys, and of mammalian and pituitary tumors in rats [14, 22 to 25] and embryotoxicity in pregnant rats [26 to 28]. ^{211}At administration of 4×10^4 Bq (1 μCi) per gram body weight was reported to be lethal for rats (MLD 60) [15]. Experiments dealing with the suppression of immunoreactions of transplanted organs by ^{211}At labeled antigens were partially successful in primates [29 to 33], cf. also [34 to 36]. Attempts to transport ^{211}At via labeled precursors of DNA into the cells of malignant tumors did not show much success, most probably due to the adsorption of astatinated compounds to larger cell clusters such as macrophages [3, 37 to 40]. New attempts to incorporate more specific ^{211}At labeled drugs, especially monoclonal antibodies, into tumors were more successful [41 to 51].

Recent experiments revealed a specific action of ^{211}At in the form of radiocolloids with tellurium against malignant tumors [52 to 56]. Another interesting application of ^{211}At α radiation is the autoradiography of labeled biomolecules and cells [57, 58]. One of the most drastic administrations was the injection of up to 8×10^9 Bq (220 mCi!) into the anterior segments of eyes of rhesus monkeys [59]. ^{211}At doses $\leqq 25$ mCi did not cause significant permanent ocular damage, doses $\geqq 50$ mCi produced intensive corneal edema, anterior synechias, iritis, and radiation cataracts. In adequate doses, the thyroid cells were destroyed totally. Epithelial cysts, however, resisted the destruction by medium doses.

Administration of ^{211}At to man was only undertaken in a very few experiments in the early stage of astatine research [7, 10, 15, 60]. The doses delivered to patients, suffering from diseases of the thyroidal gland, were not reported. But, from the context one might estimate them to about 4×10^5 Bq (10 μCi). No injuries to the patients were observed. When taking into account the weight of a man, these ^{211}At radioactivities are somewhat too small to induce any permanent damage. However, the experiments with humans have never been repeated again. Measurements of the distribution of astatine compounds in mice and rats showed, besides the preferential uptake in the thyroidal gland, large fractions of ^{211}At in stomach, liver, and intestine and finally the kidneys. The distribution does not always depend on the chemical form, cf. [12, 37, 38, 40, 50, 61, 62]. This might be due to adsorption or complex formation of astatine with proteins and macrophages, partially also due to the rupture of the bonds of astatine with the organic substrates [64 to 68]. Administration of high radioactivities of ^{211}At for therapeutical treatment of humans in amounts exceeding 4×10^6 Bq (100 μCi) bears some radiological risks for many important organs. It should only be undertaken if the distribution and excretion of the ^{211}At compound applied has been carefully checked in appropriate animal experiments; not only in mice and rats, but also in larger animals such as, e.g., monkeys. Borras and Gorson calculated the adsorbed dose in the blood of rats as $D=1.98\times10^{-8}\pm0.32$ Gy/Bq kg rat mass [63], cf. also [27]. A determination of absorbed doses delivered to the organs of mice [62] was extrapolated to man and compared with the predictions of the International Commission on Radiological

References on pp. 154/6

Protection (ICRP) [69]. The doses extrapolated from the mice data to man, assuming a body mass of 70 kg and a quality factor of 20 for α particles, yield dose equivalents of $\sim 0.5 \times 10^{-8}$, $\sim 1.2 \times 10^{-8}$, $\sim 0.9 \times 10^{-8}$ and $\sim 2.6 \times 10^{-8}$ Sv for blood, kidneys, testes, and thyroid, respectively. The values for blood, kidneys, and testes are in line with the dose equivalent of 1.1×10^{-8} Sv predicted by ICRP for all tissues [63, 69]. However, the value of the thyroid is ~240 times greater than predicted. It should be emphasized here that it is certainly not easy to extrapolate from mice to man only on the base of weight, neglecting the differences in rates of metabolism of ^{211}At.

The European Communities classify ^{211}At into class II of radiotoxicity, i.e., "high radiotoxicity", together with, e.g., ^{22}Na, ^{36}Cl, ^{60}Co, ^{90}Sr, ^{125}I, ^{224}Ra, ^{232}Th and ^{236}U, cf. [72]. Other isotopes of astatine are not listed. The International Commission on Radiological Protection recommended in 1981 limits for the intake of ^{207}At and ^{211}At by radiation exposed workers and derived concentrations in air for a working time of 40 h per week; see Table 7/3, taken from [69].

Table 7/3
Recommendations of the International Commission on Radiological Protection (ICRP) Concerning the Limit of Annual Intake of ^{207}At and ^{211}At (by workers (ALI), and the derived concentration in air for a 40 h working week (DAC), taken from [69]).

radio-isotope		ingestion (oral)	inhalation class D	inhalation class W
^{207}At	ALI, Bq (μCi)	2×10^8 (5400)	1×10^8 (2700)	8×10^7 (2200)
	DAC, $Bq \cdot m^{-3}$ ($\mu Ci \cdot m^{-3}$)	—	4×10^4 (1.1)	3×10^4 (0.8)
^{211}At	ALI, Bq (μCi)	5×10^6 (135)	3×10^6 (81)	2×10^6 (54)
	DAC, $Bq \cdot m^{-3}$ ($\mu Ci \cdot m^{-3}$)	—	1×10^3 (0.027)	8×10^2 (0.022)

Here a biological half-life of 10 days in the human body was assumed. The values for the shorter-lived and less radiotoxic ^{207}At ($T_{1/2} = 1.8$ h) are higher by a factor of about 40 than those for ^{211}At. ^{207}At seems to have been selected from the other astatine isotopes because it is frequently studied in nuclear physics. From the viewpoint of radiotoxicity and frequent production it would, however, be more appropriate to recommend limiting values for ^{210}At.

A summary of the Council Directives of the European Committies concerning the annual personal intake of ^{211}At and derived concentrations in air for 2000 working hours per year is listed in Table 7/4 for the years 1959, 1976, 1980 and 1984 [70 to 73]. Included are the corresponding values of the directives for radioprotection of the Federal Republic of Germany from 1976 [74]. Two sets of data are reported: for workers exposed to radiation (occupational exposure) and for members of the general public[1]. These show that the values for the latter group are lower by a factor of 10 than those for workers, cf. ingestion data of [69]. A gradual liberalization of the directives has taken place from 1959 to 1984. This may be due to a general tendency to reduce the limits for certain short-lived α emitters whose radiation danger had been overestimated in the past. It may also reflect the increased knowledge of ^{211}At chemistry and radiology, and, thus, the improved control of the radiation risks. ^{207}At first shows up in the 1984 directives [73].

[1] The Council directives of 1984 follow the ICRP recommendations of 1981.

Table 7/4
European Health Physics Data for Astatine.

year, Ref.	iso-tope	state a)	exposed workers: max. annual incorporation by inhalation in Bq (μCi)	exposed workers: max. radio-activity of ambient air b) in Bq · m^{-3} (μCi · m^{-3})	members of public: max. annual incorporation by inhalation in Bq (μCi)	members of public: max. annual incorporation by ingestion in Bq (μCi)
1959 [70]	^{211}At	—	—	2×10^{1} (0.0005)	—	—
1976 (German) [74]	^{211}At	—	2×10^{3} (0.053)	—	2×10^{3} (0.053)	1.5×10^{3} (0.041)
1976/80 [71, 72]	^{211}At	s i	6.7×10^{5} (18) 3.2×10^{6} (87)	2.6×10^{2} (0.007) 1.1×10^{3} (0.03)	6.7×10^{4} (1.8) 3.2×10^{5} (8.7)	5.2×10^{4} (1.4) 2.1×10^{6} (58)
1984 [73]	^{211}At	D W	3×10^{6} (81) 2×10^{6} (54)	1×10^{3} (0.027) 8×10^{2} (0.022)	3×10^{5} (8.1) 2×10^{5} (5.4)	5×10^{5} (14)
1984 [73]	^{207}At	D W	1×10^{8} (2700) 8×10^{7} (2200)	4×10^{4} (1.1) 3×10^{4} (0.81)	1×10^{7} (270) 8×10^{6} (220)	2×10^{7} (540)

a) s = soluble; i = insoluble; D, W = different storage periods in lungs upon inhalation, cf. [72]. — b) For 2000 working hours per year.

Health physics groups of institutions beginning with astatine work are in general not well prepared for an appropriate control. It is the relatively weak radiation (^{211}Po X-rays) which makes the detection of ^{211}At and especially the total body measurement relatively difficult. The chemical properties of that carrier-free, heavy radioisotope (adsorption, easy oxidation, formation of radiocolloids, reactions with any kind of proteins, complex formation, etc.) impede a proper measurement of excretions such as urine. The scientist going to be involved in astatine work should feel obliged to inform and discuss the problems of detection with his colleagues concerned with health physics prior to experiments. Surveillance of the Jülich groups of astatine workers from 1972 to 1981 (7 persons) did not yield any violation of the limits for uptake. There were no severe cases of contamination of personnel, rooms, or equipment. Transport of amounts of 50 mCi ^{211}At was carried out regularly each week during years from Karlsruhe to Jülich nuclear research center over a distance of 300 km. No severe case of personnel injury by ^{211}At was reported from the astatine research groups. The accidental uptake of ^{211}At in 10 μCi amounts (R.D.N.) did not have any negative consequence.

The positive experiences were due to a very careful handling of ^{211}At. In most of the typical experiments, ^{211}At radioactivities produced in the cyclotron, ranged from 4×10^{8} to 4×10^{9} Bq (10 to 100 mCi). Demounting of the cyclotron target, transport and opening of the sample in the laboratory were subjected to the normal health physics precautions. Attention had to be paid to the moment of dissolution, degassing, or evaporation of the Bi metal target. These processes were carried out in the Jülich laboratory in lead protected fume hoods. Two operators wearing security overalls with hoods, gas masks, and two layers of gloves (thick rubber and thin plastic) shared the work. They were controlled by a health

References on pp. 154/6

physics technician. When the astatine was in an aqueous solution or separated into several fractions, normal radiological precautions were sufficient. During the experiments especially with nonaqueous samples of astatine compounds, safety goggles should be carried in order to avoid damage to the eyes by α particles. Any vessel containing astatine in solution, irrespective whether aqueous or organic, should be closed tightly. The carrier-free ^{211}At disappears from solutions easily in the form of HAt, labeled volatile organic compounds or addition complexes with organics [75, 76], cf. also [4]. Contamination by ^{211}At and decay products can be removed by normal washing with soap or detergents. The application of iodine compounds, e.g. KI, as a kind of carrier did not improve decontamination substantially. More severe is the contamination by ^{210}At since the long-lived decay product ^{210}Po is adsorbed firmly to the skin and the walls of glass and quartz vessels. This is another reason why the formation of ^{210}At is deliberately avoided.

References:

[1] L.J. Jardine (Phys. Rev. [3] C **11** [1975] 1385/91). — [2] J.K. Basson, J.C. Shellabarger (Radiat. Res. **52** [1956] 502/14). — [3] K. Rössler, W. Tornau, G. Stöcklin (J. Radioanal. Chem. **21** [1974] 199/209). — [4] G.W.M. Visser, E.L. Diemer (Radiochim. Acta **33** [1983] 145/51). — [5] S.G. Prussin, J.M. Hollander (Nucl. Phys. A **110** [1968] 176/92).

[6] L.J. Jardine, A.A. Shihab-Eldin (Nucl. Phys. A **244** [1975] 34/44). — [7] J.G. Hamilton, M.H. Soley (J. Appl. Phys. **11** [1940] 314). — [8] J.G. Hamilton, M.H. Soley (Proc. Natl. Acad. Sci. U.S. **26** [1940] 483/9). — [9] J.G. Hamilton (J. Appl. Phys. **12** [1941] 440/60). — [10] J.G. Hamilton (Radiology **39** [1942] 541/72).

[11] J.G. Hamilton, C.W. Asling, W.M. Garrison, K.G. Scott, D. Axelrod-Heller (Proc. Soc. Exptl. Biol. Med. **73** [1950] 51/3). — [12] J.G. Hamilton, C.W. Asling, W.M. Garrison, K.G. Scott (in: H.H. Anderson, D.B. Taylor, E.L. Way, University of California Publications in Pharmacology, Vol. 2, No. 21, Univ. Calif. Press, Berkeley 1953, pp. 283/343). — [13] P.W. Durbin, J.G. Hamilton, M.W. Parrott (Proc. Soc. Exptl. Biol. Med. **86** [1954] 369/71). — [14] J.G. Hamilton, P.W. Durbin, M.W. Parrott (J. Clin. Endocrinol. Metab. **14** [1954] 1161/78). — [15] J.G. Hamilton, P.W. Durbin, M.W. Parrott (Proc. 2nd Radioisotope Conf., Oxford 1954, Vol. 1, pp. 219/31).

[16] P.W. Durbin, M. Johnston, M.W. Parrott, N. Jeung, J.G. Hamilton (UCRL-2551 [1954]). — [17] C.J. Shellabarger, J.T. Godwin (J. Clin. Endocrinol. Metab. **14** [1954] 1149/60). — [18] C.J. Shellabarger, P.W. Durbin, M.W. Parrott, J.G. Hamilton (Proc. Soc. Exptl. Biol. Med. **87** [1954] 626/9). — [19] R.W. Watts (Proc. Soc. Exptl. Biol. Med. **89** [1955] 220/2). — [20] J.G. Hamilton, P.W. Durbin, C.W. Asling, M.E. Johnston (Proc. 1st Intern. Conf. Peaceful Uses At. Energy, Geneva 1955, Vol. 10, pp. 175/81).

[21] P.W. Durbin, C.W. Asling, M.E. Johnston, J.G. Hamilton (Anat. Record **129** [1957] 17/37). — [22] P.W. Durbin, C.W. Asling, M.E. Johnston, P.W. Parrott, N. Jeung, M.H. Williams, J.G. Hamilton (Radiat. Res. **9** [1958] 378/97). — [23] C.W. Asling, P.W. Durbin, M.E. Johnston, M.W. Parrott (J. Clin. Endocrinol. Metab. **64** [1959] 579/85). — [24] K. Yokoro, A. Kunii, J. Furth, P.W. Durbin (Cancer Res. **24** [1964] 683/8). — [25] G. Ueda, T. Mori (Am. J. Pathol. **51** [1967] 601/8).

[26] C. Borrás, R.L. Brent, R.O. Gorson, J.F. Lamb (Recent Advances in Nuclear Medicine Proc. 1st World Congr. Nucl. Med. Radioisotope Assoc., Tokyo 1974, pp. 335/9). — [27] C. Borrás-Amoedo (Diss. Univ. Barcelona, Spain, 1974). — [28] C. Borrás, G.O. Gorson, R.L. Brent (Phys. Med. Biol. **22** [1977] 118). — [29] J.A. Smit, J. Little, J.A. Myburgh (Liver Proc. Intern. Liver Conf. Spec. Ref. Africa, Cape Town 1973, pp. 209/10). — [30] J.A. Smit, J.A. Myburgh, R.D. Neirinckx (Clin. Exptl. Immunol. **14** [1973] 107/16).

[31] J.A. Myburgh, J.A. Smit (Transplant. Proc. **5** [1973] 597/600). — [32] R.D. Neirinckx, J.A. Myburgh, J.A. Smit (Radiopharmaceuticals and Labelled Compounds, IAEA, Vienna 1973, Vol. 2, pp. 171/81). — [33] C. Aaij, W.R.J.M. Tschroots, L. Lindner, T.E.W. Feltkamp (Intern. J. Appl. Radiat. Isotop. **26** [1975] 25/30). — [34] S.J. Simonian, B. Wainer, A. Friedman, W. Wung, M. Zalutsky, P.V. Harper Jr., F.W. Fitch, F.P. Stuart (Fed. Proc. Fed. Am. Soc. Exptl. Biol. **35** [1976] 334). — [35] A.M. Friedman, M.R. Zalutsky, W. Wung, F. Buckingham, P.V. Harper Jr., G.H. Scherr, B. Wainer, R.L. Hunter, E.H. Appelman, R.M. Rothberg, F.W. Fitch, F.P. Stuart, S.J. Simonian (Intern. J. Nucl. Med. Biol. **4** [1977] 219/24).

[36] M.R. Zalutsky, A.M. Friedman, F.C. Buckingham, W. Wung, F.P. Stuart, S.J. Simonian (J. Labelled Compounds Radiopharm. **13** [1977] 181/2). — [37] M. Persigehl, K. Rössler (AED-CONF-75-193-078 [1975]). — [38] G.J. Meyer (Diss. Univ. Köln, FRG, 1977; JUEL-1418 [1977]). — [39] K. Rössler, G.J. Meyer, G. Stöcklin (J. Labelled Compounds Radiopharm. **13** [1977] 27). — [40] K. Rössler, G.J. Meyer, L.E. Feinendegen, G. Stöcklin (Nuklearmedizin **2** [1978] 199).

[41] I. Brown (Intern. J. Appl. Radiat. Isotop. **33** [1982] 75/6). — [42] I. Brown (J. Labelled Compounds Radiopharm. **19** [1982] 1389/91). — [43] I. Brown (Radiochem. Radioanal. Letters **53** [1982] 343/50). — [44] I. Brown (in: C. Raynaud, Nuclear Medicine and Biology, Vol. 1, Pergamon, Paris 1982, pp. 166/9). — [45] A.T.M. Vaughan, W.J. Bateman, G. Brown, J. Cowan (Intern. J. Nucl. Med. Biol. **9** [1982] 167/71).

[46] A.T.M. Vaughan, W.J. Bateman, D.R. Fisher (Intern. J. Radiat. Oncol. Biol. Phys. **8** [1982] 1943/6). — [47] W.J. Bateman, A.T.M. Vaughan, G. Brown (Proc. 7th Intern Congr. Radiat. Res., Amsterdam 1983, Abstr. No. D1-03). — [48] W.J. Bateman (Diss. Univ. Birmingham 1983). — [49] I. Brown, R.N. Carpenter, J.S. Mitchell, S.F. Russell (Proc. 7th Intern Congr. Radiat. Res., Amsterdam 1983, Abstr. No. D4-05). — [50] I. Brown, R.N. Carpenter, J.S. Mitchell (Intern. J. Appl. Radiat. Isotop. **35** [1984] 843/7).

[51] A. Harrison, L. Royle (Intern. J. Appl. Radiat. Isotop. **35** [1984] 1005/8). — [52] C.R. Harris, S.J. Adelstein, S.J. Ruth, A.P. Wolf (Radiat. Res. **74** [1978] 590). — [53] W.D. Bloomer, W.H. McLaughlin, R.D. Neirinckx, S.J. Adelstein, R.R. Gordon, T.J. Ruth, A.P. Wolf (Science **212** [1981] 340/1). — [54] S.J. Adelstein, W.D. Bloomer (J. Labelled Compounds Radiopharm. **19** [1982] 1387/8). — [55] W.D. Bloomer, S.J. Adelstein (Radioakt. Isotope Klin. Forsch. **15** [1982] 227/34).

[56] W.D. Bloomer, W.H. McLaughlin, R.A. Milius, R.W. Atcher, S.J. Adelstein, S. Mirzadeh, R.M. Lambrecht, A.P. Wolf (Proc. 7th Intern. Congr. Radiat. Res., Amsterdam 1983, Abstr. No. D4-04). — [57] J.S. Mitchell, I. Brown, R.N. Carpenter (Proc. 7th Intern. Congr. Radiat. Res., Amsterdam 1983, Abstr. No. D4-20). — [58] J.S. Mitchell, I. Brown, R.N. Carpenter (Experientia **39** [1983] 337/9). — [59] R.N. Shaffer (Ann. J. Ophthalmol. **37** [1952] 183/97). — [60] J.G. Hamilton, P.W. Durbin, M.W. Parrott (Proc. Soc. Exptl. Biol. Med. **86** [1954] 366/9).

[61] G.W.M. Visser, E.L. Diemer, C.M. Vos, F.M. Kaspersen (Intern. J. Appl. Radiat. Isotop. **32** [1981] 913/7). — [62] A. Harrison, L. Royle (Health Phys. **46** [1984] 377/83). — [63] C. Borrás, R.O. Gorson (Med. Phys. **4** [1977] Paper 21.1). — [64] G.W.M. Visser, E.L. Diemer, F.M. Kaspersen (Intern. J. Appl. Radiat. Isotop. **31** [1980] 275/8). — [65] G.W.M. Visser, F.M. Kaspersen (Intern. J. Nucl. Med. Biol. **7** [1980] 79).

[66] G.W.M. Visser, E.L. Diemer, F.M. Kaspersen (Intern. J. Appl. Radiat. Isotop. **32** [1981] 905/12). — [67] G.W.M. Visser, E.L. Diemer, F.M. Kaspersen (J. Labelled Compounds Radiopharm. **18** [1981] 127). — [68] K. Berei, L. Vasáros (in: S. Patai, Z. Rappoport, The Chemistry of Functional Groups, Suppl. D, Wiley, New York 1983, Chapter 10, pp. 405/40). — [69] International Commission on Radiological Protection 1981 (ICRP Publ. No. 30 III [1981] 102/3; Ann.

ICRP **6** [1981] Pt. 2/3). — [70] Government of Federal Republic of Germany (Richtlinien zur Festlegung der Grundnormen für den Gesundheitsschutz der Bevölkerung und der Arbeitskräfte gegen die Gefahren ionisierender Strahlung, Feb. 20, 1959, ABL-EG-2 No. 11 [1959]).

[71] European Communities, Euratom (Off. J. Europ. Comm. **19** No. 187 [1976]; Euratom Publ. No. 76-579 [1976]). — [72] European Communities, Euratom (Off. J. Europ. Comm. **23** No. 246 [1980]; Euratom Publ. No. 80-836 [1980]). — [73] European Communities, Euratom (Off. J. Europ. Comm. **27** No. L265-4 [1984]). — [74] Government of Federal Republic of Germany (Verordnung über den Schutz vor Schäden durch ionisierende Strahlen vom 13. Oktober 1976, Bundesgesetzblatt, Pt. I, No. 125 [1976] 2905/95). — [75] G.J. Meyer, K. Rössler (Radiochem. Radioanal. Letters **25** [1976] 377/89).

[76] A. Cavallero (Diss. Univ. Cathol. Louvain, Belgium, 1981, pp. 1/102; will be published 1985 as report JUEL).

8 Astatine in Biology and Nuclear Medicine

Arnulf Seidel
Institut für Genetik und für Toxikologie von Spaltstoffen
Kernforschungszentrum Karlsruhe
Karlsruhe, Federal Republic of Germany

8.1 Biological Behavior of ^{211}At

8.1.1 Routes of Absorption

After oral administration to rats, astatine is absorbed rapidly from the gastrointestinal tract and accumulates in the thyorid gland at a rate similar to that observed after intravenous or intraperitoneal injection [1]. A study of ^{211}At uptake into the thyroid from patients suffering from thyroid diseases was performed by Hamilton et al. [1]. Although the interpretation of the results is difficult (as is frequently the case with human studies) it seems likely that astatine uptake from the human gastrointestinal tract is also rapid. The uptake of the astatine analog ^{131}I from the gastrointestinal tract of man is, of course, much better known from many investigations in nuclear medicine. It can be assumed that more than 90% of ^{131}I is absorbed within two hours after oral administration. A very similar behavior is shown by pertechnetate (references in [2]). As already demonstrated for other nuclides, the absorption of ^{211}At will probably be influenced by the status of the individual (fasted or nonfasted) as well as by the chemical form in which it is incorporated. The first biological study performed with astatine (at that time called ekaiodine) had already shown that this element is also absorbed rapidly from a subcutaneous deposition site [3]. Its transcutaneous absorption has not been investigated experimentally. However, it can be assumed by analogy with iodine, that some resorption through the intact skin can occur. When inhaled, iodine is rapidly absorbed through the lung, most probably the same holds for ^{211}At, but experimental data are lacking.

8.1.2 Distribution and Retention

8.1.2.1 Organ Distribution and Retention

The blood clearance of ^{211}At is very rapid. After intravenous injection into rats, only a few percent of the injected dose remained in the blood after one hour [4], this is very similar to ^{131}I. Pertechnetate also shows a very rapid disappearance rate from blood (references in [2]).

As already observed in 1940 by Hamilton et al. only the thyroid plays a major role in the accumulation of ^{211}At [3, 5] and the behavior of this element compares well with that of ^{131}I. The same author published a more detailed comparison of the metabolism of iodine and astatine in rats with some data from monkeys [4]. **Fig. 8-1**, p. 158, shows the time dependence of the ^{211}At concentration in some rat organs during the first 24 hours after intravenous injection. Clearly, the thyroid is the main depository organ, its ^{211}At concentration being two orders of magnitude or more greater than in other organs, except for the stomach. A very similar result was obtained by Borras [6] with pregnant rats, who calculated that the absorbed radiation dose in the thyroid was about 2000 times that for muscle. The difference between the concentration of ^{131}I in the thyroid and the other organs is even higher than in the case of ^{211}At. A relatively high uptake by stomach was also reported for pertechnetate (see [2]), which was due to the excretory function of the gastric mucosa. Some fecal excretion of ^{211}At occurs, but it is unknown, from which source (salivary gland, gastrointestinal mucosa, bile) it arises.

Fig. 8-1

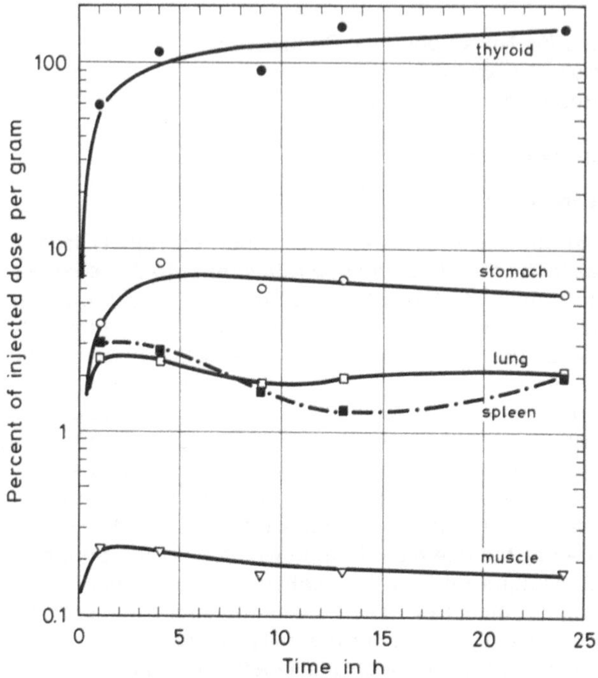

Uptake of i.v. injected ^{211}At by rat organs [4].

Like iodine and pertechnetate, astatine can be concentrated in the mammary gland and excreted via the milk [7]. Three hours after injection, the ^{211}At concentration in the mammary gland of lactating rats reached ~0.8% per gram. At four hours after injection, 10 to 15% of the maternal ^{211}At dose was passed to the litter.

The results obtained with a few monkeys confirmed the unique role of the thyroid in the metabolism and effects (see Section 8.2.1, p. 162) of astatine [4]. The uptake of astatine by the monkey thyroid was, however, higher than by rat thyroid (9 and 20% at 19 hours as compared to 2% at 24 hours for rats). In their study with oral administration of ^{211}At to man, Hamilton et al. [1, 8] also observed an unexpectedly higher accumulation in thyroid than predicted from the rat experiments. Possibly, primate thyroid has a relatively higher capacity for ^{211}At accumulation than rodent thyroid. It may be interesting to compare the kinetics of iodine and astatine accumulation in the thyroid as well the development of the relative concentrations in the organs (=ratio concentration in the organ/concentration in blood plasma). **Fig.** 8-**2** and Table 8/1 show clearly that the concentration of astatine in thyroid, though relatively high, is much lower than that of iodine. In contrast, the relative ^{211}At concentration in other organs is much higher than the relative iodine concentration. As shown by radioautography the microdistribution of ^{211}At in rat thyroid is somewhat irregular with a tendency to accumulate more in the colloid of smaller, centrally located follicles [4]. The presence of ^{211}At could also be demonstrated autoradiographically in the epithelia and lumen of mammary glands of rats [7]. This finding may have some importance for the interpretation of the increased incidence of mammary tumors in rats after ^{211}At injection (see Section 8.2.2, p. 167).

Fig. 8-2

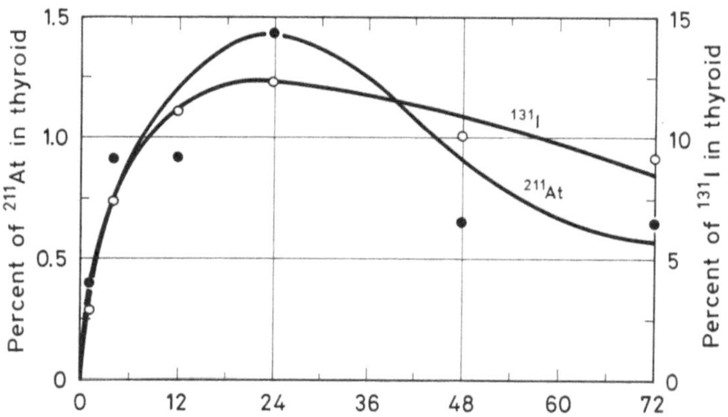

Comparison of the accumulation of i.v. injected ^{131}I and ^{211}At by the thyroid gland of the rat [4].

Table 8/1
Comparison of the Relative Concentrations of ^{211}At and ^{131}I in the Tissues of the Rat [4].

	^{211}At	^{131}I
thyroid	388	14880
stomach	33	4
lung	13	0.6
spleen	12	0.5
muscle	1	0.2

8.1.2.2 Excretion

The excretion of ^{211}At from the body occurs mainly via the urine (**Fig.** 8-**3**, p. 160). The data for fecal excretion shown in Fig. 8-3 may possibly be too high, since contamination of the feces with urine could not be ruled out during these experiments. As can be seen from this figure, the elimination of ^{131}I is much faster than of ^{211}At. This reflects the considerably longer residence time of ^{211}At in the organs of the rats, where about 60% of the injected dose were found after 24 hours. As already mentioned, another route of excretion would be salive and gastric fluid, but this pathway of elimination may be masked by intestinal reabsorption. ^{211}At can also be eliminated via the milk. When injected into pregnant animals, ^{211}At is transferred across the placenta to the embryos [6, 9]. This compares well with results for iodine, where toxic effects can be observed in the fetuses.

8.1.2.3 Factors Influencing ^{211}At Accumulation by the Thyroid Gland

The accumulation of ^{211}At in the thyroid gland is enhanced or impaired by different factors. Pretreatment with thyroid stimulating hormone [3] as well as an iodine deficient diet causes increased uptake (**Fig.** 8-**4**, p. 160, and Table 8/2, p. 161). On the other hand,

Fig. 8-3

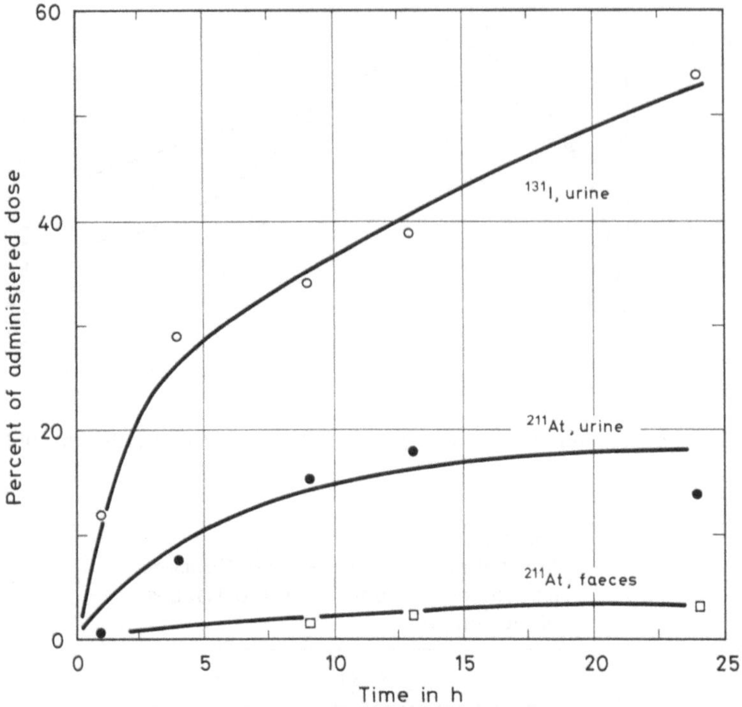

Excretion of i.v. injected ^{131}I and ^{211}At by the rat [4].

Fig. 8-4

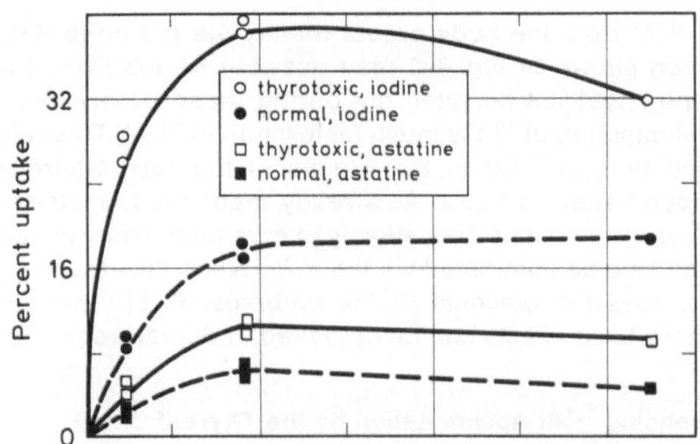

Uptake of ^{131}I and ^{211}At by the thyroid gland of normal and thyreotoxic rats [3].

Table 8/2
The Effect of the Administration of Stable Iodine and a Low-Iodine Diet upon the Uptake of Astatine by the Thyroid Gland of the Rat. (The uptake of radio-iodine by the thyroid gland after administration of stable iodine is included for comparison) [4].

	percent in thyroid gland	range	number of rats
1 mg KI given with ^{211}At			
control	1.74	1.00 to 2.46	16
treated	0.45	0.37 to 0.53	5
1 mg KI daily 6 × followed by ^{211}At			
control	0.84	0.57 to 1.11	5
treated	0.10	0.06 to 0.14	10
low-iodine diet before ^{211}At			
control	0.97	0.46 to 1.48	5
treated	2.43	1.35 to 3.51	10
1 mg KI daily 6 × followed by ^{131}I			
control	18.9	14.9 to 25.2	5
treated	0.42	0.27 to 0.60	5

Table 8/3
Effect of 8 Daily Subcutaneous Injections of Thyroxine (230 $\mu g \cdot kg^{-1} \cdot d^{-1}$) on 18-Hour Thyroidal Uptake of ^{131}I or ^{211}At, Body Weight and Thyroid Weight of 55-Day-Old Female Sprague-Dawley Rats. Last thyroxine injection given 4 h before i.v. administration of isotope [10].

treatment	number of rats	mean body wt beginning in g	mean body wt at sacrifice in g	mean thyroid weight in mg	mean ± S.E.[a] thyroid uptake in % of administrated isotope	mean ± S.E.[a] thyroid conc. in % isotope/g wet tissue
thyroxine + ^{131}I	8	121	133	9.8	0.73 ± 0.06	75.4 ± 6.6
control + ^{131}I	10	124	148	12.3	6.00 ± 0.51	502 ± 46
thyroxine + ^{211}At	10	114	124	9.6	0.068 ± 0.001	6.9 ± 0.5
control + ^{211}At	9	119	141	9.7	0.305 ± 0.01	32.0 ± 4.3

[a] S.E. = standard error.

the ^{211}At content of the thyroid was lower than in controls after pretreatment with thyroxine and KSCN [10, 11] (Table 8/3 and 8/4, p. 162). While these effects are similar to those of the iodine metabolism, an essential difference exists with regard to the action of thiouracil [5] or propylthiouracil [12] and the influence of posttreatment with KSCN [10]. The thiouracils decrease the iodine uptake whereas the ^{211}At uptake was increased (Table 8/5, p. 162). KSCN has little influence when iodine is already incorporated into the thyroid gland, but it reduces the ^{211}At content to about 50% of control when given 16.5 hours after the nuclide (Table 8/4, p. 162). The results could be explained by assuming that ^{211}At is trapped in the thyroid like other monovalent anions (see [2]), but that it is bound within the gland in a different manner to iodine, which forms I-C bonds in thyroxine. Visser et al. [13] concluded from the results described above and from their own experiments that astatine is bound most probably through At-S bonds.

References on pp. 176/8

Table 8/4
Effects of a Single or Double Subcutaneous Injection of KSCN on Thyroidal Uptake of ^{211}At in 130-Day-Old Female Sprague-Dawley Rats [10].

interval between ^{211}At i.v. injection and sacrifice in h	interval between KSCN injection and sacrifice in h	number of rats	mean body weight in g	mean thyroid weight in mg	mean ± S.E.[a] thyroid uptake in % of administrated ^{211}At	mean ± S.E.[a] thyroid concentration in % ^{211}At/g wet tissue
1	none	5	252	20.5	0.31 ± 0.02	15.60 ± 1.35
1	2.5	5	254	20.3	0.07 ± 0.01	3.19 ± 0.13
4	none	5	246	16.8	0.32 ± 0.03	18.95 ± 1.36
4	5.5	5	245	16.7	0.10 ± 0.01	5.85 ± 0.45
18	none	6	248	19.9	0.40 ± 0.06	20.00 ± 2.16
18	19.5	6	244	18.8	0.45 ± 0.08	23.08 ± 3.34
18	19.5 to 5.5	6	255	18.4	0.16 ± 0.02	8.67 ± 0.52
18	1.5	6	248	21.0	0.22 ± 0.02	10.37 ± 1.12

[a] S.E. = standard error.

Table 8/5
Effect of Treatment (for 11 d before sacrifice) with Thiouracil or Propylthiouracil on the Thyroidal Uptake of ^{131}I and ^{211}At by Rats.

nuclide	treatment	thyroidal uptake	n	Ref.
^{131}I	control	7.3 ± 1.5	5	[12]
	propylthiouracil[a]	1.1 ± 0.2	10	[12]
	control	3.1	5	[5]
	thiouracil[b]	1.7	5	[5]
^{211}At	control	0.2 ± 0.003	5	[12]
	propylthiouracil[a]	3.8 ± 0.3	10	[12]
	control	1.2	4	[5]
	thiouracil[b]	3.8	4	[5]

Arithmetic means ± standard error, n = number of rats per group.
[a] Sacrifice 21 h after nuclide injection. – [b] Sacrifice 1.5 h after nuclide injection.

8.2 Biological Effects of Incorporated ^{211}At

8.2.1 Acute and Subacute Effects

The short term mortality in rats after intravenous injection of ^{211}At was determined by Hamilton et al. [1] and compared with the effects of ^{131}I. The results of a probit analysis of the 10 and 60 day mortality are shown in **Fig. 8-5**. By extrapolation, the "median lethal dose 60 days" was estimated to be 1.12 μCi/g for ^{211}At and 85 μCi/g for ^{131}I (3.7×10^7 Bq/kg and 3×10^9 Bq/kg, respectively). The animals died with signs of severe haemorrhage. The body weight of the animals receiving 4×10^7 Bq/kg ^{211}At decreased rapidly and remained lower than that of controls (**Fig. 8-6**).

Fig. 8-5

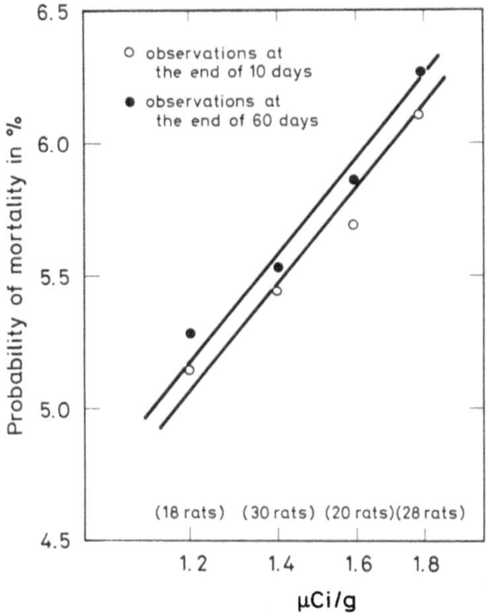

Probit analysis of the 10- and 60-day mortality of female Sprague-Dawley rats after a single i.v. injection of ^{211}At. The number of rats in each group is indicated at the foot of the diagram; extrapolated MLD60 = 1.12 µCi/g [1].

Fig. 8-6

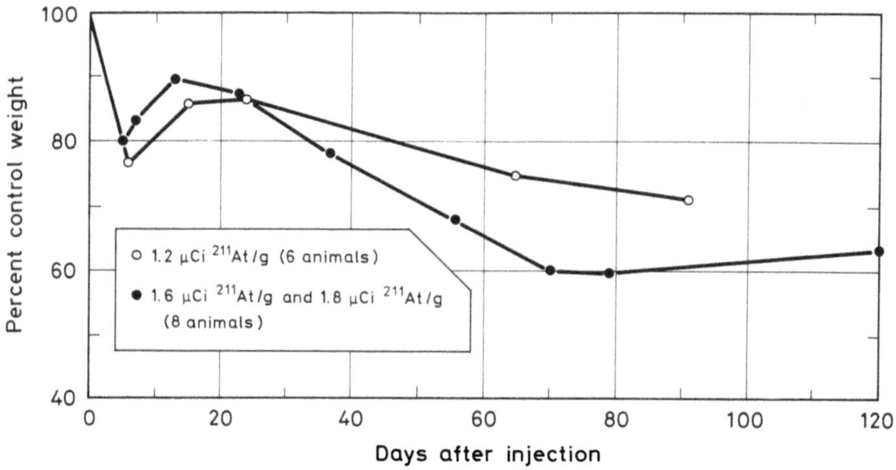

The effect of a single i.v. injection of ^{211}At on the growth of the Sprague-Dawley female rat. The weights are expressed as percent of control weight [18].

The main effects of sublethal doses of ^{211}At occur, of course, in the thyroid. The injection of ^{211}At into rats causes thyroid destruction (but not of the parathyroid) leading to almost complete disappearance [14, 19]. This is illustrated by some data on thyroid weight shown in Table 8/6, p. 164. Also shown in this table is the uptake of ^{131}I by the thyroid gland,

References on pp. 176/8

Table 8/6
Mean Weight and 24-Hour Iodine Uptake of Residual Thyroid Tissue of Rats Following ^{211}At Administration [14].

dosage	postinjection interval 6 days			40 days		1 year		
^{211}At µCi/gm body wt	number of rats	thyroid weight in mg	% ^{131}I uptake	number of rats	% ^{131}I uptake	number of rats	thyroid weight in mg	% ^{131}I uptake
controls	6	14.3±0.5	9.0±0.7	10	16.7±0.9	9	26.0±1.4	7.4±0.1
0.4	–	–	–	3	4.8±2.1	7	10.6±1.0	2.6±0.2
0.6	–	–	–	–	–	5	11.6±1.7	3.0±0.4
0.8	6	12.7±0.4	4.1±0.7	3	0.8±0.3	6	9.3±0.9	1.6±0.3
1.2	6	11.3±0.3	1.3±0.1	–	–	1		0.2
1.5	5	13.2±0.4	0.3±0.1	–	–	–	–	–
1.8	5	11.3±0.6	1.3±0.2	–	–	–	–	–

a) No thyroid tissue was found at autopsy. The ^{131}I uptake was measured on tissues en bloc.

which is a measure of its functional integrity. The functional disorders develop within six days, as indicated by the drastic decrease of ^{131}I uptake. The dose dependency of this effect was investigated 41 days after ^{211}At injection by Hamilton et al. [4], and is illustrated in **Fig. 8-7**. The minimally effective dose for reducing the ^{131}I uptake is below 2×10^6 Bq/kg ^{211}At. Histologically, progressive fibrosis and destruction of the epithelium was found, depending on the injected ^{211}At dose and the time of examination [14]. As a consequence of the "radiothyroidectomy" by ^{211}At the metabolic rate of the animals decreased rapidly to about three quarters of the control value between six and eight days [17], see **Fig. 8-8.**

Fig. 8-7

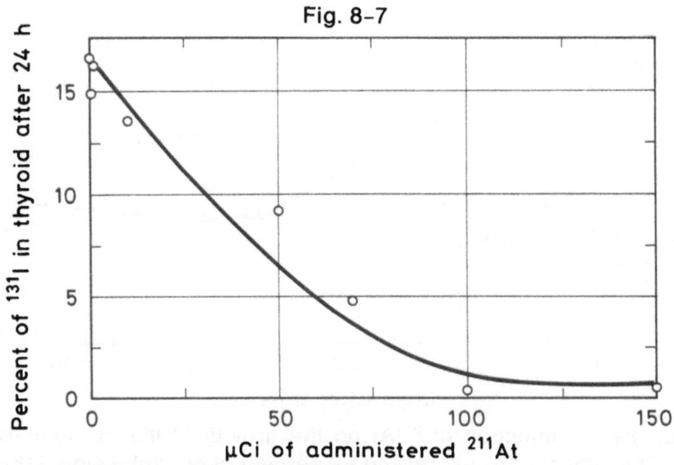

Correlation between the accumulation of radio-iodine by the thyroid gland of the rat fortyone days after the administration of different amounts of astatine. Seven groups of rats and one control group were given radio-iodine 24 h before the animals were sacrificed. The values shown are the percent of the administered radio-iodine accumulated by the thyroid gland [4].

Fig. 8-8

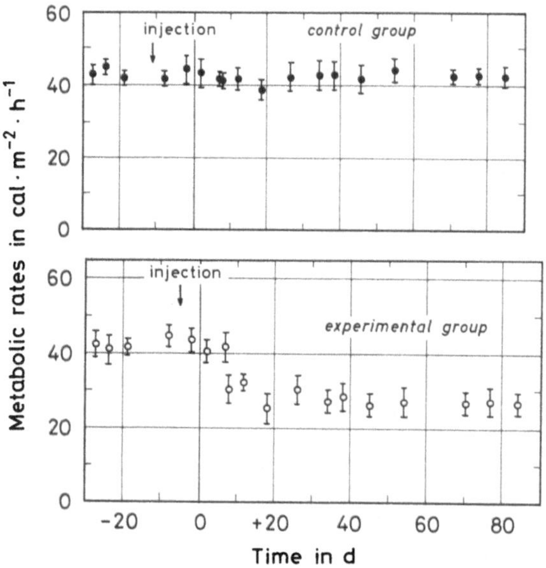

Metabolic rates of rats given a single i.v. injection of ^{211}At (0.8 μCi/g body weight) and of control animals given isotonic NaCl average values ± standard deviation are given for each group of animals [17].

Table 8/7
Effect of a Single Intraperitoneal Injection of Astatine-211 on Growth and Blood Picture of the Young Rhesus Monkey. (The monkeys were six to eight months old when injected) [1].

monkey	sex	days after injection	^{211}At dose	body wt in kg	RBC[a] millions/mm^3	WBC[b] thousands/mm^3	haemoglobin g/100 mL	differential white count	
								lymphocytes	granulocytes
3	M	358	control	3.32	5.77	14.40	13.8	81.5	18.5
6	M	358	control	3.92	7.27	9.25	15.8	92.0	8.0
2	M	358	0.42	2.92	3.52	5.75	9.5	73.5	26.5
5	M	358	μCi/g	2.89	3.53	3.30	11.2	85.0	15.0
1	F	294	0.83	2.61	4.66	7.85	11.8	84.0	16.0
4	F	294	μCi/g	3.35	3.56	7.05	10.3	68.0	32.0

a) RBC = red blood cell count. – b) WBC = white blood cell count.

References on pp. 176/8

In principle, the effects of ^{211}At injection into monkeys were the same as in rats. The growth of rhesus monkeys is affected by 0.36 μCi/g or 1.6×10^7 Bq/kg ^{211}At [18]. Table 8/7 shows that also some permanent haematologic changes occur [1]. The animals showed loss of hair and apathy, which can be attributed to diminished thyroid function. One monkey which had been injected with 1.6×10^6 Bq/kg and which exhibited distinct signs of diminished thyroid function was treated with thyroid extract in its food [4]. The signs of hypothyroidism rapidly disappeared. This indicates that the symptoms shown by the animal were due to ^{211}At-induced "thyroidectomy". Indeed, 3×10^6 Bq/kg ^{211}At injected into a rhesus monkey, caused almost complete destruction of thyroidal tissue after two years [18]. The effects of lower doses have not been investigated neither have more detailed dose-effect studies been made.

During an experiment performed to determine the RBE (relative biological effectiveness) of ^{211}At alpha particles as compared to ^{132}I beta particles in rat thyroid, Basson et al. [15] measured in more detail the dose-effect relationships for the influence of ^{211}At injection on thyroidal iodine accumulation. For ^{132}I as well as for ^{211}At a linear dose-effect function on a double-logarithmic scale (**Fig. 8-9**) was obtained with an RBE value of 2.8. This means that a beta particle dose from ^{132}I of 2.8 times greater than a corresponding alpha particle dose of ^{211}At is necessary to induce the same degree of functional damage in the thyroid gland. As pointed out by the authors, a drawback of this study was the fact that a uniform distribution of radiation dose throughout the thyroid gland was assumed, which is not the case. Presently it is believed that the RBE value for mammalian cell killing lies between 2 and 5, depending on the linear energy transfer [16].

In addition to its destructive action [24], ^{211}At injected in sublethal doses (2×10^7 Bq/kg) into rats also influences other tissues [4]. Persistent damage to bone marrow and/or lymphoid tissue, was indicated by the observation that the white blood cell count remained

Fig. 8-9

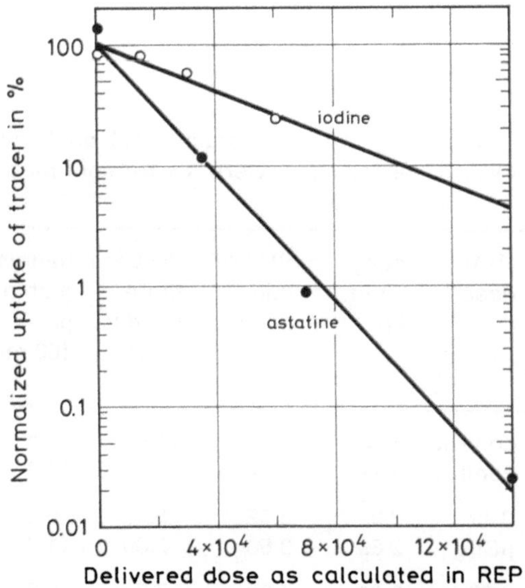

Eighteen-hour uptake of tracer ^{131}I by groups of animals which had received different doses of ^{211}At or ^{132}I. Linear graphs were drawn by the least squares method and normalized as a percentage of the no-dose uptake obtained [15].

lower than in controls. The pituitary gland shows the picture typical of thyroid deficiency and atrophy of the adrenal cortex was observed. Severe functional and gross anatomical ovarian deficiency occur, which seems to be related to direct irradiation damage (see [11]).

8.2.2 Carcinogenesis

A high incidence of mammary gland neoplasias was observed by Hamilton et al. [18] after injection of 2 to 3×10^7 Bq/kg ^{211}At into rats. A more detailed study on tumor induction by ^{211}At in rats was later published by Durbin et al. [11] for a dose level of 2×10^7 Bq/kg. The data are presented in Table 8/8. The most prominent finding was the increased incidence of mammary gland tumors and pituitary or adrenal cortical adenomas, the latter two tumors were very small in most cases. The incidence of mammary tumors in controls and ^{211}At-injected rats as a function of age is shown in **Fig.** 8-**10**, p. 168. The tumors appeared earlier and their incidence was higher after ^{211}At than in controls. Most of the mammary tumors in the ^{211}At group were malignant. The authors themselves discussed the question, whether the mammary tumor induction by ^{211}At injection was a direct radiation effect or an indirect effect resulting from irradiation of other parts of the body [11], leading to endocrine disturbances. However, no clear conclusion could be drawn, although ^{211}At concentration in the rat mammary gland has been demonstrated (see 8.1.2.1, p. 157).

Table 8/8
Number and Site of Tumors Induced in Female Sprague-Dawley Rats after 0.5 μCi of Astatine-211 per Gram Administered at 55 Days of Age [11].
(The experiment was terminated when the rats were 420 days old.)

tumor site and type	untreated controls	^{211}At-injected, untreated
number of rats autopsied	46	55
total tumors	10	107
mammary tumor-bearers (total)	8	39
pituitary adenoma [a)]	—	15
adrenal cortical [a)] adenoma	—	8
endometrial polyps	2	42
bronchiolar carcinoma	—	—
ovarian granulosa cell tumor	—	1
ovarian endometrioma	—	1
uterine squamous cell carcinoma	—	1

a) Most of the adenomata were very small and only detectable microscopically.

Tumors of the thyroid gland after ^{211}At incorporation have not been reported in the literature. At first glance, this seems to be strange, but the radiation dose to the thyroid may have been high enough to induce radiation sterilization of potentially malignant cells in most of the studies. It was possible to induce thyroid gland tumors in rats and sheep by incorporation of ^{131}I or external irradiation under certain conditions. Also in man, thyroid tumors were caused by external irradiation. On the other hand, in patients, receiving ^{131}I for diagnostic or therapeutic purposes, no excess thyroid cancer has yet been observed; a small group of normal subjects with incorporated iodine isotopes exists, in which a few thyroid cancers have occurred. Specific data for carcinogenesis by alpha-emitters in the thyroid gland are not available, but there is ample evidence for the induction of malignant

References on pp. 176/8

Fig. 8-10

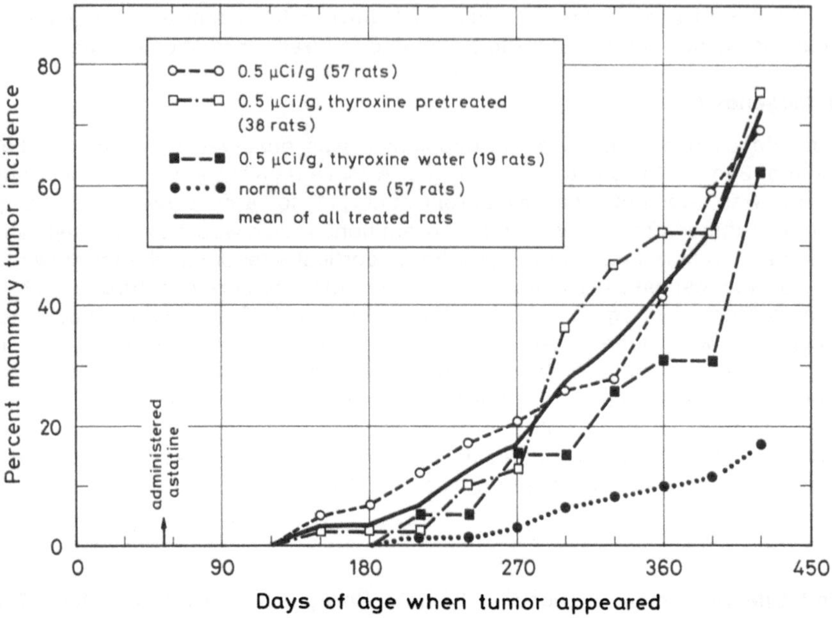

Incidence of mammary tumors in female Sprague-Drawley rats as a function of age after an i.v. injection of 0.5 μCi of ^{211}At per gram of body weight. Data are also shown for intact controls [11].

neoplasias in different tissues by incorporated alpha-emitters (^{226}Ra, ^{232}Th, ^{222}Rn) also in man. At present, risk estimates for ^{211}At and thyroid may be based on those for the tumor induction in this organ by external irradiation or ^{131}I, which have recently been reviewed by Taylor [20].

Considering the well-documented teratogenic effects of ionizing radiation it is not surprising that injection of ^{211}At into pregnant rats caused disturbances of embryonic development [9]. In addition to other effects, the incidence of malformations was increased. The authors conclude that the observed effects were qualitatively and quantitatively comparable to those produced by external irradiation with X-ray.

8.3 ^{211}At in Nuclear Medicine

The potential usefulness of ^{211}At in radiotherapy was first discussed by Hamilton et al. when they published the first biological study with ^{211}At [3]. The use of ^{211}At would offer the advantage of combining a short half-life with a very intense irradiation of circumscribed tissue areas, since the range of alpha particles in tissue is only 40 μm. Several categories of application of ^{211}At in radiotherapy have been discussed in addition to its use after organ transplantations [21 to 23]. Some considerations regarding the radiation dose can be found in [60].

It would seem logical to use ^{211}At for the treatment of thyroid disorders or neoplasms, since the thyroid gland is the main storage organ for that nuclide (see 8.1.2.1, p. 157). As described in Section 8.2.1, p. 162, the thyroid gland can be almost completely destroyed by injecting ^{211}At in large amounts, whereas the parathyroid remains unaltered. However, therapy with ^{211}At would also expose other organs to irradiation by an alpha-emitter. Until now, ^{211}At has not replaced radioactive iodine for treatment of diseases of the thyroid gland. At present, its use does not seem to be justified at least in the case of hyperthyroidism.

8.3.1 DNS Precursors Labeled with ^{211}At

Iodo-deoxyuridine is incorporated into the DNA of proliferating cells, e.g., tumor cells, like thymidine or other DNA precursors. The analogous astatinated compound might be of interest for obtaining a selective alpha irradiation of rapidly proliferating tumor cells. Attempts have been made to prepare astatinated uracil and deoxyuridine by electrophilic reactions with ^{211}AtCl but the yield was very low [24]. Another procedure uses the 5-amino derivatives of the same compounds and obtains the desired labeled molecules by formation of 5-diazonium salts [25, 26]. However, the yield of the nucleoside remains very low (3%). This led to the search for more efficient labeling procedures. After the successful preparation of 5-^{211}At-uracil by the chloromercury method, Visser et al. [27] also reported the astatination of different nucleosides and nucleotides in high yield (see 8.4.1, p. 174). The authors tested the stability of the astatopyrimidines and astatinated DNA and RNA at different pH values, temperatures and in presence of sulfite of hydrogen peroxide and concluded that the astatinated compounds were stable in vitro. However, when injected into rats, as 5-^{211}At-cytidine, 5-^{211}At-uridinemonophosphate or ^{211}At-RNA, the distribution of ^{211}At was similar to that of free ^{211}At (Table 8/9). Obviously, rapid deastatination of these compounds occurs in vivo.

Table 8/9
Distribution of Astatinated Pyrimidines in Rats (240 min after injection). Values are % injected dose/g tissue after i.v. administration (3 to 4 rats per group) [46].

tissue	5-^{211}At-UMP [a)]	5-^{211}At-cytidine	^{211}At-RNA
blood	0.3	0.2	0.2
skin (abdominal)	0.5	0.4	0.3
muscle (abdominal)	0.2	0.2	0.1
spleen	2.7	2.5	1.9
kidneys	0.8	0.9	1.7
liver	0.3	0.4	1.4
stomach (muscle)	5.2	5.8	3.0
stomach content [b)]	8.3	10.0	8.9
coecum	0.3	0.2	0.1
intestines (small)	0.7	0.6	0.5
heart	0.7	0.8	0.3
lungs	2.2	1.7	1.0
thyroid [b)]	0.7	0.7	0.5

The distribution of 5-^{211}At-deoxyuridine prepared by the diazonium salt method has been studied in normal mice and those with a malignant tumor (Sarcom 180) [26]. In normal mice no essential difference between the retention pattern of free ^{211}At and that of the labeled compounds was seen. At present, it seems unlikely that 5-^{211}At-pyrimidines are suited for therapy in man [13].

References on pp. 176/8

8.3.2 Anticancer Drugs Labeled with ^{211}At

Another way of achieving more or less selective irradiation of tumor tissue is the development of radioactive anticancer drugs with special affinity to malignant cells. In case of ^{211}At, derivatives of 2-methyl-1,4-naphthoquinol diphosphate are investigated (6-^{211}At-astato-2-methyl-1,4-naphthoquinol disodium diphosphate, ^{211}At-MNDP). The 6-^{211}At-MNDP can be prepared by the chloromercury method [28] but thermal heterogenous isotopic exchange proved to occur more quickly [29] (see 8.4.2, p. 175). The substance was injected into mice bearing a transplanted rectal adenocarcinoma [30, 31]. Its relative tissue uptake is presented in **Fig. 8-11.** (The thyroid was blocked by $KClO_4$ prior to injection.) The radioactivity is concentrated mainly in lung, tumor and spleen, but the relative specific activity in lung and spleen is a factor of 4 lower than that of ^{211}At. According to the authors, the molecule is stable in vitro and in vivo. Recently, therapeutic results obtained with this substance have been published [32]. Mice bearing a transplanted rectal adenocarcinoma were treated with ^{211}At-MNDP when the tumor was 100 to 1000 mm^3, by a single intraperitoneal injection. **Fig. 8-12** shows that a significant percentage of the mice was still alive 6 weeks after treatment, whereas all of the untreated control mice had died within 11 days. Autoradiographic

Fig. 8-11

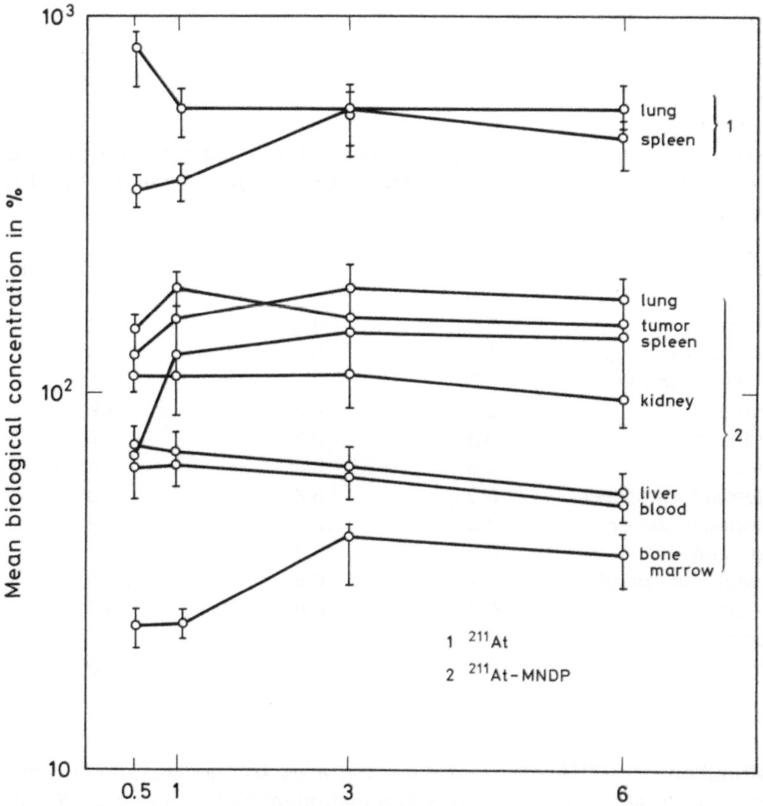

Tissue distribution of 6-^{211}At-astato-MNDP in mice (each point 4 mice; mean biological concentration: 100% × specific activity of organ/specific activity of the whole body) [31].

Fig. 8-12

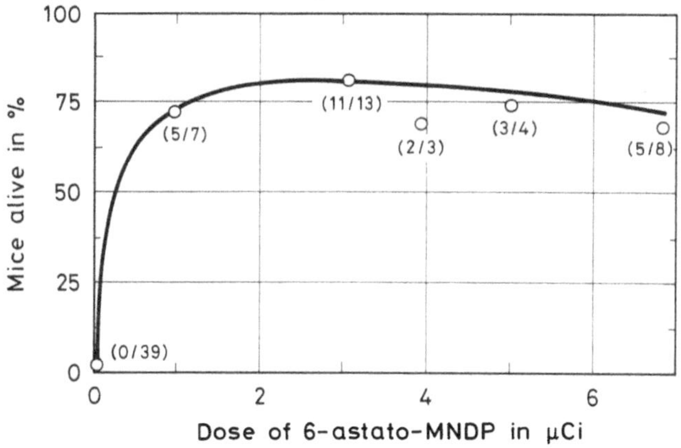

Graph showing percent mice still alive at ≧6 weeks after treatment plotted against dose of 6-^{211}At-astato-MNDP [32].

studies have shown that the compound is localized in the nuclei of tumor cells, but not of cells from normal colon [33] or bone marrow (see also [61]). Considerable uptake by specific cells in liver and spleen, however, was observed.

8.3.3 Colloids Labeled with ^{211}At

Several concepts have been proposed for local irradiation by ^{211}At of special types of malignant diseases, like malignant ascites or surface spreading tumors in the abdominal cavity. Amongst these, the use of ^{211}At-Te colloids has been tested experimentally [34 to 37, 62] including radiobiological studies on cell cultures [38]. To prepare the radioactive colloid, ^{211}At in 2 N nitric acid was added to a suspension of tellurium particles with a diameter of 2 to 25 µm. The experiments were performed with mice bearing an ascitic form of an ovarian tumor. This tumor can be transplanted and leads to death if no treatment is given. Different doses of ^{211}At were injected 24 hours after the inoculation of the mice with the tumor cells. The data are expressed as "change in median survival". The **Fig. 8-13**, p. 172, shows that the median survival time increases in proportion to the ^{211}At dose administered up to 25 and 50 µCi level, higher doses causing death within a few days. The animals receiving 25 or 50 µCi could be regarded as cured, no tumorous material was found, when they were autopsied at 200 days or later. The results can also be expressed as a "cellular survival function" (**Fig. 8-14**, p. 172) from which the authors calculated a so-called "D_{37}" (1.6 µCi ^{211}At per cell). The efficacy of ^{211}At has been compared with that of another colloid labeled with a beta-emitter (chromium-^{32}P-phosphate), with the Te colloid without ^{211}At or with decayed ^{211}At-tellurium colloid (**Fig. 8-15**, p. 172). Only the ^{211}At colloid was curative. As a side effect of treatment with ^{211}At-tellurium colloid, histopathological changes in the thyroid of the mice cured from malignant ascites were observed, which probably reflects dissociation of the compound. If ^{211}At colloids were indeed used for therapy, the thyroid could be blocked (see Section 8.1.2.3, p. 159, and 8.3.2, p. 170). A single report dated 1950 [39] concerning the application of an ^{211}At solution into the eyes of monkeys and of a man suffering from an epithelial cyst in the anterior part of the eye has been published. The nuclide was rapidly resorbed from the injection site into the thyroid of the monkeys.

Fig. 8-13

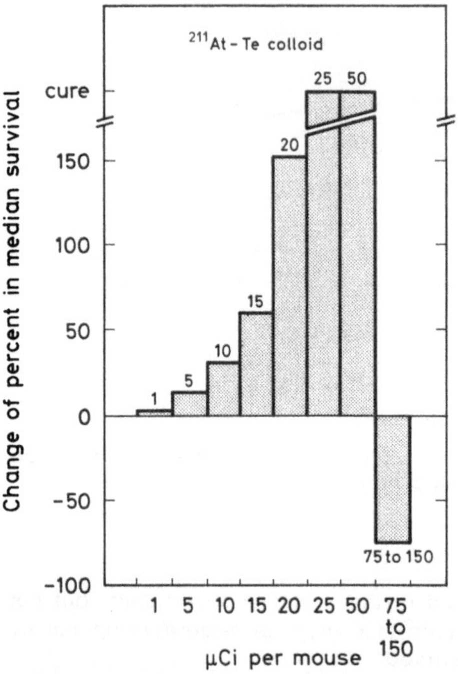

Results of radiocolloid therapy on experimental malignant ascites in mice, expressed as the percentage of change in median survival. Each experimental group contained 10 to 12 mice; experiments were performed three times. Nonradioactive tellurium colloid <2 μm in size is uniformly lethal in 3 days, presumably the result of pulmonary insufficiency [41].

Fig. 8-14

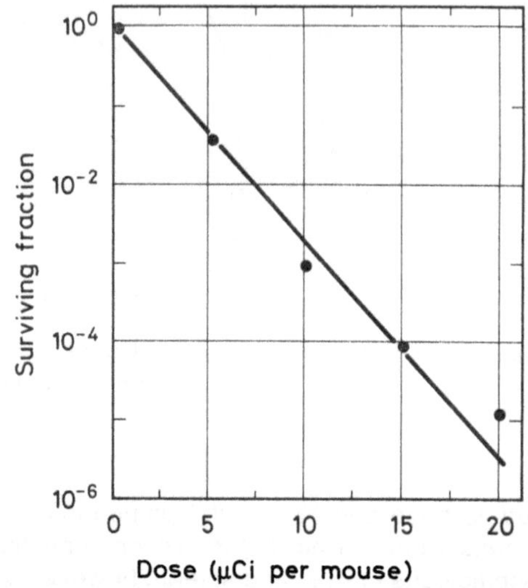

Results of ^{211}At colloid therapy expressed as cellular survival fraction. The dose to reduce survival to 0.37 is 1.6 μCi [35].

Fig. 8-15

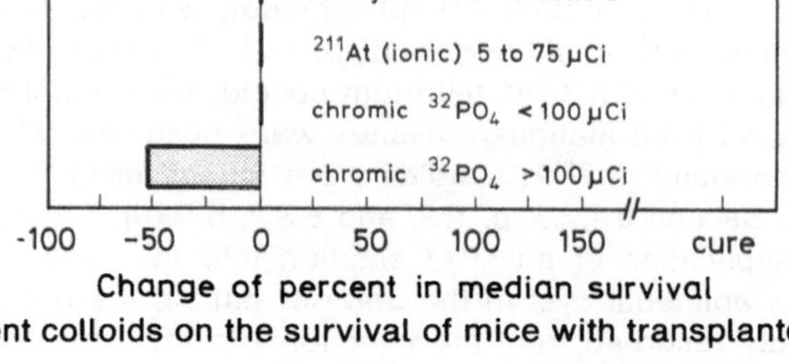

Influence of different colloids on the survival of mice with transplanted tumor [34].

No curative effect on the cyst in the human eye was observed and it was stated that this mode of treatment was not to be recommended.

8.3.4 Use of ^{211}At in Tumor Immunology and Transplantation

Tumor cells possess specific cell surface antigens, which can be recognized immunologically by the corresponding antibodies [40]. Such an antibody coupled with ^{211}At would specifically be bound to certain tumor cells, which consequently might be killed selectively by the local alpha irradiation [22]. Antitumor antibodies have been labeled with iodine, but not yet with ^{211}At [41]. The lectin Concanavalin A is bound to specific receptors on cell membranes. Vaughan et al. [42] were able to label this lectin with ^{211}At and to bind the nuclid via Concanavalin A to the cell membranes to human leukemia cells. The cytotoxic action of ^{211}At is shown in form of the survival curve in **Fig. 8-16**, which, thus, represents an example of cell killing by specifically bound ^{211}At.

Compounds labeled with ^{211}At may be directed against immuncompetent cells instead of tumor cells. Therefore, the use of ^{211}At-labeled compounds has also become interesting in transplantation science. Thus, a series of papers appeared describing the labeling of various types of proteins with different procedures and the testing of the stability of the ^{211}At-protein bound (see also Section 8.4.3, p. 176).

Proteins used were streptokinase, tuberculin and phytohaemagglutinine [43], hemocyanine [44], bovine serum albumin [45, 46], immunoglobulin [47], and the aminoacid tyrosine [48, 49]. In an in vitro system, Smit et al. [43] were able to inactivate selectively sensitized lymphocytes by incubation with suitable ^{211}At-labeled antigens. Lymphocytes from volunteers with hypersensitivity against streptokinase and tuberculin were incubated with the respective radioactive antigen. Autoradiographic examination showed that only a small, i.e., the sensitized fraction of the lymphocytes bound radioactivity. After incubation with ^{211}At-labeled streptokinase or tuberculin, only the lymphocyte fraction which was sensitive

Fig. 8-16

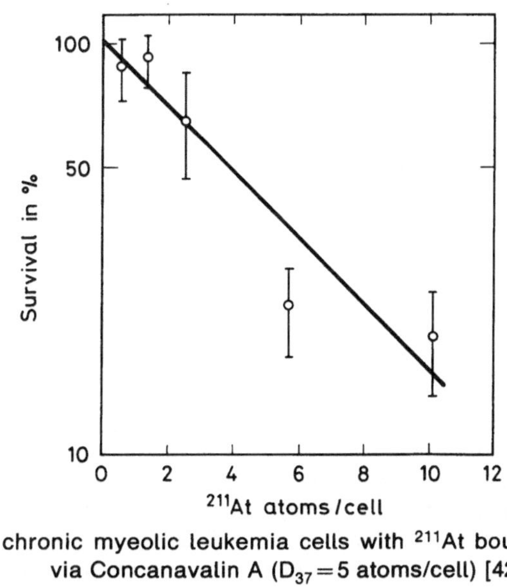

Survival of human chronic myeolic leukemia cells with ^{211}At bound to the cell membrane via Concanavalin A ($D_{37}=5$ atoms/cell) [42].

References on pp. 176/8

Table 8/10
Distribution of Free ^{211}At Astatide and ^{211}At-Labeled Bovine Serum Albumin (BSA) in Rats (240 min after i.v. injection 3 rats per group) [46] (in % injected dose/g tissue).

	^{211}At astatide	^{211}At BSA		^{211}At astatide	^{211}At BSA
blood	0.2	0.3	liver	0.2	0.3
skin	0.2	0.6	stomach (muscle)	4.9	6.1
spleen	2.3	2.7	stomach (content)	8.1	8.5
kidneys	0.7	1.0	thyroid [a)]	3.1	4.1

[a)] % of injected dose.

to the respective antigen was eliminated. With respect to the clinical use, the main problem with ^{211}At-labeled proteins is their in vivo stability. Labeled rabbit immunoglobulin has been found to be unstable under in vitro conditions [47]. Visser et al. [46] compared the tissue distribution of electrophilically astatinated BSA in rats and observed a retention pattern very similar to that of free $^{211}At^{-}$ (Table 8/10). Since the nuclide also behaved like the free form after injection of ^{211}At antigens into mice in vivo deastatination appears to be a serious disadvantage for the clinical use of ^{211}At proteins. The nature of the ^{211}At protein bound is still subject to controversy [46, 50 to 52].

An attempt to destroy lymphocytes in order to inhibit rejection after liver transplantation was made [53, 54] by injection of ^{211}At-labeled donor lymphocytes. Baboon donor lymphocytes were labeled by electrooxidation and injected after liver transplantation into the baboon recipient. The mean survival of the untreated group was 11+3 days, of the treated group 36+6 days, the difference being statistically significant ($p \leq 0.01$). All animals died eventually of rejection, which indicates that the sensitive recipient lymphocyte population was not completely destroyed. In connection with transplantation problems it should be mentioned finally that the use of bone marrow seeking radiocolloids containing alpha-emitters has been suggested for the elimination of the recipients bone marrow cells, whereby the radiation dose to other organs would be minimized [55]. Pertinent results are, however, not yet available.

8.4 Preparation of Biologically Interesting ^{211}At Compounds

In the following paragraphs, only an outline of the most frequently used procedures and principles can be given. For more detailed information, Chapter 4 and the original papers must be consulted. Some aspects of the chemistry of astatine have been discussed by Visser [63].

8.4.1 Aminoacids and Pyrimidines

Visser et al. [13] described the preapration of 5-^{211}At-cytidine and 5-^{211}At-uridinemonophosphate by astatination of the corresponding chloromercury compounds. Of these compounds, 10 μmol were suspended in 0.5 mL of aqueous NaOAc buffer, pH = 5. The ^{211}At activity in dilute NaOH was added followed by 2×10^{-2} μmol of KI_3 after one minute.

After stirring for 30 min at room temperature, the precipitated HgI_2 was dissolved by addition of excess KI. The astatinated compounds were isolated by two step chromatography. First, the unreacted ^{211}At and HgI_4 were removed by chromatography over DEAE-Sephadex using 0.9% NaCl as eluant. Then, unreacted chloromercury compounds were separated by chromatography over reduced thiosepharose 6B using 0.9% NaCl as eluant. A radiochemical purity of 95% was demonstrated by thin-layer chromatography.

Another mechanism for astatination was used by Meyer et al. [56] for uracil. The reaction started by dissolving 0.1 mmol of 5-aminouracil in 200 µL 7 N HCl at about 50 °C followed by cooling to −5 °C. After addition of 50 µL 2.5 N $NaNO_2$ and 2 to 3 mg of urea, up to 1 mCi of ^{211}At in 50 µL 0.4 N $NaSO_3$ solution were added. The reaction mixture was heated for a few minutes to 80 °C, then cooled down to 0 °C and the precipitate filtered. The filtrate, which contained most of the radioactivity, was further analyzed by high pressure liquid chromatography. The reaction is shown in **Fig.** 8-**17.** Also ^{211}At-desoxyuridine was prepared by a similar reaction [26]. Astatinated phenylalanines like ^{211}At-tyrosine, ^{211}At-4-methoxyphenylalanine and ^{211}At-phenylalanine were prepared by Visser et al. [49] through a reaction with the chloromercury compounds similar to that described above. The radiochemical yields were 60 to 80% and purification obtained by chromatography over DEAE-Sephadex. Attempts to prepare ^{211}At-tyrosine by exchange in the solid state (starting with iodotyrosine and ^{211}At astatide) showed only a poor radiochemical yield (1 to 5%) [49].

Fig. 8-17

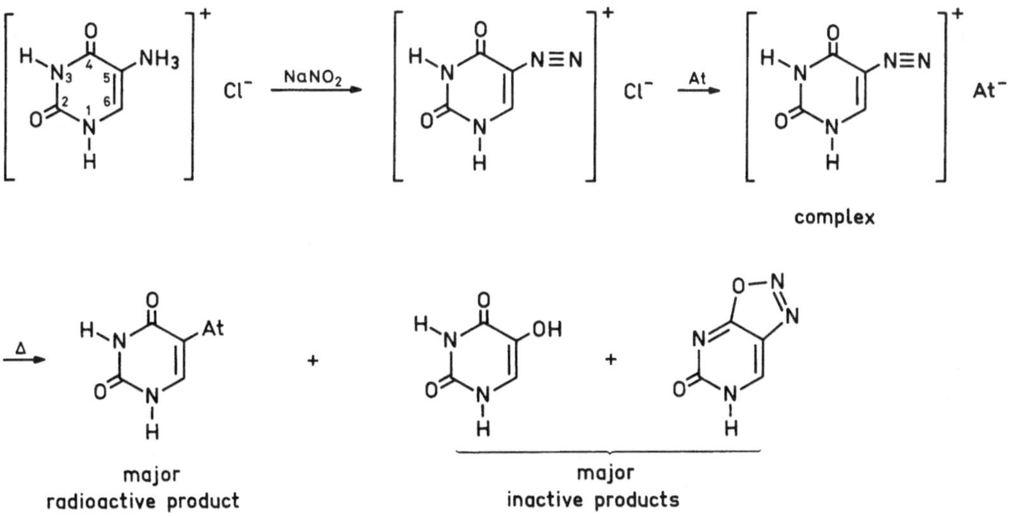

Astatination of uracil [56].

8.4.2 Antitumor Drugs

The antitumor drug 6-^{211}At-2-methyl-1,4-naphthoquinol bisphosphate (MNDP) can be prepared via the 6-chloromercury compound or by thermal heterogenous exchange. For the chloromercury method [28], an ethanolic suspension of 50 µmol 6-chloromercury-1,4-naphthoquinone was refluxed with 200 µCi ^{211}At and 55 µmol ICl and the reaction product (II) (see reaction scheme, **Fig.** 8-**18**, p. 176) was precipitated by KI. The naphthoquinone was reduced by $NaBH_4$ and product (III) phosphorylated by addition of $POCl_3$. After evaporation to dryness, the residue was dissolved in CH_2Cl_2-H_2O saturated LiOH solution. The separated aqueous phase (after stirring at pH 8 to 11 for 0.25 h) contained 80% of the remaining radioactivity, the major product was the desired astatinated naphthoquinol bisphosphate, which could be purified by thick-layer chromatography on cellulose. In order to improve the overall efficiency and to shorten the time needed for preparation, the same author proposed the method of thermal heterogenous exchange, starting with 6-iodo-methyl-

References on pp. 176/8

Fig. 8-18

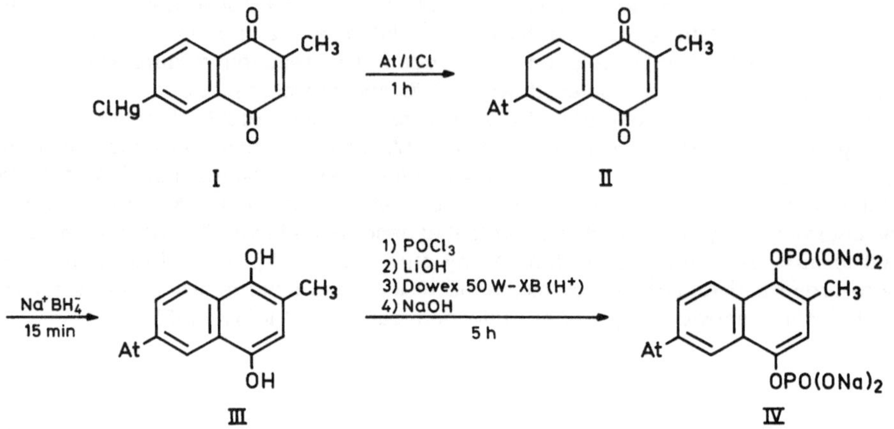

Synthesis route for 6-^{211}At-astato-2-methyl-1,4-naphthoquinol bis (disodium phosphate) [28].

1,4-naphthoquinol bis-dilithiumphosphate [29]. After addition of an ^{211}At solution to the iodo compound dissolved in distilled water, the mixture was evaporated to dryness at 170 °C. After cooling and dissolving in distilled water ^{211}At was removed by chromatography on DEAE-Sephadex and further analysis was performed by thin-layer chromatography on silica gel.

8.4.3 Proteins

Bovine serum albumin and RNA have been prepared by the chloromercury method (see preceeding chapter) [13]. A different method, which possibly leads to compounds of greater in vivo stability [13], is the conjugation of the protein with p-^{211}At-benzoic acid. This has been used for Concanavalin [42], bovine serum albumine [45] and an "antibody type" protein [57]. In principle ^{211}At is added to a diazonium sulfate salt solution of p-amino-benzoic acid [45] and after reaction the ^{211}At-para-benzoic acid is recovered in the diethylether fraction used for extraction. This fraction is evaporated to dryness and a mixed anhydride generated by addition of tributylamine and isobutylchloroformate. This mixed anhydride is added dropwise to a bovine serum albumin solution. The ^{211}At-labeled protein is purified by chromatography on Sephadex G-25. Electrolytic labeling of proteins was attempted for streptokinase, tuberculin and phytohaemagglutinine [43], hemocyanine [44] and a rabbit immunoglobulin [47]. Basically, the proteins are mixed with ^{211}At in sodium phosphate buffer. Electrolytic labeling is performed in a platinum crucible which serves as the anode, and the solution is separated from the cathode by a dialysis membrane. In the case of the immunoglobulin, rapid in vivo deastatination was observed [47]. For bovine serum albumin, haemoglobin and lactoglobuline astatination by the chloramine-T method [58, 59] has been described [46].

References:

[1] J.G. Hamilton, P.W. Durbin, M.W. Parrott (Radioisot. Conf. Proc. 2nd Intern. Conf., Oxford 1954, Vol. 1, pp. 219/31, Lect. AC1/AC12; C.A. **1955** 6471). – [2] A. Seidel (Gmelin Handbook "Technetium" 1982, pp. 271/314). – [3] J.G. Hamilton, M.H. Soley (Proc. Natl.

Acad. Sci. U.S. **26** [1940] 483/9). – [4] J.G. Hamilton, C.W. Asling, W.M. Garrison, K.G. Scott (Univ. Calif. Publ. Pharmacol. **2** [1953] 283/343). – [5] C.J. Shellabarger, J.T. Godwin (J. Clin. Endocrinol. Metab. **14** [1954] 1149/60).

[6] C. Borras (Phys. Med. Biol. **22** [1977] 118 [Abstr. only]). – [7] C.W. Asling, P.W. Durbin, M.E. Johnston, M.W. Parrott (Endocrinology [Baltimore] **64** [1959] 579/85). – [8] J.G. Hamilton, P.W. Durbin, M.W. Parrott (Proc. Soc. Exptl. Biol. Med. **86** [1954] 366/9). – [9] C. Borras, R.L. Brent, R.D. Gorson, J.F. Lamb (Recent Advan. Nucl. Med. Proc. 1st World Congr., Tokyo 1974, pp. 335/9; C.A. **83** [1975] No. 24414; CONF-740926-2 [1974] 1/5; COO-3268-5 [1974] 1/5; N.S.A. **30** [1974] No. 29781; C.A. **82** [1975] No. 66749). – [10] C.J. Shellabarger, P.W. Durbin, M.W. Parrott, J.G. Hamilton (Proc. Soc. Exptl. Biol. Med. **87** [1954] 626/9).

[11] P.W. Durbin, C.W. Asling, M.J. Johnston, M.W. Parrott, N. Jeung, M.H. Williams, J.G. Hamilton (Radiat. Res. **9** [1958] 378/97). – [12] P.W. Durbin, J.G. Hamilton, M.W. Parrott (Proc. Soc. Exptl. Biol. Med. **86** [1954] 369/71). – [13] G.W.M. Visser, E.L. Diemer, C.M. Vos, F.M. Kaspersen (Intern. J. Appl. Radiat. Isotop. **32** [1981] 913/7). – [14] J.G. Hamilton, P.W. Durbin, C.W. Asling, M.E. Johnston (Proc. 1st Intern. Conf. Peaceful Uses At. Energy, Geneva 1955, Vol. 10, pp. 175/81; C.A. **1958** 5613). – [15] J.K. Basson, C.J. Shellabarger (Radiat. Res. **5** [1956] 502/14).

[16] M.E. Elkind, G.F. Whitmore (The Radiobiology of Cultured Mammalian Cells, Gordon and Breach, New York 1967, p. 218). – [17] R.W.E. Watts (Proc. Soc. Exptl. Biol. Med. **89** [1955] 220/2). – [18] J.G. Hamilton, P.W. Durbin, M.W. Parrott (J. Clin. Endocrinol. Metab. **14** [1954] 1161/78). – [19] P.W. Durbin, C.W. Asling, M.E. Johnston, J.G. Hamilton (Anat. Record **129** [1957] 17/37). – [20] D.M. Taylor (J. Radioanal. Chem. **65** [1981] 195/208).

[21] G. Samson (Nuklearmedizin Suppl. [Stuttgart] No. 12 [1974] 506/8). – [22] A.M. Friedman (in: R.P. Spencer, Therapy in Nuclear Medicine, Grune & Stratton, New York 1978, pp. 139/44). – [23] K. Berei, L. Vasáros (KFKI-1981-10 [1981] 1/74, 53; INIS Atomindex **12** [1981] No. 609573). – [24] K. Rössler, W. Tornau, G. Stöcklin (J. Radioanal. Chem. **21** [1974] 199/209). – [25] G.J. Meyer, K. Rössler, G. Stöcklin (Gen. Meeting GdCh [Ger. Chem. Soc.], Cologne 1975; AED CONF-75-404-029 [1975] 1/9; INIS Atomindex **7** [1976] No. 257238).

[26] K. Rössler, G.J. Meyer, G. Stöcklin (J. Labelled Compounds Radiopharm. **13** [1977] 271). – [27] G.W.M. Visser, E.L. Diemer, F.M. Kaspersen (J. Labelled Compounds Radiopharm. **18** [1981] 799/807). – [28] I. Brown (Intern. J. Appl. Radiat. Isotop. **33** [1982] 75/6). – [29] I. Brown (Radiochem. Radioanal. Letters **53** [1982] 343/50). – [30] I. Brown (Proc. 3rd World Congr. Nucl. Med. Biol., Paris 1982, Vol. 1, pp. 166/9).

[31] I. Brown (J. Labelled Compounds Radiopharm. **19** [1982] 1389/91). – [32] I. Brown, R.N. Carpenter, J.S. Mitchell, S.F. Russel (Proc. 7th Intern. Congr. Radiat. Res., Amsterdam 1983, Lect. D4-05). – [33] J.S. Mitchell, I. Brown, R.N. Carpenter (Proc. 7th Intern. Congr. Radiat. Res., Amsterdam 1983, Lect. D4-20). – [34] S.J. Adelstein, W.D. Bloomer (DOE-EV-04115-4 [1979/81] 54/64; INIS Atomindex **12** [1981] No. 615556). – [35] W.D. Bloomer, W.H. McLaughlin, R.D. Neirinckx, S.J. Adelstein, P.R. Gordon, T.J. Ruth, A.P. Wolf (Science **212** [1981] 340/1).

[36] S.J. Adelstein, W.D. Bloomer (J. Labelled Compounds Radiopharm. **19** [1982] 1387/8). – [37] W.D. Bloomer, S.J. Adelstein (Radioakt. Isot. Klin. Forsch. **15** [1982] 227/34). – [38] C.R. Harris, S.J. Adelstein (Radiat. Res. **74** [1978] 590). – [39] R.N. Shaffer (Trans. Am. Ophthalmol. Soc. **50** [1952] 607/27). – [40] C.W. Parker (Pharmacol. Rev. **25** [1973] 325/42).

[41] S.J. Adelstein, M. Zalutsky, W. Bloomer (DOE-EV-04115-T3 [1981] 1/17; INIS Atomindex **13** [1982] No. 696346). – [42] A.T.M. Vaughan, W. Batteman, J. Cowan (J. Radioanal.

Chem. **64** [1981] 33/9). – [43] J.A. Smit, J.A. Myburgh, R.D. Neirinckx (Clin. Exptl. Immunol. **14** [1973] 107/16). – [44] C. Aaij, W.R.J.M. Tschroots, L. Lindner, T.E.W. Feltkamp (Intern. J. Appl. Radiat. Isotop. **26** [1975] 25/30). – [45] A.M. Friedman, M.R. Zalutsky, E.H. Appelman (Intern. J. Nucl. Med. Biol **4** [1977] 219/24).

[46] G.W.M. Visser, E.L. Diemer, F.M. Kaspersen (Intern. J. Appl. Radiat. Isotop. **32** [1981] 905/12). – [47] A.T.M. Vaughan, J.H. Fremlin (Intern. J. Nucl. Med. Biol. **5** [1978] 229/30). – [48] A.T.M. Vaughan, J.H. Fremlin (Intern. J. Appl. Radiat. Isotop. **28** [1977] 595/8). – [49] G.W.M. Visser, E.D.L. Diemer, F.M. Kaspersen (Intern. J. Appl. Radiat. Biol. **30** [1979] 749/52). – [50] G.W.M. Visser, F.M. Kaspersen (Intern. J. Nucl. Med. Biol. **7** [1980] 79).

[51] A.T.M. Vaughan (Intern. J. Nucl. Med. Biol. **7** [1980] 80). – [52] G.W.M. Visser, E.L. Diemer (J. Labelled Compounds Radiopharm. **18** [1981] 127). – [53] R.D. Neirinckx, J.A. Myburgh, J.A. Smit (Radiopharm. Labelled Compounds Proc. Symp., Copenhagen 1973, Vol. 2, pp. 171/80). – [54] J.A. Smit, J. Little, J.A. Myburgh (Liver Proc. Intern. Conf. Spec. Ref. Africa, Cape Town 1973, pp. 209/10). – [55] W.D. Bloomer, W.H. McLaughlin, R.A. Milius, R.W. Atcher, S.J. Adelstein, S. Mirzadeh, R.M. Lambrecht, A.P. Wolf (Proc. 7th Intern. Congr. Radiat. Res., Amsterdam 1983, Lect. D4-04).

[56] G.-J. Meyer, K. Rössler, G. Stöcklin (J. Labelled Compounds Radiopharm. **12** [1976] 449/58). – [57] A.T.M. Vaughan (Intern. J. Appl. Radiat. Isotop. **30** [1979] 576/7). – [58] K. Krohn, L.C. Knight, J.F. Harwig, M.J. Welch (Biochim. Biophys. Acta **490** [1977] 497/505). – [59] P.J. Fraker, J.C. Speck (Biochim. Biophys. Res. Commun. **80** [1978] 849/57). – [60] A. Harrison, L. Royle (Health Phys. **46** [1984] 377/83).

[61] J.S. Mitchell, I. Brown, R.N. Carpenter (Experientia **39** [1983] 337/9). – [62] W.D. Bloomer, W. McLaughlin, R.M. Lambrecht, R.W. Atcher, S. Mirzadeh, J.L. Madara, R.A. Milius, M. Zalutsky, S.J. Adelstein, A.P. Wolf (Intern. J. Radiat. Oncol. Biol. Phys. **10** [1984] 341/8). – [63] G.W.M. Visser (INIS-mf-8634 [1982] 1/129; C.A. **100** [1984] No. 80773).

9 Irradiation and Self-Irradiation of Astatine Compounds

Kurt Rössler
Institut für Chemie 1 (Nuklearchemie)
Kernforschungsanlage Jülich
Jülich, Federal Republic of Germany

9.1 General Aspects

The decay schemes of the two most important isotopes ^{210}At and ^{211}At have been described in the preceding chapters, cf. also [1 to 3]. It is generally agreed that the radiation effects of ^{211}At are mainly due to the energetic α particles. For a more detailed treatment of irradiation and self-irradiation effects it is, however, necessary to consider the different kinds of radiation emitted in the decay of ^{210}At and ^{211}At somewhat more closely. The largest contribution to radiation damage of the environment in which astatine decays is, besides the effects of the α particles themselves the α emission induced recoil of the decay products, i.e., Bi and Pb isotopes. Table 9/1 lists the recoil energies of the various decay products of ^{210}At and ^{211}At. It shows, that the α recoil of ^{207}Bi and ^{207}Pb can contribute to the effects of α radiation only in the case of ^{211}At. The fact that the decay of ^{210}At leads primarily to the long-lived ^{210}Po, which slowly emitts the alphas, hinders an effective participation in the immediate radiolysis. An exception may be for long storage times in solids. Thus, ^{211}At is by far more important for radiolysis effects than ^{210}At.

Table 9/1
Recoil Energies of Decay Products of ^{210}At and ^{211}At.

radioisotope	decay mode	recoil atom	recoil energy E_r in keV
^{210}At	0.17% α	^{206}Bi	1.07
	99.83% ε	^{210}Po	0 a)
^{210}Po	100% α	^{206}Pb	1.05
^{211}At	41.9% α	^{207}Bi	1.14
	51.8% ε	^{211}Po	0 a)
^{211}Po	100%α	^{207}Pb	1.44
^{207}Bi	100%ε+β^+	^{207}Pb	max. 2.3×10^{-2} a)

a) If Auger effect occurs, low energy recoil energy may arise from Coulomb repulsion.

Electron capture processes do not contribute directly to recoil. However, some recoil energy may arise from Coulomb repulsion stimulated by the Auger effect. The total energy will certainly not surpass a few eV to some 10 eV and is negligible in this context. The effects of the β^+ emission are also small. The direct radiolysis by the positrons emitted and the X-rays of polonium must be added to that of α particles and recoil. Since these radiations affect a much wider portion of the medium in which the decay occurs, they will not contribute too much to the local effects.

Table 9/2, p. 180, lists the mean ranges of α particles and recoil atoms from the decay of ^{211}At estimated for three different materials: H_2O, MgO and Fe, according to [4, 5]. The mean penetration of alphas ranges from 12 to 48 μm, i.e., all α particles are stopped within

Table 9/2
Mean Ranges $\bar{r}$ of α Particles and Recoil Atoms in the Decay of ^{211}At.

projectile	energy in MeV	$\bar{r}$ in μm in H_2O	in MgO	in Fe	estimated from
$\alpha(^{211}At)$	5.867	33	18	12	[4]
$\alpha(^{211}Po)$	7.450	48	27	20	[4]
^{207}Bi	1.14×10^{-3}	$\approx2\times10^{-1}$	$\approx6\times10^{-2}$	$\approx3\times10^{-2}$	[5]
^{207}Pb	1.44×10^{-3}	$\approx2.5\times10^{-1}$	$\approx8\times10^{-2}$	$\approx4\times10^{-2}$	[5]

a radius of about 20 to 60 μm. The much lower energetic recoil particles are stopped within a radius ranging from 30 to 250 nm. Both kinds of projectiles show different collision characteristics. The light α particles with their relatively high kinetic energies mainly undergo inelastic (electronic) collisions and induce excitation and ionization. The medium energy, heavy recoils undergo mainly elastic collisions and will induce atomic defects, especially displacements in solids. The linear energy transfer for both kinds of radiation is about 0.17 keV · nm^{-1} for the α particles and about 0.5 keV · nm^{-1} for ^{207}Bi and ^{207}Pb in H_2O.

9.2 Irradiation Effects of Aqueous Solutions

Any kind of energetic radiation produces radicals in H_2O such as $H^{\cdot}$, $OH^{\cdot}$, $O_2H^{\cdot}$, and in addition solvated electrons and H_2O_2. In not too concentrated solutions these radicals govern the radiolysis of substrates and self-radiolysis of astatine compounds in dilute solutions. Since, in general, the reactivity of the oxidizing species is higher than that of the reducing ones, oxidation will occur. This is the reason that At^- in a nonreducing medium is gradually converted to At^0, cf. Chapter 7.3. At^0 can attack organic solvents and substrates or slowly disappear by evaporation out of the reaction vessel upon standing. The easy oxidation of At^- is the reason that storage in the dark is often recommended in order to avoid the formation of radicals by light [6]. The higher the astatine radioactivity in a given volume, the more carefully redox conditions have to be controlled in order to preserve the desired chemical state. The direct action of the radiation on the substrate will only take place at higher concentrations of the latter. Bond breakage, especially of freshly formed carbon-astatine bonds, may be the consequence. Since most of the modern work with astatine is done in rather dilute solutions, radiation and self-radiolysis effects are in general neglected, cf., e.g. [7]. A systematic study of the effects is hitherto lacking. It would have been interesting to follow the changes in the extremely concentrated solutions used in the early nuclear medical experiments, cf., e.g. [8]. Recent studies on the stability of astatine labels in biomolecules do not report any bond breakage due to radiolytic effects in vitro and in vivo [9, 10].

9.3 Irradiation Effects in Organic Solutions and Solids

Even less is known for organic solutions and solids than for aqueous solutions. It is assumed that the latter systems are much more stable against radiolysis. Less stable radicals are formed in organics than in water. Scavenging effects may be operative in order to stop chain reactions of the radicals. Polymerization has been observed under high doses: organic residues in the vessel in which the bismuth targets had been dissolved after activation became brown. In solids there are recombination reactions of defects at least at ordinary temperature. Decoloration of glasses has been observed for regions where astatine had been deposited [6]. Work with weighable amounts (some 10 ng) of ^{211}At seems, however, to be very difficult with respect to self-radiolysis and temperature.

9.4 Radiation Effects in Biological Environment

The health physics aspects of the α radiation of ^{211}At has been dealt with in Chapter 7.4 on "Radiation Risks, Health Physics and Protection", together with a more general treatment of the radiobiological and nuclear medical applications. For a long time the aim of astatine chemists and medical staff has been to concentrate astatine in malignant cells for a radiation therapeutical treatment by the localized α emission. The simple inorganic compounds finally proved to be inadequate for a selective concentration in specific organs, with the exception of the thyroid gland. A nonselective application such as injection of astatine into the malignant tissue, i.e., into the water between the cells, only increases the general radiation risks. The astatine is finally transported to liver, kidneys and stomach and can not act specifically. Thyroidal glands have been treated with some success. However, the thyroidea and the lymphatic tissues were severely injured and tumors were induced in other parts of the body when applying doses $\geq 1.2 \times 10^4$ Bq (0.3 μCi) per g body weight to rats, for details cf. [11 to 15] and Chapters 7.4 and 8, p. 150 and p. 157. An example of a successful transport of simple species to malignant tumors and therapeutical action in close contact to the tissue has been reported for At radiocolloids with tellurium [16 to 19]. The idea of using organic compounds of ^{211}At and labeled biomolecules and their precursors as a vehicle for transporting the astatine to specific sites was born in 1955 [20, 21] and reviewed in 1972 [22, 23]. It was finally applied to problems of immunosuppression and cancer treatment, cf., e.g. [10, 24, 25] and Chapter 7.4. Numerous compounds have been labeled with ^{211}At because of to skill of organic chemists. There may be a chance of selective α therapy in the near future, cf. Chapter 7.4.

In these experiments the radiation acts in direct contact with or even upon incorporation into to the cells of the malignant tissue. Here the recoil nuclei offer a specific chance for introducing nearby strand or double-strand breaks in the DNA of cell nuclei. For some specific problems, the effects of the recoils may be even more important than those of the α particles, since the first ones produce more atomic damage with a somewhat higher LET. The latter ones produce less displacements, but more electronic damage in a wider region.

Last but not least, a specific application of ^{211}At in biochemistry and biophysics is autoradiography of labeled molecules and cells [26, 27], following an idea proposed by Adelstein [28]. The α particle track can be followed individually. It is possible to locate the exact position of the α emitter inside a cell or even in genes by microscopy of etched plates.

References:

[1] L.J. Jardine (Phys. Rev. [3] C **11** [1975] 1385/91). — [2] S.G. Prussin, J.M. Hollander (Nucl. Phys. A **110** [1968] 176/92). — [3] L.J. Jardine, A.A. Shihab-Eldin (Nucl. Phys. A **224** [1975] 34/44). — [4] U. Littmark, J.F. Ziegler (Handbook of Range Distributions for Energetic Ions in All Elements, Pergamon, Elmsford, N.Y., 1980, pp. 1/493). — [5] K. Rössler (Gmelin Handbook "Uranium" A6, 1983, pp. 135/64).

[6] E.H. Appelman (UCRL-9025 [1960] 1/113; N.S.A. **14** [1960] No. 11501). — [7] G.W.M. Visser, E.L. Diemer, F.M. Kaspersen (Intern. J. Appl. Radiat. Isotop. **32** [1981] 905/12). — [8] R.N. Shaffer (Am. J. Ophthalmol. **37** [1952] 183/97). — [9] A.T.M. Vaughan, W.D. Bateman, D.R. Fisher (Intern. J. Radiat. Oncol. Biol. Phys. **8** [1982] 1943/6). — [10] A. Harrison, L.Royle (Intern. J. Appl. Radiat. Isotop. **35** [1984] 1005/8).

[11] J.G. Hamilton, P.W. Durbin, M.W. Parrott (J. Clin. Endocrinol. Metab. **14** [1954] 1161/78). — [12] P.W. Durbin, C.W. Asling, M.E. Johnston, P.W. Parrott, N. Jeung, M.H. Williams, J.G. Hamilton (Radiat. Res. **9** [1958] 378/97). — [13] C.W. Asling, P.W. Durbin, M.E.

Johnston, P.W. Parrott (J. Clin. Endocrinol. Metab. **64** [1959] 579/85]. — [14] K. Yokoro, A. Kunii, J. Furth, P.W. Durbin (Cancer Res. **24** [1964] 683/8). — [15] G. Ueda, T. Mori (Am. J. Pathol. **51** [1967] 601/8).

[16] W.D. Bloomer, W.H. McLaughlin, R.D. Neirinckx, S.D. Adelstein, R.R. Gordon, T.J. Ruth, A.P. Wolf (Science **212** [1981] 340/1). — [17] S.J. Adelstein, W.D. Bloomer (J. Labelled Compounds Radiopharm. **19** [1982] 1387/8). — [18] W.D. Bloomer, S.J. Adelstein (Radioakt. Isotop. Klin. Forsch. **15** [1982] 227/34; C.A. **99** [1983] No. 18894). — [19] W.D. Bloomer, W.H. McLaughlin, R.A. Milius, R.W. Atcher, S.J. Adelstein, S. Mirzadeh, R.M. Lambrecht, A.P. Wolf (Proc. 7th Intern. Congr. Radiat. Res., Amsterdam 1983, Paper D4-04; CONF-830710 [1983] D4-04). — [20] W.L. Hughes, J. Klinenberg (BNL-367 [1955] 1/59, 2/3; N.S.A. **10** [1956] No. 3143).

[21] W.L. Hughes, E. Smith, J. Klinenberg (BNL-406 [1956] 1/56, 44/5; N.S.A. **11** [1957] No. 12). — [22] K. Rössler, W. Tornau, G. Stöcklin (AED-CONF-72-350-002 [1972]; Abstr. Meeting GDCh Fachgr. KRS/Ges. Nukl. Med., Freiburg, FRG, 1972). — [23] K. Rössler, W. Tornau, G. Stöcklin (J. Radioanal. Chem. **21** [1974] 199/209). — [24] J.A. Smit, J.A. Myburgh, R.D. Neirinckx (Clin. Exptl. Immunol. **14** [1973] 107/16). — [25] I. Brown, R.N. Carpenter, J.S. Mitchell (Intern. J. Appl. Radiat. Isotop. **35** [1984] 843/7).

[26] J.S. Mitchell, J. Brown, R.N. Carpenter (Proc. 7th Intern. Congr. Radiat. Res., Amsterdam 1983, Paper D4-20; CONF-830710 [1983] D4-20). — [27] J.S. Mitchell, J. Brown, R.N. Carpenter (Experientia **39** [1983] 337/9). — [28] S.J. Adelstein (Personal Communication 1980).

10 Chemical Behavior and Compounds of Astatine

10.1 General Aspects of the Chemistry of Astatine

S.H. Eberle
Universität und Kernforschungszentrum Karlsruhe
Institut für Radiochemie
Karlsruhe, Federal Republic of Germany

Introduction

In spite of its limited availability and the many difficulties connected with the experimental handling of astatine there is a continuous interest in the chemistry of the element since its discovery in 1940 [1]. As a result, a relatively large number of compilations or summaries exist on its chemical behavior, e.g.:

G.L. Johnson, R.F. Leininger, E. Segrè, Chemical Properties of Astatine, J. Chem. Phys. **17** [1949] 1/10.
A.H.W. Aten, T. Doorgeest, U. Hollstein, H.P. Moeken, Analytical Chemistry of Astatine, Analyst [London] **77** [1952] 774/8.
E.K. Hyde, The Present Status of Element 85 and 87, J. Phys. Chem. **58** [1954] 21/6.
K.W. Bagnall, Chemistry of the Rare Radioelements, Butterworths, London 1957, pp. 97/118.
E.H. Appelman, The Radiochemistry of Astatine, NAS-NS-3012 [1960] 1/33; N.S.A. **14** [1960] No. 20201.
A.H.W. Aten, The Chemistry of Astatine, Advan. Inorg. Chem. Radiochem. **6** [1964] 207/23.
A.K. Lavrukhina, A.A. Pozdyakov, Analiticheskaya Khimiya Tekhnetsiya, Prometiya, Astatina i Frantsiya, Nauka, Moscow 1966, pp. 1/307; Analytical Chemistry of Technetium, Promethium, Astatine, and Francium, Ann Arbor-Humphrey, Ann Arbor, Mich., 1970, pp. 1/316 or Israel. Progr. Sci. Transl., Jerusalem 1969.
E.H. Appelman, Astatine, MTP [Med. Tech. Publ. Co.] Intern. Rev. Sci. Inorg. Chem. Ser. One **3** [1972] 181/98; C.A. **76** [1972] No. 120925.
W.A. Chalkin [Khalkin], E. Herrmann, Anorganische Chemie des Astats, Isotopenpraxis **11** [1975] 333/40.
W.A. Chalkin [Khalkin], E. Herrmann, J.W. Norseev, I. Dreyer, Der gegenwärtige Stand der Chemie des Astats, Chemiker-Ztg. **101** [1977] 470/81.

The thesis of E.H. Appelman, Diss. Univ. California 1960, pp. 1/131, published as UCRL-9025 [1960] 1/131; N.S.A. **14** [1960] No. 11501, is a rather fundamental source on the experimental methods and for many important results. The literature up to this date is summarized in "The Radiochemistry of Astatine" referenced above. Modern experimental techniques in astatine chemistry at the time of printing of this volume are in a series of articles by R. Dreyer and I. Dreyer. The most recent one is reference [2].

For a number of reasons the present knowledge of the chemistry of astatine is rather incomplete and uncertain. It is the least known element below atomic number 100. One consequence is the difficulty of treating its chemical behavior in the normal way, that is, divided according to its valency state. In many cases of experiments in solution it is not quite clear what was the true valence state or how was the purity of valency state. Therefore, it was decided to present the solution chemistry grouped according to the investigation methods rather than according to the valency. But, whenever possible, the relevant information is given. A summary regarding the valency states is included in this chapter and, therein, further experimental procedures for preparing chemically uniform solutions are outlined.

10.1.1 Valency States and Summary of the Chemical Behavior of Astatine

No stable isotopes of astatine are known and none can be expected. There are several isotopes with half-lives of hours, but ^{211}At with a half-live of 7.2 h is normally used for chemical experiments. The isotope emits α particles in 42% of its degradations and undergoes electron capture in 58%. Therefore it can be measured easily and specifically by either its α radiation or the X-rays of 80 keV. ^{211}At is formed by irradiation of ^{209}Bi with ^{4}He particles of 20 to 28 MeV through an ^{4}He, 2n reaction.

General Chemical Nature

Astatine is the higher homologue of iodine and resembles this element in its general chemical nature but with a more pronounced metallic character.

Important general features of the chemical properties are the high volatility which is advantageous for the purification of the element, the tendency to adsorb onto surfaces, especially metallic surfaces, the photosensitivity in solution, the high extractability into organic solvents in certain valency states, and the easy formation of organic compounds by substitution reactions. It appears to be oxidized to positive valency states easier than iodine.

Another general feature of astatine chemistry is the complexity of its behavior during experiments. Experiments must be interpreted very cautiously because of the low concentration to which one is limited (normally $<10^{-8}$ M), the thermodynamic possibility to exist

Fig. 10-1

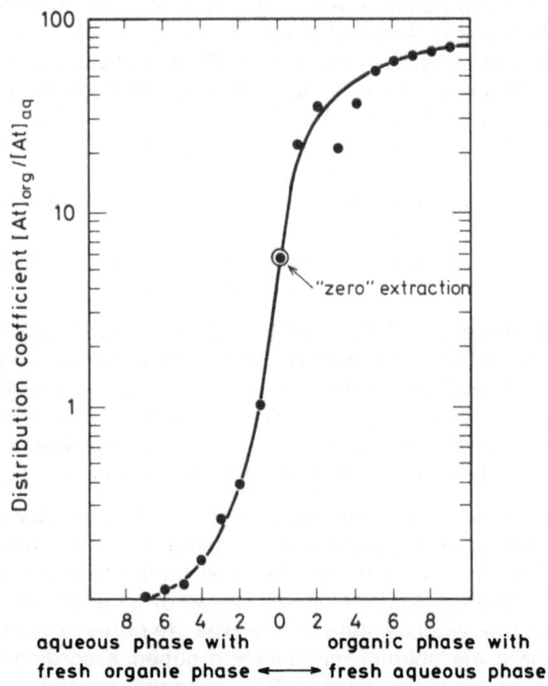

Result of the consecutive equilibration of the aqueous and organic phase of a "zero extraction" of astatine from 0.01 M HNO_3 by benzene with fresh portions of the respective other phase. Astatine was originally in the aqueous phase [3].

simultaneously in different oxidation states and the obviously high reactivity with impurities. In many cases, the results are intelligible but of poor reproducibility and far yield from 100% of the expected value. It is rather difficult to obtain reliable physicochemical data therefore, other conclusions about the meaning of the results cannot be excluded fully. An example is given in **Fig.** 10-1 for the results of successive extractions of astatine from 0.01 M HNO_3 by benzene. Both phases of the "zero equilibration" were treated several times with fresh portions of the other phase. The zero value of the distribution coefficient is 5.7 but a limiting distribution coefficient of 70 was reached after equilibrating the organic phase 9 times with fresh astatine containing aqueous phase. Consecutive extractions of the "zero aqueous phase" with fresh benzene did not remove all astatine but left about 1% of the activity as "unextractable". Appelman [3] concluded from these results that there were three forms of astatine present. The main fraction was At^0. The others were possibly "reduced" At and At bound to impurities. He said that sometimes the results are "uninterpretable – not to say exceedingly frustrating" [4]. An important factor is the presence of invisible suspended material (dust) in the solutions, onto which adsorption may occur. Johnson et al. [5] were able to improve the consistency of their experiments remarkably by centrifuging the solutions and removing any fine solids by this.

Basic Physicochemical Data

Physicochemical data of astatine have not been measured up to now directly but were estimated from those of the other halogens by various extrapolation methods. There are two exceptions, the detection of two light absorption lines of gaseous astatine at 2162.25 and 2244.01 Å by McLaughlin [6] and the gas chromatographic determination of the boiling point 230 °C, by Otozai, Takahashi [7]. Table 10/1 is an internally consistent set of relevant data, indicating astatine to have a melting point well above iodine and similar to bismuth and polonium. The boiling point is low compared to both elements.

Table 10/1
Physicochemical Properties of the Halogens and Extrapolations for Astatine [8].

halogen X	At	I	Br	Cl	F
covalent radius in Å	1.44	1.33	1.14	0.99	0.64
electronegativity according to Pauling	2.3	2.5	2.8	3.0	4.0
first ionization potential in eV	9.5	10.45	11.84	13.1	17.42
second ionization potential in eV	17.0	19.4	19.1	23.7	34.8
radius of X^+ in Å	1.22	1.13	0.96	0.84	–
electron affinity in eV	2.7	3.12	3.34	3.68	4.0
radius of X^- in Å	2.3	2.16	1.96	1.81	1.33
boiling point of HX in °C	–20	–35	–67	–84	20
boiling point of element in °C	309 [a)]	184	59	–34	–187
melting point of element in °C	244	113	–6	–101	–218
dissociation energy of X_2 in kcal/mol	28	36	46	58	72

a) Experimental value of Otozai, Takahashi. [7] is 230 °C.

Valency States and Redox Potentials

There is evidence for the existence of six valency states: At^{-I}, At^0, At^I, At^{III}, At^V and At^{VII}. Appelman has proposed the following redox potential scheme at room temperature

References on pp. 208/9

Fig. 10-2

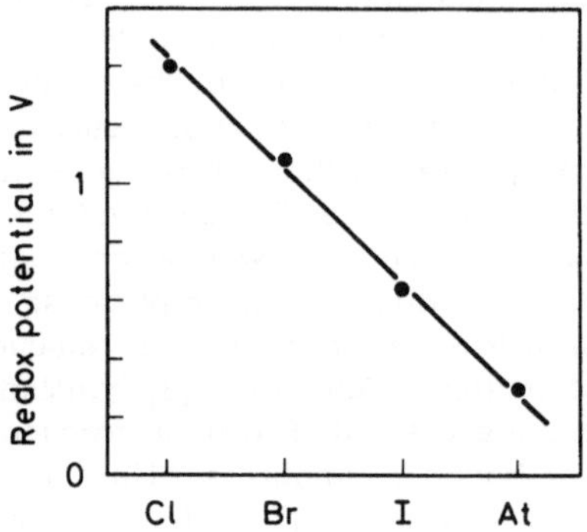

Redox potentials of the X^-/X_2 couples.

and 0.1 M acid [3, 9]:

$$At^- \underset{}{\overset{0.3V}{\rightleftharpoons}} At^0 \overset{1V}{\rightleftharpoons} At^{+X} \overset{1.5V}{\rightleftharpoons} At^{V} \overset{>1.6V}{\rightleftharpoons} At^{VII}$$

He based his results on the following assumptions: only At^0 is extracted by CCl_4, At^-, At^V, and At^{III} will coprecipitate exclusively with insoluble iodides, iodates, and periodates, respectively, and will not be removed from the precipitates by washing them with the aqueous phase or organic solvents.

The redox potentials for the reaction $2X^- \rightleftharpoons X_2 + 2e^-$ of the halogens in **Fig.** 10-**2** show a linear decrease going from chlorine to astatine and identify the latter behaving as a member of the homologue row of halogens. G.-J. Meyer has determined the following redox potentials at 80 °C: At/At^0 0.23 V, At^0/At^+ 1.09 V [10]. The identification of the valency states was made mainly by extraction and coprecipitation, some chromatographic and electrophoretic evidence has also been obtained. A fundamental fact for the assignment of charge and valency is that astatine is extractable by CCl_4 at a redox potential in the region realized by the Fe^{II}/Fe^{III} couple but nonextractable under both strong reducing and strong oxidizing conditions.

A short summary of the redox and extraction behavior in which the recently discovered trivalent state is not included is presented as **Fig.** 10-**3**.

Astatide Ion

The astatide anion, At^- is formed by reduction of any of the higher valency states with reducing agents of medium strength like SO_2, but not with ferrous ion [5]. Dreyer et al. [44] have prepared "pure" At^- by mixing 0.1 mL of a solution of thermochromatographically deposited At^0 in water with 0.02 mL 70% hydrazine hydrate and heating the mixture to 100 °C for 5 minutes. In contrast, trivalent iron oxidizes At^- to At^0 as does iodine. The At^- anion is characterized by almost complete coprecipitation with AgI, TlI and $PbCl_2$ and inextractability into organic solvents [5]. Hydrogen astatide HAt was identified by mass spectrometry [12] and as an emanation product from uranyl stearate [13]. Johnson et al. [5] have shown that At^{-I} accompanies the distillate if its solution in 0.5 M H_2SO_4 + 0.05 M Fe^{II} is boiled. This behavior is expected for HAt.

One of the rare cases of measured physicochemical data of the ion is the ratio of diffusion coefficient D(I):D(At) = 1.41:1 as given by Johnston et al. [14].

Fig. 10-3

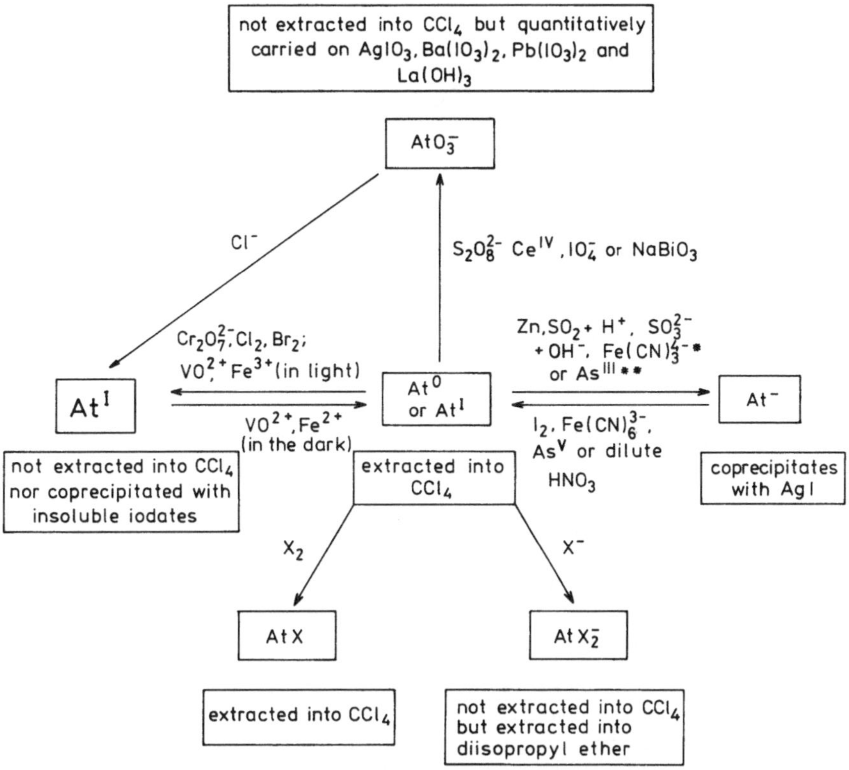

* Low acidity and ionic strength $[Fe(CN)_6^{4-}] \gg [Fe(CN)_6^{3-}]$ $(E < +0.4\ V)$.
** Low acidity and ionic strength $[As^{III}] \gg [As^{V}]$ $(E < +0.3\ V)$.

Aqueous oxidation and reduction reactions of astatine [11].

As a short note description of the chemical properties, one may say that there is no fundamental difference between At^- and I^-. As a typical result of the comparison of halide ions, **Fig. 10-4**, p. 188, shows the retention volumes of an anion exchange separation plotted against the inverse ionic radius. There is a regular variation. From this and other results, one can judge that the astatide ion behaves as expected from its position in the group of halogens.

A presumably specific reaction of the astatide is the adsorbability onto tellurium from acid solution. This has been observed in the form of coprecipitation with tellurium from 3 M $HClO_4$ upon reduction with SO_2 [1, 16] and the sorption onto crystalline tellurium from 0.1 M $SnCl_2$ in 3 M HCl [17]. Polonium is not coprecipitated but adsorbed. A separation can be achieved from the crystalline tellurium by selective elution of the astatine with 2 M NaOH. Iodine is neither coprecipitated nor adsorbed. This behavior was used for the separation of the At from proton irradiated thorium [17, 18].

Zerovalent Astatine

The so-called At^0, is the form usually found when astatine is produced from bismuth metal or bismuth oxide, separated from the target material by evaporation and condensation

References on pp. 208/9

Fig. 10-4

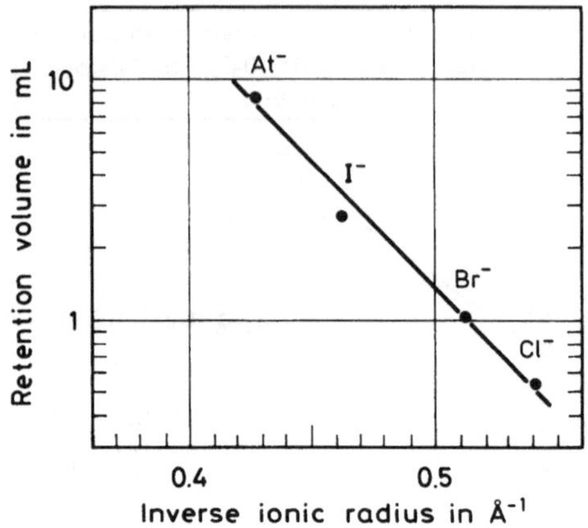

Separation of carrier free radiohalides on an Aminex A27 column with 1 M $NaNO_3$+0.1 M Na_2SO_3 (data of [15]).

and dissolved from the condensation spot by dilute acid under nonoxidizing and nonreducing conditions, e.g., with dilute nitric acid. It may also be prepared by reduction of higher valency states with Fe^{II} or I^- and by oxidation of At^{-I} with Fe^{III} or I_2 (see Fig. 10-2, p. 186). At^0 seems to be highly photosensitive and is oxidizied to a higher valency state by Fe^{III} solution upon illumination but not in the dark [3, 9, 25].

The true chemical nature of At^0 is not quite clear. It was not found in time-of-flight mass spectrometry [12] which gave evidence for At^+, HAt^+, CH_3At^+, AtI^+, $AtBr^+$ and $AtCl^+$ ions. The authors themselves conclude: "Our failure to observe At_2 does not allow us to draw any(!) definite conclusions about the equilibrium $2At \rightleftharpoons At_2$". Further scepticism comes from a consideration of the concentration dependence of the dissociation reaction mentioned [9, 19]. There are, on the other hand, results which at least convinced the researchers of the existence of At_2 in their experiments. Merinis et al. [20] from thermochromatography and Otozai, Takahashi [7] from gas chromatographic investigation concluded the presence of At_2. As an example of the type of results are shown in **Fig.** 10-**5** the radio gas chromatograms of carrier-free and carrier containing ^{131}I and ^{211}At. There is one peak which can be assigned without doubt to At_2. The boiling points derived from the experiments are: At_2 230 °C, IAt 213 °C and I_2 184 °C. Golovkov et al. [21] have identified the molecular ion At_2^+ in a plasma ion source. Therefore, it seems now more justified to assume the existence of At_2 than to neglect it.

In the presence of other halogens except fluorine, the zerovalent astatine must be assumed to exist as interhalogen species AtI, AtBr or AtCl. All of them have been found in the mass spectrum [4] and by thermochromatography [20].

The most informative and characteristic property of At^0 is its solubility in both hydrophobic and hydrophylic organic solvents like benzene, carbon tetrachloride, n-heptane, isopropylether [4], and even in water. The solution in water is stable during the practical lifetime [5]. The organic solvents mentioned extract At^0 from neutral to acidic aqueous solutions with distribution coefficients in the range 10 to 100. Astatine condensates on glass walls can be completely dissolved in benzene [22, 23] and astatine vapor is "absorbed" by water [24]. It is volatile at 200 °C from concentrated sulfuric acid [24]. From these and other reactions, one must conclude that At^0 behaves as a nonmetal.

Fig. 10-5

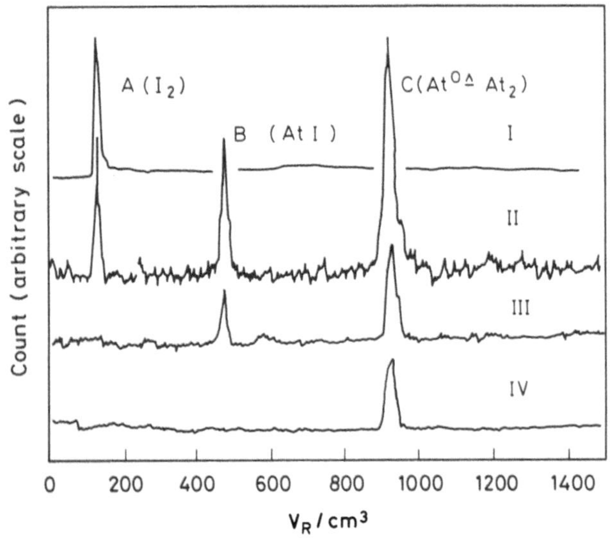

Radio gas chromatogram of astatine and iodine at 120 °C with He as carrier gas [7]. I) $^{131}I_2$ (carrier-free), α-ray+β-ray. — II) $^{131}I_2$ (containing carrier I_2)—At^0, α-ray+β-ray. — III) $^{131}I_2$ (containing carrier I_2)—At^0, α-ray. — IV) At^0 (carrier-free), α-ray+β-ray.

At^0 becomes unextractable by all organic solvents when its acidic aqueous solution is rendered alkaline. The extractability is restored if the solution is reacidified within a short time [5]. This effect and the coprecipitation of At from the alkaline solution with AgI was explained by a reversible disproportionation at pH > 13 [5] according to:

$$At_2 + H_2O \rightleftharpoons H^+ + At^- + AtOH$$

An upper limit for the equilibrium constant of the analogous reaction for the AtI species to form AtOH and I^- was set by Appelman as 10^{-11} [25].

A property of practical interest is the high tendency of At^0 to be "collected" by solid surfaces from both the gaseous state and from solution. It is easily deposited from the gaseous state onto metallic surfaces, especially Pt, Au, Ag [3, 26]. This was used to isolate elemental ^{211}At from an irradiated bismuth target by evaporation from the molten metal at 560 °C and adsorbing the radionuclide onto a silver foil at 230 to 370 °C [22]. Strong adsorption from aqueous solution onto glass walls [3], noble metals [16, 27, 35, 64], copper and bismuth [29] has been reported. The astatine valency state was not always clear in the experiments.

Monovalent Astatine

A species is formed by oxidation of At^0 with dichromate in perchloric acid solution (e.g., 2 M $HClO_4$ + 0.01 M $H_2Cr_2O_7$ at 100 °C for 20 to 30 min) which is not extractable by CCl_4 [25, 27]. It carries a positive charge according to electromigration [30 to 32], electrolysis [5], and cation exchange experiments [33, 34]. The latter indicated a charge of +1 in 0.3 M to 1.5 M perchloric acid. At^I is also obtained by oxidation of At^0 with H_2O_2 [19] and chlorine [9].

References on pp. 208/9

The nature of the monovalent astatine cation was derived mainly from its ion exchange behavior, adsorbability onto hydrated titanium hydroxide [8], adsorbability onto oxidized platinum surfaces [35], electrophoretic mobility [31, 36, 37], and extraction behavior [19]. Besides having a positive charge, it behaves differently than the normal univalent cations like Tl^+. The example in **Fig.** 10-**6** shows that the positive charge exists only at pH$\leqq$4 [8]. Further evidence for the special properties of the At^I species is the extraction with ethers from strong HCl solution [38], the decrease of sorption onto a cation exchanger with increasing HCl concentration in an HNO_3/HCl mixture of almost constant total hydrogen ion concentration [12], and the electrophoretic movement as an anion in the presence of halide anions Cl^-, Br^- and I^- [36, 37]. These observations were interpreted by the formation of 1:1 and 1:2 complexes of the astatine with the halide ions.

Many authors have stated that the monovalent astatine cannot be regarded as the simple ion At^+. Chalkin [Khalkin] et al. [39] preferred the notation $At(\theta)^+$, and discussed the species as a "hydrated" cation. As a result of several observations, including the establishment of the molecular ions AtO^+, $AtOH^+$ and $At(OH_2)^+$ with a plasma ion source [21], most of the recent publications formulate the species present in noncomplexing acid media as $At(OH_2)^+$ [8, 23, 31, 40]. Visser, Diemer [19] assume the following dissociation scheme:

$$At(OH)_2^- \xrightleftharpoons{\text{pH } 10} AtOH^0 \xrightleftharpoons{\text{pH } 4} At(OH_2)^+$$

The anionic species $At(OH)_2^-$ may also be considered as AtO^-, formed by dissociation of the hypoastatic acid AtOH. Milanov et al. [31] derived the dissociation constant of $At(OH_2)^+$ from the variation of the electrophoretic mobility with pH in the intervall pH 0.63 to 1.68. They found $K=0.032\pm0.005$ at 25 °C and ionic strength 0.4.

A very pronounced property of the monovalent astatine is the ability to form complexes with inorganic anions and organic polar molecules. Several authors have demonstrated complex formation with halide ions in weakly and strongly acidic solution [33, 36 to 38, 40], with the pseudohalide ions SCN^- and CN^- [14] and with nitrate, sulfate and dichromate [33]. The stoichiometry At:X equals 1:1 and 1:2, respectively, for the halides and sulfates and 1:1 for the nitrate. Formation constants have been established showing an increase in

Fig. 10-6

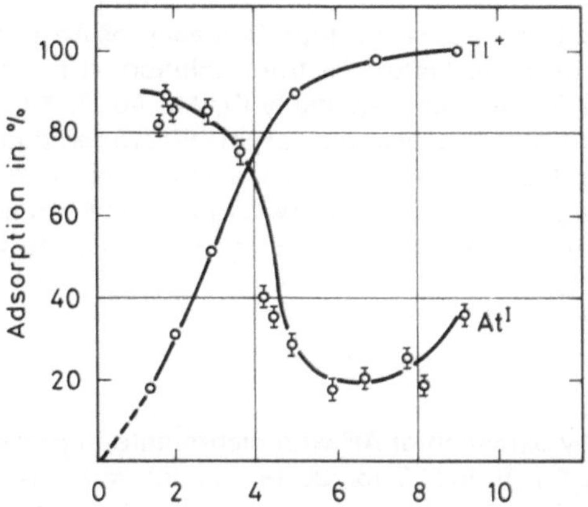

Adsorption of At^I and Tl^+ onto titanium hydroxide [8].

Fig. 10-7

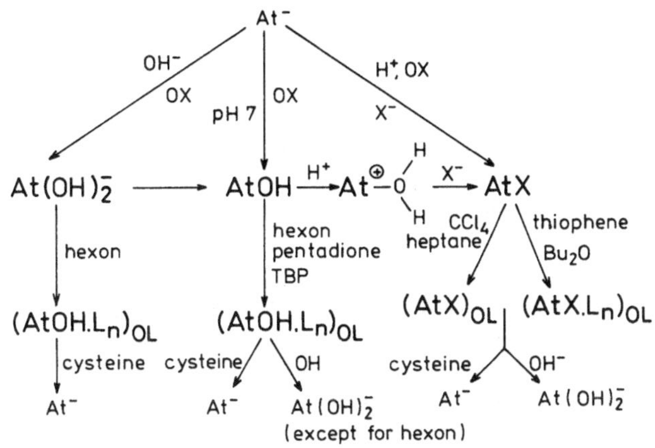

Extraction and complex formation of At^{I} (OL = organic phase) [19].

stability in the sequence

$$I > SCN > Br > Cl > Cr_2O_7 > SO_4 > NO_3$$

Visser, Diemer [19] have shown that monovalent astatine is readily extracted by oxygen containing solvents. A summary of their results given in **Fig.** 10-**7** indicates that species of both types, AtOL and AtXL, are proposed. X represents a halogen (Cl, Br) and L represents a solvent molecule. The order of increasing extraction ability is thiophene < tributylphosphate < pentanedione < hexone.

Trivalent Astatine

In his fundamental paper of 1961 on the oxidation states of astatine in aqueous solution Appelman [9] discusses the possibility that the intermediate oxidation state between At^0 and At^V obtained may be either At^I or At^{III} or even varying mixtures of them. It was generally accepted that At^+ is formed with $Cr_2O_7^{2-}$ as the oxidant. In 1978, Dreyer et al. [32] observed that upon oxidation of astatine with 0.02 M $Na_2S_2O_8$ in 0.5 M $HClO_4$ at 100 °C a second cationic species results which was identified as At^{III}. This cation could also be obtained by treating an aqueous solution of At^0 with XeF_4. The discrimination between the two oxidation states was made by electrophoretic mobility in 0.02 M $HClO_4$. At^I produced by dichromate oxidation had $(1.9 \pm 0.2) \times 10^{-4}$ $cm^2 \cdot s^{-1} \cdot V^{-1}$ and At^{III} produced by peroxydisulfate oxidation has $(3.3 \pm 0.2) \times 10^{-4}$ $cm^2 \cdot s^{-1} \cdot V^{-1}$ (all mobilities measured at 25 °C).

In neutral or alkaline medium the cation transforms into a nonmoving species which is converted into an anion if the solution is heated to 100 °C for one hour. The latter had an electrophoretic mobility of 4.6×10^{-4} $cm^2 \cdot s^{-1} \cdot V^{-1}$, which is different than that of At^-, AtO_3^- and AtO_4^-.

The authors conclude from these observations that the cation belongs to trivalent astatine and the species in acid solution is AtO^+ or $At(OH)_2^+$. The "hydrolysis" in neutral or alkaline medium is supposed to proceed in a two-step process as follows:

$$AtO^+ + OH^- + H_2O \rightleftharpoons At(OH)_3$$

$$At(OH)_3 + OH^- \rightleftharpoons AtO_2^- + 2H_2O$$

References on pp. 208/9

The first step takes place almost immediately upon neutralization but the second step proceeds rather slowly and requires higher temperatures.

Complex anions are formed with the estimated composition $AtOX_2^-$ (X=Cl, Br, I) in 0.02 M HCl and HBr and in 0.015 M $HClO_4$+0.01 M KI. Putting the solution of the chloro and bromo complex in an electrolyte consisting of 0.02 M $(NH_4)_2CO_3$ and 0.015 M NH_4OH yielded a different anion which on the basis of its mobility (1.4×10^{-4} $cm^2 \cdot s^{-1} \cdot V^{-1}$) is assumed to be AtO_2^- [37]. Visser, Diemer [19] supposed that At^{III} is extracted from 0.5 M S_2O_8 solution by dibutylether as an etherate.

Pentavalent Astatine

The At^V state was assumed by Appelman [27] to be achieved upon oxidation with alkaline hypochlorite and with periodate, persulfate or Ce^{IV} in acidic solution. He said that the reactions are frequently sluggish at room temperature and heating as high as 100 °C is "sometimes" necessary to obtain complete oxidation. In the light of recent findings for At^{III} some of the older results require reinvestigation, but Dreyer et al. [42] showed that at least the oxidation by 2 M KOCl in 1 M KOH yields At^V. Furthermore, At^V is obtained if a solution of At^{VII} is acidified.

Pentavalent astatine was shown to behave as an anion, having a different ionic mobility than the other anions of astatine. Table 10/2 shows that the At^V species produced by KOCl is clearly different from the most possible interferences AtO_2^- and AtO_4^-. It coprecipitates with $AgIO_3$, $Ba(IO_3)_2$ and $Pb(IO_3)_2$ [3, 13]. From the properties investigated up to now, there is no doubt that pentavalent astatine is present in aqueous solution as the astatate anion AtO_3^-.

Table 10/2
Electrophoretic Mobilities of Iodine and Astatine in Different Oxidation States [32, 41].

anion	$u \cdot 10^4$ in $cm^2 \cdot s^{-1} \cdot V^{-1}$
At^-	6.1
AtO_2^-	4.6
AtO_3^-	2.3
AtO_4^-	3.4
I^-	8.0
IO_3^-	4.0
IO_4^-	4.9

0.04 M KNO_3 pH 6 25 °C

Heptavalent Astatine

Khalkin et al. [43] reported in 1970 the formation of At^{VII} as perastatate anion AtO_4^- upon treatment of 1 mL of an At solution in 1 M NaOH with 80 mg XeF_2 at 100 °C. The resulting astatine species cocrystallized with KIO_4 and $CsIO_4$, and showed the same paper electrophoretic behavior as IO_4^-, i.e., no migration at a field of 50 V $\cdot$ cm^{-1} [42]. The distribution coefficients between solution and crystals were 0.068$\pm$0.021 for KIO_4 and 9.31$\pm$0.08 for $CsIO_4$ and exhibited no dependency on the amount of precipitate. These results have been confirmed by Dreyer et al. [41, 42], who showed further that the perastatate species is formed neutral to alkaline solution either with XeF_2 or more convenient with 0.01 M KIO_4 [42]. The AtO_4^- is decomposed upon acidification to pH<1 and heating to 90 °C within 5 to 10 min to AtO_3^- [42]. These authors were able to add more evidence for the perastatate ion by measuring the electrophoretic mobility in a capillary tube. The results are included in Table 10/2 and allow a clear differentiation of perastatate and astatate ion.

Interhalogen Compounds

Astatine has a high tendency to form interhalogen compounds, as does iodine. The first indication for this was made by Neumann [38] who interpreted the high extractability of astatine from 4 to 10 M HCl by the formation of the $AtCl_2^-$ complex. The binary compounds AtCl, AtBr and AtI are formed if At^0 comes in contact with Cl_2, Br_2 or I_2 [4, 20]. Appelman [25] has shown that upon extraction of astatine solutions containing an alkali halogenide with carbon tetrachloride containing free halogen, the compounds AtCl, AtBr and AtI are extracted and the anions AtC_2^-, $AtBr_2^-$ and AtI_2^- form in the aqueous phase. The initial valency state was not mentioned by Neumann [38] and Appelman [25] but it should have been either At^0 or At^{-I}. Dreyer et al. [37, 44] have obtained the same halide species by the reaction of HCl, HBr and $HClO_4$+KI (0.02 M) with At^+. They assume the reaction

$$Ac^+ + 2X^- \rightleftharpoons AcX_2^-$$

and were able to characterize the individual species by their electrophoretic mobility. They further showed the formation of the same interhalide anions starting with At^-, presumably via oxidation by the dissolved oxygen of the solution. Furthermore, the pseudohalogen compounds $At(SCN)_2^-$ and $At(CN)_2^-$ have been detected [45].

The formation constants of the ZI_2^- species increases remarkably from Z=Cl to At. **Fig.** 10-**8** indicates that astatine has the highest coordination tendency among the halogens.

Organic Astatine Compounds

A substantial number of organic compounds of astatine have been synthesized and were examined mainly using radio gas chromatography and high pressure liquid chromatography to determine physicochemical properties like boiling point, evaporation enthalpy,

Fig. 10-8

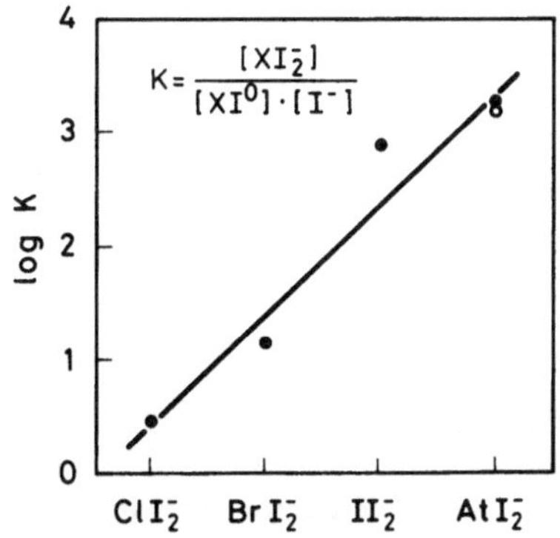

Stability constants of interhalogen complex anions [25, 36].

References on pp. 208/9

carbon–astatine dissociation energy and acid dissociation constants. It is justified to say that the reliability of the knowledge of the organic reactions of astatine is presently superior to that of the inorganic reactions.

The following methods have been applied for the synthesis of aliphatic and aromatic astatine derivatives: homogeneous and heterogeneous halogen exchange using At^- [46], electrophilic halogen exchange using At^+ or At(Cl,Br) [26, 33, 47], reactions of diazonium compounds with At^- [46, 48] and recoil astatination [49, 50]. Labeling of proteins with astatine was achieved by electrooxidation and by chemical oxidation with hydrogen peroxide, both reactions were carried out in phosphate buffer [51]. The reported radiochemical yields reach 80%. In general methods suitable for synthesis of organic iodine compounds may be successfully applied for the preparation of organic astatine compounds [46].

Samson, Aten [46] irradiated triphenylbismuth two hours in the ^{4}He beam of a cyclotron (0.01 µA) for recoil astatination. Berey et al. [52] used the electron capture reaction ^{211}Rn(ε)^{211}At for generating the recoil atoms. The synthesis consisted simply in "condensing" ^{211}Rn onto the organic substrate, e.g. chlorobenzene, in a vacuum apparatus. Radioactive yield was about 14% for chlorine replacement and 3 to 4% for hydrogen replacement.

Electrophilic substitution takes place with At^I, either using an $HClO_4$ solution of the radionuclide [23] or the interhalogen compound AtCl [53]. In both cases the astatine is supposed to react as At^+ cation. Vasaros et al. [23] state that the substitution occurs only in acidic solution where $At(OH_2)^+$ exist and much less in neutral solution where $AtOH^0$ is dominating. The rather slow reaction was explained by a dehydration step prior to the substitution $At(OH_2)^+ \rightarrow At^+ + H_2O$.

The astatine atom in organic compounds exerts a less inductive effect than its lower homologues. One example of this is the sequence of dissociation constants of the monohalogen acetic acid (**Fig.** 10-**9**). Investigating the acidity of a number of halogenated benzoic acids, phenols and amines, Visser et al. [55] concluded, that the field effect of astatine is weaker than that of iodine, whereas the resonance effect is about the same. In general the variation of the properties of halogen organic compounds seem to vary rather regularly going from chlorine to astatine.

Fig. 10-9

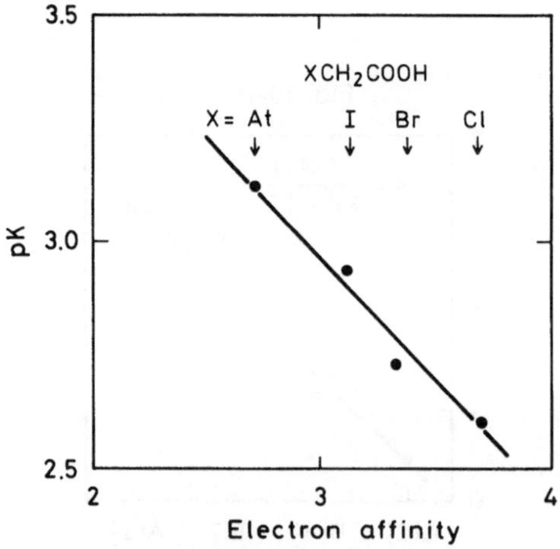

Acid dissociation constants (pK) of monohalogen acetic acids, astatine data of [54].

10.1.2 Isotopes for Chemical Investigation

Suitable Isotopes and Radioactivity

The isotopes ^{209}At, ^{210}At and ^{211}At, with half-lives of several hours have been used for chemical experiments in most cases (Table 10/3). They are usually produced by bombarding either a bismuth target with accelerated ^{4}He particles, or thorium or uranium with high energy protons. The latter reaction proceeds through a spallation, and yields a complex mixture of radionuclides from which the astatine has to be separated.

Table 10/3
Astatine Isotopes Used for Chemical Experiments.

isotope	half-life	radiation in %	Ref.
^{209}At	5.5 h	ε 95.9	[9, 12, 29]
		α 4.1	
^{210}At	8.1 h	ε 99.83	[9, 12, 29]
		α 0.17	[43]
^{211}At	7.21 h	ε 58.06	[2, 3, 5, 6, 10, 57]
		α 41.94	[43]

ε = electron capture

The isotope normally used is ^{211}At, which undergoes 42% α decay and 58% electron capture. The α particles emitted of 5.94 MeV energy are easily measured. It is also possible to measure the 80 keV γ radiation (K_α line) emitted due to electron capture. The ^{211}Po daughter disintegrates with a half-life of 0.5 s, emitting an α particle of 7.43 MeV. Thus, one obtains one α particle per ^{211}At decay.

Depending on energy control of the ^{4}He beam varying amounts of ^{210}At are formed besides the ^{211}At. It seems better suited for experiments becauce of its longer half-life, but this isotope decays by electron capture to ^{210}Po, an α emitter of 134 d half-life. Thereby the astatine activity becomes self-contaminated with an α emitter, although its radiation intensity is low due to its long half-life. A further contamination may come from the fact that in many cyclotrons the helium contains a large number of deuterons leading to ^{210}Po by the d,n reactions [56].

Isotopically pure ^{211}At is accessible by electron capture of ^{211}Rn which can be isolated in radiochemically pure form from irradiated thorium by gas chromatography [8].

Experiments with astatine are restricted to the tracer scale because of its short half-life and the low yield that can be achieved with the nuclear reactions necessary. A typical total activity produced is 370 MBq that is equal to about 10 mCi or 5 ng or 2×10^{-11} mol. The largest single batch produced seems to be 50 ng [12]. The highest concentration applied in experiments seem to be 10^{-8} M, normally solutions $\leq 10^{-10}$ M in astatine have been used. Even if larger amounts of astatine were available, chemical operations with them would be quite impossible because of the high energy output. One mg of ^{210}At represents an activity of 2000 Ci [11] and its α energy is equal to 1.4 cal/s. The initial self-heating rate of a solution of 1 g/L would be of the order of 1 °C per second, or in other words it would become boiling within minutes.

The tendency of astatine to concentrate in the thyroid and its volatility makes it particularly dangerous [4]. The dose rate of pure ^{211}At is reported to be 4 mrad/h per 1 mCi (37 MBq)

References on pp. 208/9

at 10 cm distance [24]. But the irradiated Bi targets have a much higher dose rate depending on irradiation intensity, target construction and particle energy. Beyer et al. [57] reported 15 mrem/h at 10 cm distance for their Bi target irradiated 1 to 5 h with 28 MeV and 5 μA. Meyer, Roessler [24] found 50 rad/h at 10 cm distance for their target assembly after 3 to 6 h irradiation with 28 to 30 MeV and 20 μA.

Thus precautions customary in the handling of highly radioactive substances must be observed and even for tracer amounts adequate venting is recommended. ^{210}At is additionally hazardous because of its hard gamma ray and its α-emitting ^{210}Po daughter [4].

Production of ^{211}At Activity by ^{4}He, 2n Reaction

Metallic bismuth was most frequently used as target material, see e.g. [4, 16, 22, 24, 38, 51, 57, 58] but in some cases Bi_2O_3 was preferred [46].

The formation of ^{211}At and ^{210}At upon ^{4}He irradiation of ^{209}Bi is strongly energy dependant. As shown in Fig. 3-26, p. 64, the cross section for the α,2n reaction becomes substantial at about 20 MeV and that for the α,3n reaction at about 29 MeV. The threshold energy for the α,4n reaction leading to ^{209}At was reported as 34 MeV [4]. Based on the energy dependance of isotope formation, the optimal α energy for achieving pure ^{211}At and high yield is 28 MeV [24]. Neumann [38] obtained specific yields of 4 mCi/μA · h ^{210}At and 3.1 mCi/μA · h ^{211}At.

Every experimenter has to produce the At activity specifically for his own purpose because of the short half-life. Therefore a multitude of target constructions and irradiation conditions have been applied. Table 10/4 summarizes references which contain enough information to be regarded as a procedure description.

Table 10/4
Production of ^{211}At by He Irradiation of Bismuth.

	Bi target	irradiation	yield	Ref.
a)	thickness in mg/cm²	energy in MeV	MBq/μA · h	
b)	area in cm²	current in μA	total MBq	
c)	support	time in h	—	
a)	120	29	20	[5, 12]
b)	?	30 to 40	5000	
c)	Al	8 to 10	—	
a)	100 to 220	29	22	[3, 4]
b)	≅10	12 to 30	800	
c)	Au	1 to 4	—	
a)	500	33	7 to 15	[51, 58]
b)	1.75	2 to 4	≅40	
c)	Cu	3 to 4	—	
a)	200	28 to 30	≅18	[10, 24]
b)	0.7	20	≦2000	
c)	Cu	6	—	

Table 10/4 [continued]

	Bi target	irradiation	yield	Ref.
a)	thickness in mg/cm^2	energy in MeV	MBq/μA · h	
b)	area in cm^2	current in μA	total MBq	
c)	support	time in h	—	
a)	110 [a)]	28	20 to 30	[57]
b)	20	5	370	
c)	Cu	5	—	
a)	150 [a)]	36	110	[22]
b)	6	15	≦2000	
c)	Cu	0.17?	—	

[a)] The "pathway thickness" given is the equivalent thickness for the beam angle applied.

The target device used consists normally of a metallic support foil made of gold, aluminium or in most cases, copper, to which the bismuth is fused or vacuum-condensed (Table 10/4). Intensive cooling is required to remove the heat generated by the slowing down of the α particles in the target and the support. Meyer [10] reports a heat generation of about 50 W/cm^2 for a 28 MeV/20 μA beam. Cooling is achieved by water flowing along the backside of the target whereas the upper (beam incident) side was kept under vacuum or He gas under reduced pressure. As an example, **Fig.** 10-**10** shows the target construction which was used by Doberenz et al. [22]. In this case, the Bi layer was about 100 mg/cm^2 thick, but because of the 45° beam incident angle, the pathway thickness is equal to 150 mg/cm^2. This target head was later used directly for isolation of ^{211}At by evaporation. Most often, experimenters preferred to separate the support and the bismuth before processing it.

The yield depends on the energy and the target thickness. Lindner et al. [58] have shown that there is a decrease of the specific yield with increasing current (**Fig.** 10-**11**, p. 198). The mean value is 18 MBq/μAh, which is equal to 0.5 mCi/μAh. The highest reported yields are 110 MBq/μAh of $^{211}At+^{210}At$ using 36 MeV ^4He particles [22] and 148 MBq/μAh of ^{211}At with 46 MeV ^4He particles [38]. The content of ^{210}At in the activity produced by 36 MeV ^4He ions is not given, but it should be substantial. The ^{211}At is more than 99.9% pure for 28 MeV ^4He particles [51, 57].

Fig. 10-10

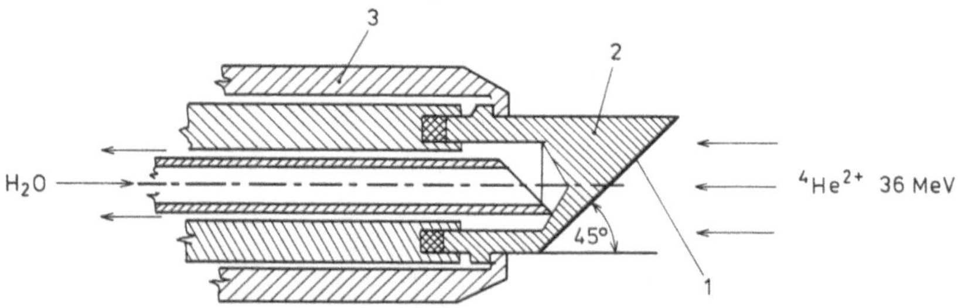

Principal scheme of the target device for the irradiation of bismuth used by Doberenz et al. [22]. 1) Bismuth layer 110 mg/cm^2. — 2) Copper support. — 3) Target tester.

Fig. 10-11

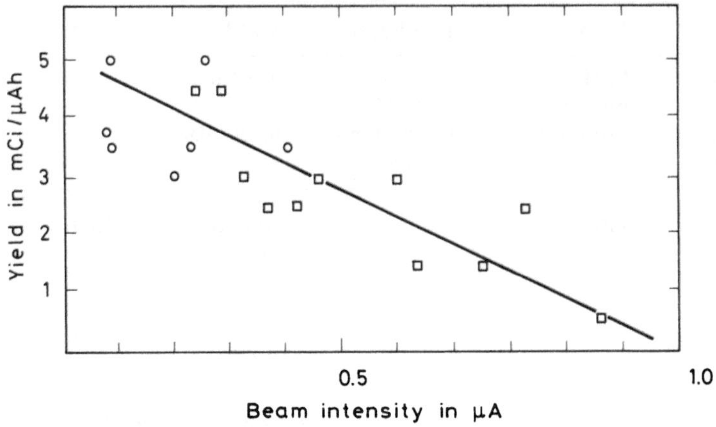

Yield in mCi/μAh of the reaction $^{209}Bi(\alpha, 2n)^{211}At$ as a function of beam intensity (irradiation with 32 MeV(○) and 50 MeV(□)) [58].

Production of $^{211/209}$At Activity by Proton Irradiation

The use of a high energy proton accelerator seems to be a rather extravagant undertaking at a first glance. But, astatine chemists have no other choice than to use the accelerator accessible to them.

Astatine isotopes have been produced by 660 MeV proton irradiation of lead, bismuth, thorium and uranium [18, 59]. In the first two cases they are formed by secondary (Li, xn) reactions with a cross section of 0.1 and 10 μbarn, respectively. The total cross section for formation of spallation and fission products is about 1 μbarn. Therefore, the astatine activity amounts to only 10^{-5} to 10^{-7} of the total activity.

With thorium and uranium the astatine is formed directly by spallation. The cross section has been reported to be 10 mbarn for ^{211}At and 15 mbarn for ^{209}At with 660 MeV protons [18]. Lefort et al. [60] found 0.21 mbarn for ^{211}At and 1.8 mbarn for ^{210}At with 160 MeV protons.

Table 10/5
Relative Yields of Astatine Isotopes from a Uranium Target Irradiated with 660 MeV Protons [59].

mass number	relative yield
211	0.7
209	1
208	0.4
207	0.5
206	0.3
205	0.14
204	0.09
203	0.02

Vakhtel et al. [59] used uranium metal plates 1 mm thick, weighing 5±1 g. These were irradiated for 20±10 min in the inner proton beam of the 660 MeV synchrocyclotron of the United Institute of Nuclear Research at Dubna. The target had a total activity equal to 10 g of radium after irradiation. The relative yields of astatine atoms from a uranium target are given in Table 10/5, showing a maximum for ^{209}At and the presence of 8 astatine isotopes. Despite the high radioactive contamination, it was possible to obtain a 99.95% pure astatine activity by a highly selective purification process based on the adsorption onto metallic tellurium [17].

10.1.3 Isolation and Purification of Astatine from Irradiated Targets

The discoverers of astatine [1] isolated the new element from a bismuth target by heating the target to its melting point in a glass tube and transporting the radionuclide to a water-cooled condensing plate by a helium stream. This dry distillation method was used subsequently by many researchers in a number of variants [3, 5, 16, 22, 24, 51, 58, 59]. Other principles are the dissolution of the target in nitric acid followed by extraction of the astatine by diisopropylether [18, 61], the adsorption from the hydrochloric acid solution onto metallic tellurium [17], the coprecipitation from nitric acid with tellurium [18] and extractive distillation from sulfuric acid solution [10, 24]. Appelman [4] has described in detail and tested several procedures published up to 1960. He reported yields ranging from 10 to 80%. It seems that since that time some progress has been made, but up to now no method has proved to be the best one.

Dry Distillation Procedure of Parrot et al. [62]

This procedure was originally developed for preparing astatine solutions for biological studies. It was also described and tested by Appelman [4]. The main feature consists of the distillation of the astatine in a nitrogen stream of 0.2 mm Hg pressure.

The procedure is carried out in the device made of quartz shown in **Fig.** 10-**12**. The target plate of about 7×2 cm was cut into two pieces and these placed in a small quartz boat, which is inserted into the large quartz tube at C in Fig. 10-12. Liquid nitrogen is put in the cold finger reservoir and around the trap. In order to facilitate collection and to prevent the (irreversible) adsorption of the astatine on the glass of the cold finger, a

Fig. 10-12

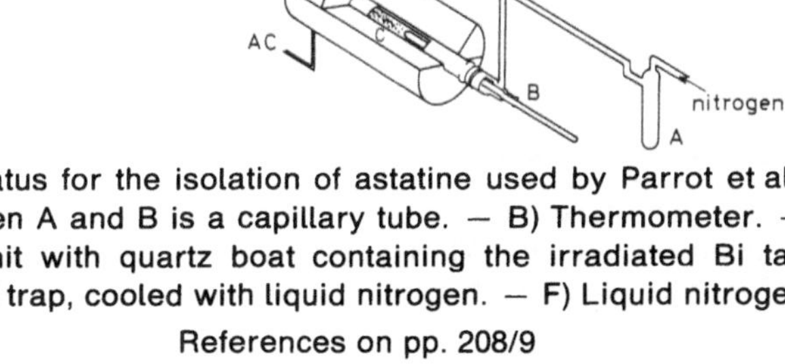

Dry distillation apparatus for the isolation of astatine used by Parrot et al. [62]. A) Trap A, the connection between A and B is a capillary tube. — B) Thermometer. — C) Quartz tube in electric heating unit with quartz boat containing the irradiated Bi target. — D) Cold finger. — E) Cold trap, cooled with liquid nitrogen. — F) Liquid nitrogen reservoir.

References on pp. 208/9

thin film of ice is deposited on it by soft breathing. The entire system is evacuated. When a vacuum of less than 200 microns is obtained, an inert streaming gas — helium or nitrogen — is introduced through the capillary between A and B in Fig. 10-13. Two pounds of pressure is maintained on the tank side in order to permit a steady flow of gas over the bismuth to the cold finger. Neither water nor oxygen is permitted to enter the system, because it has been found that these tend to increase colloid formation.

The furnace is turned on, and the target is brought rapidly to 700 °C as determined by a thermometer inserted in B of Fig. 10-13. Below this temperature, the yield of ^{211}At is low and above a brown to black layer of bismuth may distill onto the cold finger and cause great loss of activity in the subsequent steps. The quartz tube between the furnace and the finger is flamed periodically to prevent adsorption on the walls of the tube. The temperature is held for about 40 min. Then the furnace is turned off and opened, and the apparatus is allowed to cool to 100 °C. The vacuum pump is turned off. When the system has returned to atmospheric pressure, the cold finger is carefully lifted straight up to prevent any of the ice from flaking off. The finger is washed with an "isotonic saline" containing 5 mg Na_2SO_3/mL. This solution is placed in an ultracentrifuge for 20 min at 30000 rpm. The supernatant is the product solution.

Appelman [4] has tested this procedure and obtained astatine yields of about 10%. Garrison et al. [18] used a flow of nitrogen at 0.01 to 0.001 Torr and heated the bismuth to only 425 °C. Heating time was 20 min, no yield data have been given. Lower temperatures were also used successfully by Lindner et al. [58]. Aaij et al. [51] applied He as carrier gas and heated sequentially to 250 and 500 °C during a 90 min period. The latter two research groups connected the outlet of the distillation furnace directly to a gas bubbler containing 1.5 mL of either phosphate buffer containing 0.005 mol of Na_2SO_3 or 0.05 M NaOH. Both solutions were >90% efficient in trapping the astatine. Aaij et al. [51] conclude the NaOH absorber solution is more reproducible and at least 99% efficient. The ^{211}At activity collected was 1 to 2 mCi.

Oxidative Distillation Procedure of Vakhtel et al. [59]

This procedure was used for the processing of proton-irradiated uranium by several researchers, e.g. [37, 41, 42]. It was originally designed by Bayar et al. [28] as a general approach for the thermochromatographic isolation of radioactive elements characterized by the evaporation of volatile oxides from the target. An oxygen containing carrier gas is used for achieving this.

The apparatus used is shown in **Fig. 10-13**, together with a typical result. The column was made of quartz and had a 4 mm diameter and 300 mm length. A temperature drop from 800 down to 20 °C was maintained along the tube. The carrier gas was a 9:1 mixture of O_2 and He. With a flow rate in the range of 100 mL/min, reaction time was only 10 min at the optimum temperature in the reaction zone of 850 ± 30 °C. The astatine deposited at 80 ± 10% onto the preoxidized platinum foil in the temperature zone 45 to 100 °C. The major part of the undistilled astatine remained in the target material. Decontamination from iodine and bromine was achieved by a silver foil positioned in the temperature zone 425 to 475 °C. The level of contamination of the astatine with other radionuclides present in the target is given as less than 0.5%.

Vacuum Dry Distillation Procedure of Meyer, Roessler [24]

The vacuum distillation of astatine from the bismuth target irradiated with ^{4}He ions was first described by Johnson et al. [5]. They heated the 40 mg/cm^2 Bi on an Al target to its melting point (271 °C) and collected the astatine in a 4 mm U tube cooled with liquid nitrogen.

Fig. 10-13

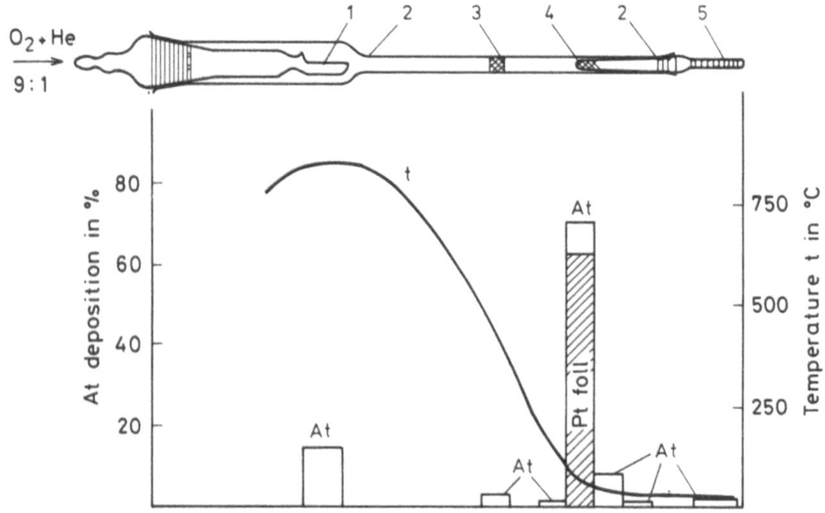

Thermochromatographic device for the oxidative distillation of astatine from proton irradiated uranium [59]. 1) Quartz boat with target. — 2) Quartz gas thermochromatographic column. — 3) Silver filter, 50 μm thick, length 1 to 1.5 cm, area 20 to 30 cm^2. — 4) Platinum foil 30×3 mm previously oxidized in pure oxygen at 800 to 900 °C. — 5) Adsorber with activated charcoal vented to a trap (not shown).

Yields of up to 95% were reported with less than 0.001% polonium contamination. Appelman [4] was not able to repeat the result with 500 to 1000 mg/cm^2 targets. The astatine did not distill below 600 °C. It is possible that the difference in thickness of the bismuth was in some way responsible for the difference in results.

Mayer, Roessler [24] successfully used the apparatus shown in **Fig.** 10-**14**, p. 202. The target (≅0.14 g Bi+0.3 g Cu) is placed within a quartz tube fitted to a vacuum line (10^{-3} Torr). A capillary trap and a charcoal trap both cooled by liquid nitrogen are attached to it. The oven is heated for $^1/_2$ h at 450 to 500 °C, then the temperature is raised to 600 °C and the Bi begins to sublime. It condenses at the colder parts of the quartz tube whereas At distills off and is trapped to 60 to 80% in the capillary tube, most probably as At^0. Depending to the size of the metal target piece, 15 to 35% of the ^{211}At is retained by the sublimed Bi. Five percent of the radioactivity is adsorbed on the cooling jacket.

Simple inorganic forms can easily be obtained by rinsing the trap with reducing, neutral or oxidizing aqueous solutions. For preparation of carrier-free interhalogen compounds [10] the capillary trap is closed against the pump and about 10 μmol of $Cl_2 \cdot Br_2$ or I_2 from a special supply tube are frozen on the At. Subsequently, the trap is warmed and kept for 5 min at room temperature to allow the quantitative formation of AtX. The capillary trap is cooled to dry ice temperature in the case of AtCl and AtBr. The excess of Cl_2 or Br_2 is distilled off into the charcoal trap (still at liquid nitrogen temperature) whereas AtX sticks to the walls of the capillary trap. In the case of AtI it is necessary to keep the trap at room temperature for removal of excess I_2. The whole procedure takes up to 2 h. The trap is removed from the vacuum line after flooding with argon. One hundred μL of a liquid organic substrate can be reacted with the AtX directly. The trap can also

References on pp. 208/9

Fig. 10-14

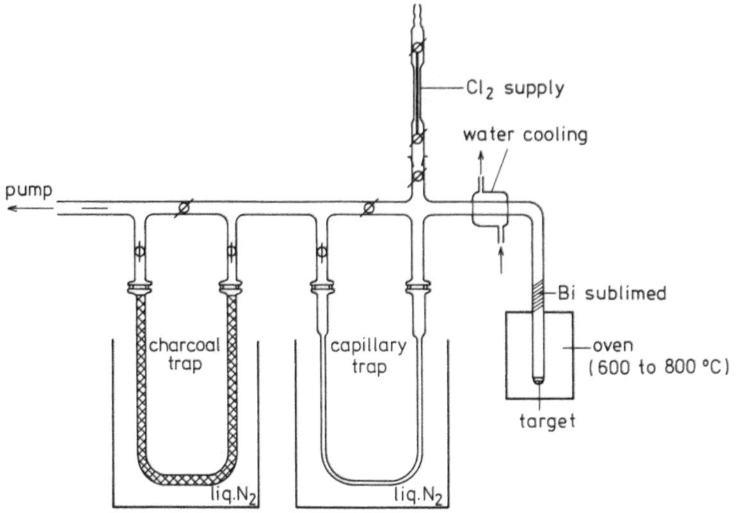

Vacuum line for dry distillation of astatine from ^4He irradiated bismuth targets, cf. pp. 104, 202 [24].

be rinsed by a 1 M aqueous solution of HX (X = Cl, Br, I) to yield AtX_2^- ions. The overall yield in the capillary was 55 to 75% and in the rinse solutions 40 to 50%.

Radioactive contamination of the produced astatine with ^{210}Po, ^{209}Po, ^{207}Po and radioactive isotopes of Bi, Ga, Zn and Ge was less than 0.01%.

Two Stage Dry Distillation Procedure of Doberenz et al. [22]

This procedure was developed for the preparation of astatine of high specific activity and free from inorganic or organic impurities. It is based on the special target construction described in Fig. 10-10, p. 197.

The irradiated target head was placed into a cylindrical stainless steel vessel with a teflon cover (**Fig.** 10-**15**). A glass tube, the lower part of which had a cylindrical funnel form with an inside diameter exceeding the diameter of the target head by 1 to 1.5 mm, was passed through the Teflon cover into the center of the steel vessel. Helium was passed through the steel vessel at a rate of 100 mL/min for 10 to 15 min to remove the air and then the latter was placed into a cylindrical furnace and heated up to 560 °C. The helium flow in the narrow clearance between the glass funnel and target head excluded the diffusion of astatine into the steel vessel and transported it into the glass tube, where astatine was adsorbed onto a silver foil. For this purpose, two silver foil stripes of 0.01 × 5 × 50 mm size were placed into the glass tube and were separated from each other with quartz wool. The temperature of zones containing the silver foils was 370 to 230 and 230 to 130 °C, respectively. After 10 min heating of the target head, astatine began to volatilize from bismuth, and within 15 min after the beginning of distillation, activity on silver foils in the glass tube stopped increasing. All the astatine removed from the gas flow was found on the first half of the lower silver foil strip. The activity of the upper (control) foil strip was lower by a factor of 10^4 than that of the lower (sample) one. Apparently, in the selected distillation conditions, astatine almost completely volatilized from the molten bismuth. γ-spectral analysis of the bismuth remainder did not show any line characteristic for astatine isotopes.

Fig. 10-15

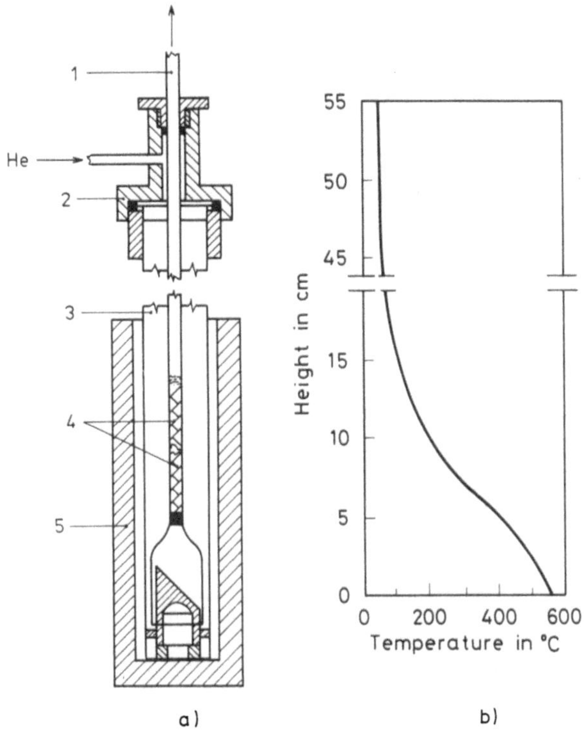

Equipment for the distillative isolation of astatine from molten bismuth targets (a) and temperature gradient within it (b) [22]. 1) Glass tube with a cylindrical funnel. — 2) Teflon cover. — 3) Stainless steel vessel. — 4) Silver foil. — 5) Cylindrical furnace.

For preparing aqueous solutions from the astatine deposited onto the silver foil, the astatine is redistilled in a device shown in **Fig. 10-16**, p. 204. The silver foil is inserted into a glass ampule, the upper part of which can be heated electrically and the lower part can be cooled by water. The heating temperature was 500 °C. After 30 min distillation in the closed ampule 95 to 98% of astatine was condensed into the aqueous solution. Alkaline, neutral and acidic solutions of astatine were obtained by this method in a volume of 30 to 100 μL with a specific activity up to 1000 mCi/mL (37 GBq/mL).

The element in the solutions investigated was found in the form of astatide ions, hydrated astatine atoms, positive astatine ions and chloride complexes of astatine, respectively.

Distillation of astatine has been carried out in dry ampules when nonaqueous solutions of astatine were required for organic synthesis. Astatine was condensed on the cooled part of the closed ampule, and then washed away with organic solvent, e.g. benzene, bromobenzene or amines.

The time required for preparation of astatine by this technique was not more than 2 h. Noticeable loss of astatine during the separation and concentration was only connected with the incomplete evaporation of the element from the silver foil. It approached 5%.

Fig. 10-16

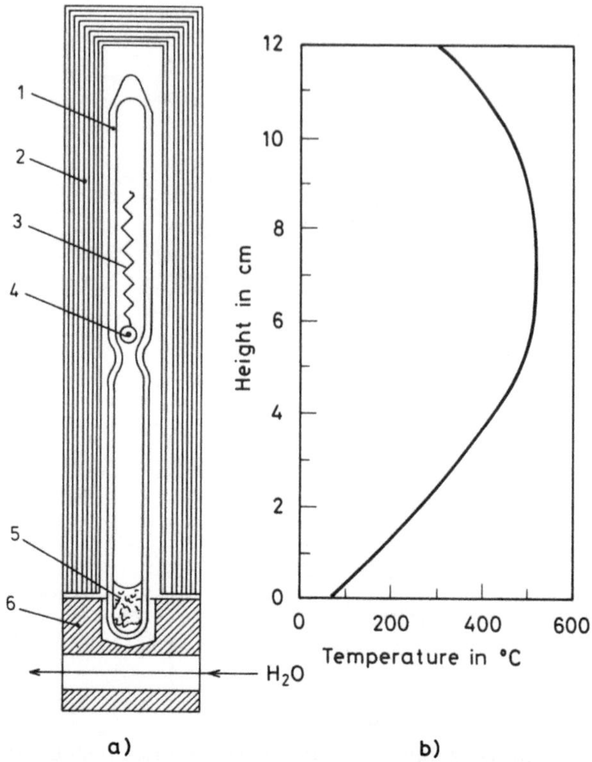

Schematic diagram of the equipment for preparation of high specific activity astatine solutions (a) and temperature gradient applied along the ampule (b) [22]. 1) Glass ampule. — 2) Cylindrical furnace. — 3) Silver foil with astatine. — 4) Bead. — 5) Aqueous solution. — 6) Water-cooled metallic disk.

Extractive Distillation Procedure of Meyer, Roessler [24]

Johnson et al. [5] have reported that astatine evaporates effectively if a solution containing 0.5 M H_2SO_4+0.05 M Fe^{II} is boiled. Later Belyaev et al. [18] reported 90% yield in the condensate when a solution 1 M in H_2SO_4 and 0.5 M in $FeSO_4$ is distilled to 85% of its volume.

This effect is the basis of an extractive distillation procedure for the isolation of astatine from irradiated bismuth targets developed by Meyer, Roessler [24]. In this the carry-over of the astatine is enhanced by simultaneous distillation of chloroform. The active mechanism is not well understood.

The target is dissolved in bulb I of the quartz apparatus shown in **Fig. 10-17** in 1 mL of concentrated H_2SO_4 at 170 °C. Sulfuric acid is used instead of HNO_3 in order to keep the solution free from water, to improve extraction and to avoid formation of nitric oxides. Ten batches of 5 mL $CHCl_3$ are added in 3 to 4 min intervals to the 150 °C hot solution under continuous stirring. The organic solvent distills vigorously and carries the At into the cold trap II, perhaps via adsorption or formation of adducts. The solution containing

Fig. 10-17

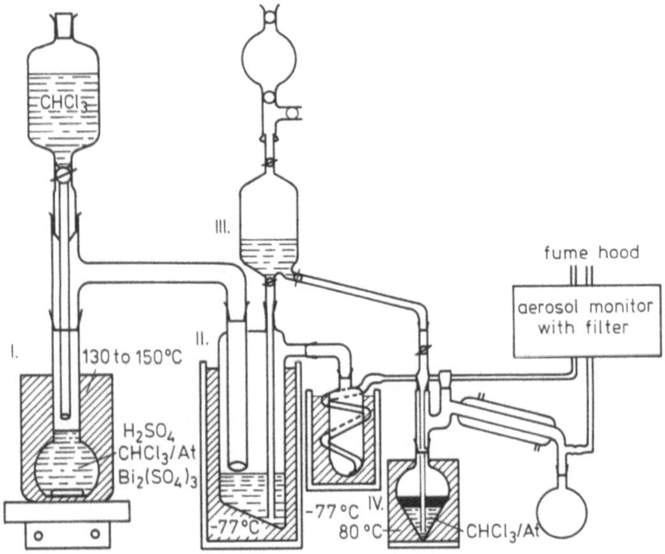

Apparatus for extractive distillation (quartz) [24]. 1) Bulb for dissolution of target and distillation. — 2) Cooling trap with security trap. — 3) Reservoir for At/$CHCl_3$ solution. — 4) Second extractive distillation.

70 to 75% of the At is transferred with the aid of a siphon into the reservoir III, from where it is slowly added to bulb IV containing 2 mL of an aqueous solution of 0.2 M Na_2SO_3 at 80 °C. While the $CHCl_3$ distills through the water layer (50 mL in 1 to 1.5 h) At is reduced to At^- and 90% stays in the aqueous fraction. A solution of 1 M HNO_3 with 10^{-3} M $Cr_2O_7^{2-}$ leads to 90% oxidation to the At^+ state, whereas pure water retains only 70 to 80% most probably as At^0. Specific activities in the order of 10 mCi/mL can easily be reached.

Special care was taken to avoid contamination or incorporation of the dangerous At during processing. The outlets of the quartz apparatus as well as the air in the fumehood where it is placed are continuously controlled by aerosol monitors. Although the concentration of ^{211}At in the air never exceeded 1×10^{-9} µCi · cm^{-3} (MPC value: 7×10^{-9} µCi · cm^{-3}), the workers were protected by wearing closed overalls, gas masks and threefold gloves.

Extraction Coprecipitation Procedure of Belyaev et al. [18]

The method is based on the successive application of two wet separation steps, the extraction from acid solution with diisopropylether and the coprecipitation with tellurium from acid solution upon reduction. It was successfully applied to proton irradiated bismuth and thorium targets requiring higher decontamination factors than ^4He irradiated bismuth targets.

The extractability of astatine from concentrated HCl solution was first reported by Neumann [38] and used, e.g., by Neirinckx, Smit [61] and Barton et al. [63] for the separation from irradiated bismuth. The simple procedure of Neirinckx was to dissolve the target in a minimum of concentrated nitric acid, cool the solution, make it 8 M in HCl and extract for a minute with 20 mL of diisopropylether. After two washings of the organic phase with

References on pp. 208/9

20 mL of concentrated HCl and 20 mL water containing 250 mg of $NH_2OH \cdot HCl$, the main part of the astatine is recovered with 20 mL of water containing 1 g $NH_2OH \cdot HCl$.

The coprecipitation of astatine with tellurium in 3 M HCl upon treatment with SO_2 was reported by the discoverers of the element [1]. They also mentioned that At is not coprecipitated with tellurium from alkaline solution with Sn^{II}. This was used, e.g., by Lefort et al. [64] for the isolation of astatine from proton bombarded thorium. These authors give a rather detailed description of their procedure and report an overall yield of 40 to 50%.

The recommended procedure for the two-step separation of Belyaev et al. is as follows: One gram of the target (bismuth, thorium) was dissolved in 5 mL concentrated nitric acid by heating in a flask equipped with a reflux condenser. Forty mL of 8 M HCl, saturated with chlorine, were added to the nitrate solution. The extraction was carried out, with 60 mL of diisopropylether, in an extractor equipped with a mechanical stirrer. The organic layer was washed twice with 15 mL of 8 M hydrochloric acid. The astatine was reextracted from the ether by 40 mL of a 0.1 M solution of sodium stannite in 2 M NaOH. Ten to fifteen mg sodium tellurite, 2 to 3 mg lanthanum (as $LaCl_3$) and 1 to 2 mg sodium chloroaurate were added to the alkaline solution. The solution was separated from the precipitate by filtration through a glass filter No. 4. The precipitation of tellurium by sodium stannite was again repeated. The alkaline filtrate was acidified with 20 mL concentrated hydrochloric acid containing about 0.2 mg Te/1 mL. The precipitation of the tellurium from the acid solution was carried out with vigorous stirring. After the coagulation of the precipitate, 5 mg of tellurium were added twice. The astatine-containing tellurium precipitate was separated from the solution by centrifugation, washed with 6 M HCl, and dissolved in several drops of nitric acid. Twenty mL 6 M HCl were added to the resulting solution and the tellurium was precipitated by $SnCl_2$. After the coagulation of the precipitate, the precipitation of tellurium was repeated (5 mg). The precipitate formed was centrifugated, washed with concentrated hydrochloric acid and dissolved in 5 mL 8 M HCl through which gaseous chlorine was bubbled. The astatine was separated from the tellurium by extraction with diisopropylether. The ether layer (about 6 mL) was washed twice with 1.5 to 2 mL of 8 M HCl, and the astatine was reextracted with two 5 mL portions of water. After the extraction ~0.01 M HCl solution of radiochemically pure astatine, containing traces of the solvent, was obtained.

The procedure gets somewhat complicated in the separation of astatine formed from lead. First, it is necessary to separate the lead chloride, which is precipitated by the addition of hydrochloric acid to the nitrate solution; a final additional purification of the astatine by distillation from a sulfate solution is also necessary.

The irradiated lead (1 g) was dissolved in 5 mL 6 M HNO_3 and 20 mL of 4 M HCl were added to the resulting hot nitrate solution. The solution was cooled and separated from the precipitated $PbCl_2$ crystals by filtration through a glass filter No. 2. The solution was then acidified by the addition of 20 mL concentrated HCl, saturated with chlorine. The subsequent operations are analogous to the stages of purification in the separation of astatine from bismuth.

The additional purification by distillation is necessary because of the very low cross section of formation of astatine (10^{-31} cm^2). For this purpose, 10 mL of 1 M solution of $FeSO_4$ in 2 M H_2SO_4 were added to an equal volume of the aqueous astatine solution resulting from the separation by extraction of the tellurium. The distillation was carried out until the first appearance of crystals in the distillation flask. The distilled astatine was absorbed in a 0.5 M NaOH solution.

The average chemical astatine yields for the procedure was 56 ± 4% with Pb, 64 ± 2 with Bi and 61 ± 2% with Th targets. Measurement of the purified astatine was possible

Fig. 10-18

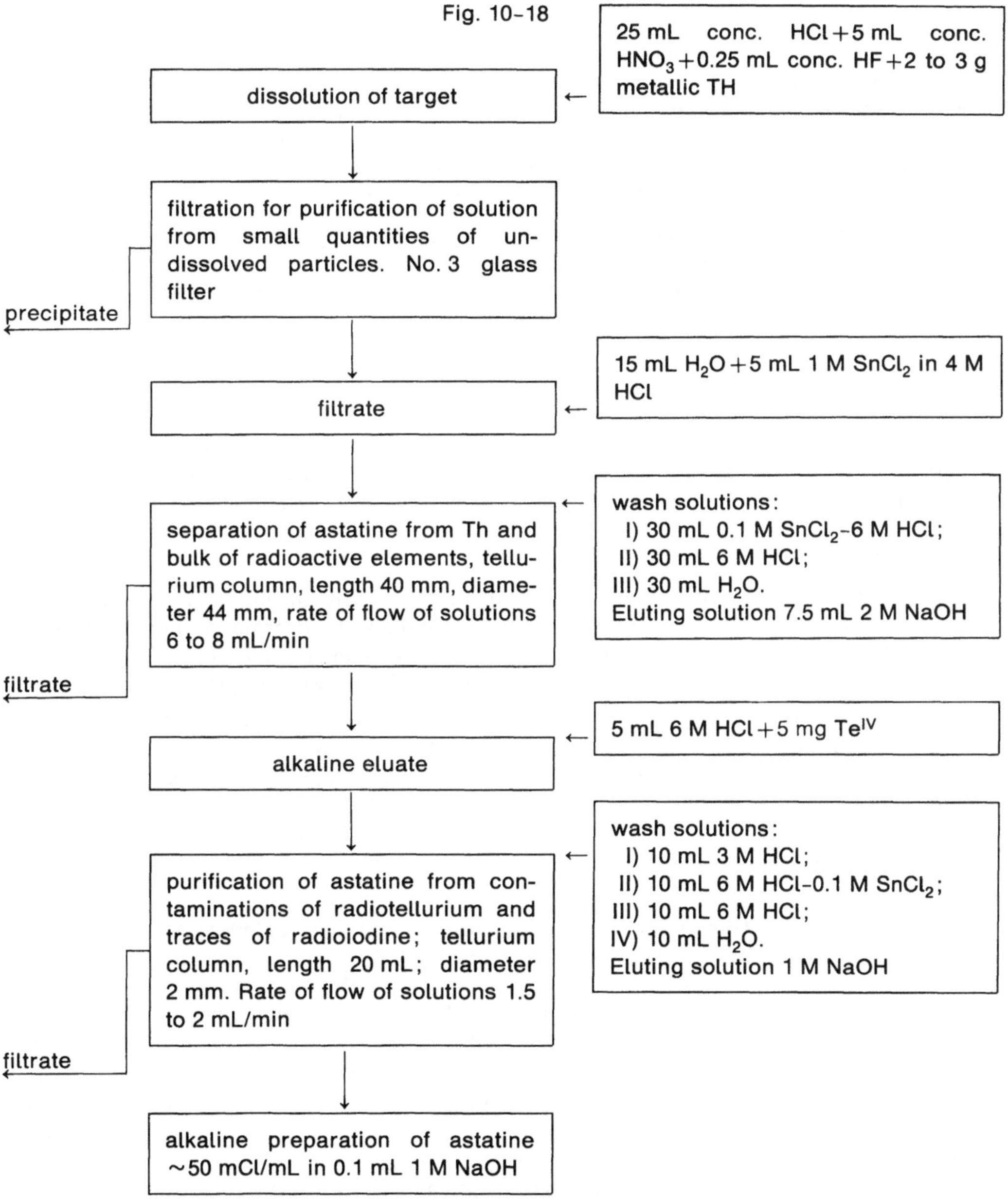

Procedure for the isolation of astatine from thorium irradiated with 660 MeV protons by tellurium columns [17].

5 to 6 h after end of irradiation. It was radiochemically pure in the sense that both α spectra and γ spectra did not indicate the presence of other radionuclides than from astatine in the case of bismuth and thorium but not in the case of lead.

References on pp. 208/9

Tellurium Adsorption Procedure of Bochvarova et al. [17]

The procedure was developed starting with the knowledge that astatine is coprecipitated with tellurium in acid media in the presence of reducing agents but not in alkaline media. It was assumed that the coprecipitation is the result of the formation of a coordination bond between the terminal atoms of the crystal chains of tellurium and the readily polarizable At^- anions, which possess four unshared pairs of electrons. Thus, the tellurium is thought to act as an anion exchanger that selectively adsorbes the astadide ion.

The procedure recommended by Bochvarova et al. for the application to proton irradiated thorium targets is schematically described in **Fig. 10-18**, p. 207. The chemical yield was $80 \pm 5\%$. The product contained about 3 µg tellurium and was obtained 5 to 7 h after the end of the irradiation. Radioactive contamination was no more than 0.04%.

References:

[1] D.R. Corson, K.R. MacKenzie, E. Segrè (Phys. Rev. [2] **58** [1940] 672/8). — [2] R. Dreyer, I. Dreyer, F. Roesch, S. Fischer (Z. Chem. [Leipzig] **23** [1983] 346/7). — [3] E.H. Appelman (UCRL-9025 [1960] 1/131; N.S.A. **14** [1960] No. 11501). — [4] E.H. Appelman (NAS-NS-3012 [1960] 1/33; N.S.A. **14** [1960] No. 20201). — [5] G.L. Johnson, R.F. Leininger, E. Segrè (J. Chem. Phys. **17** [1949] 1/10).

[6] R. McLaughlin (J. Opt. Soc. Am. **54** [1964] 965/7). — [7] K. Otozai, N. Takahashi (Radiochim. Acta **31** [1982] 201/3). — [8] W.A. Chalkin [Khalkin], E. Herrmann, J.W. Norseev, I. Dreyer (Chemiker-Ztg. **101** [1977] 470/81). — [9] E.H. Appelman (J. Am. Chem. Soc. **83** [1961] 805/7). — [10] G.-J. Meyer (JUEL-1076-NC [1974] 1/67; N.S.A. **31** [1975] No. 5663).

[11] J.A. Downs, C.J. Adams (in: J.C. Bailar, H.J. Emeléus, R. Nyholm, A.F. Trotman-Dikenson, Comprehensive Inorganic Chemistry, Vol. 2, Pergamon, New York 1973, pp. 1107/595). — [12] E.H. Appelman, E.N. Sloht, M.H. Studier (Inorg. Chem. **5** [1966] 766/9). — [13] J.R. Grover, E. Lebowitz, E. Baker (J. Inorg. Nucl. Chem. **31** [1969] 3705/20). — [14] M. Johnston, C.W. Asling, P.W. Durbin, J.G. Hamilton (UCRL-3013 [1955] 1/61, 35/6; N.S.A. **9** [1955] No. 5571). — [15] J.K. Roessler, W. Tornau, G. Stoecklin (J. Radioanal. Chem. **21** [1947] 199/209).

[16] W. Garrison, J. Gile, R. Maxwell, J. Hamilton (Anal. Chem. **23** [1951] 204/5). — [17] M. Bochvarova, Do Kim Tyung, I. Dudova, Yu.V. Norseev, V.A. Khalkin (Radiokhimiya **14** [1972] 858/65; Soviet Radiochem. **14** [1962] 889/95). — [18] B.N. Belyaev, Wang Yung-Yü, Ye.N. Sinotova, L. Nemet, V.A. Khalkin (Radiokhimiya **2** [1960] 603/13; Radiochem. [USSR] **2** [1960] 243). — [19] G.W.M. Visser, E.L. Diemer (Radiochim. Acta **33** [1983] 145/51). — [20] J. Merinis, Y. Legoux, G. Boussières (Radiochem. Radioanal. Letters **11** [1972] 59/64).

[21] N.A. Golovkov, I.I. Gromova, M. Janicki et al. (Radiochem. Radioanal. Letters **44** [1980] 67/78). — [22] V. Doberenz, Dang Duc Nang, R. Dreyer, M. Milanov et al. (Radiochem. Radioanal. Letters **52** [1982] 119/28). — [23] L. Vasaros, Yu.V. Norseev, D.D. Khan, V.A. Khalkin (Radiochem. Radioanal. Letters **54** [1982] 239/48). — [24] G.-J. Meyer, K. Roessler (Radiochem. Radioanal. Letters **25** [1976] 377/90). — [25] E.H. Appelman (J. Phys. Chem. **65** [1961] 325/31).

[26] J. Merinis, G. Boussières (Radiochim. Acta **12** [1969] 140/52). — [27] E.H. Appelman (MTP [Med. Tech. Publ. Co.] Intern. Rev. Sci. Inorg. Chem. Ser. One **3** [1972] 181/98; C.A. **76** [1972] No. 120925). — [28] B. Bayar, I. Votsilka, N.G. Zaitseva, A.F. Novgorodov (Radiokhimiya **16** [1974] 329/36; Soviet Radiochem. **16** [1974] 329/36). — [29] K.W. Bagnall (Chemistry of the Rare Radioelements, Butterworths, London 1957, pp. 1/177, 97/118). — [30] Do Kim

Tyung, I.V. Dudova, V.A. Khalkin (Radiokhimiya **14** [1972] 766/7; Soviet Radiochem. **14** [1972] 790/2).

[31] M. Milanov, V. Doberenz, V.A. Khalkin, A. Marinov (J. Radioanal. Nucl. Chem. Articles **83** II [1984] 291/9). – [32] I. Dreyer, R. Dreyer, V.A. Chalkin [Khalkin] (Radiochem. Radioanal. Letters **36** [1978] 389/98). – [33] Do Kim Tyung, I.V. Dudova, V.A. Khalkin (Radiokhimiya **15** [1973] 548/53; Soviet Radiochem. **15** [1973] 552/6). – [34] Wang Fu-Chueing, Yu.V. Norseev, V.A. Khalkin, Chao Tao-Nan (Radiokhimiya **5** [1963] 351/5; Soviet Radiochem. **5** [1963] 318/22). – [35] Yu.V. Norseev, Chao Tao-Nan, V.A. Khalkin (Radiokhimiya **8** [1966] 497/504).

[36] R. Dreyer, I. Dreyer, F. Roesch, G.-J. Beyer (Radiochem. Radioanal. Letters **54** [1982] 165/76). – [37] I. Dreyer, R. Dreyer, V.A. Chalkin [Khalkin], M. Milanov (Radiochem. Radioanal. Letters **40** [1979] 145/54). – [38] H.M. Neumann (J. Inorg. Nucl. Chem. **4** [1957] 349/53). – [39] W.A. Chalkin [Khalkin], E. Herrmann (Isotopenpraxis **11** [1975] 333/40). – [40] Yu.V. Norseev, V.A. Khalkin (J. Inorg. Nucl. Chem. **30** [1968] 3239/43).

[41] I. Dreyer, R. Dreyer, V.A. Chalkin [Khalkin] (Radiochem. Radioanal. Letters **35** [1978] 257/62). – [42] I. Dreyer, R. Dreyer, Yu.V. Norseev, V.A. Chalkin [Khalkin] (Radiochem. Radioanal. Letters **33** [1978] 291/300). – [43] V.A. Khalkin, Yu.V. Norseev, V.D. Nefedov et al. (Dokl. Akad. Nauk SSSR **195** [1970] 623/5; Dokl. Chem. Proc. Acad. Sci. USSR **190/195** [1970] 855/7). – [44] R. Dreyer, I. Dreyer, F. Roesch (Z. Chem. [Leipzig] **22** [1982] 54/6). – [45] R. Dreyer, I. Dreyer, M. Pfeiffer, F. Roesch (Radiochem. Radioanal. Letters **55** [1982] 207/14).

[46] G. Samson, A.H.W. Aten (Radiochim. Acta **13** [1970] 220/1). – [47] G.-J. Meyer (JUEL-1418 [1978] 1/107; C.A. **88** [1978] No. 71113). – [48] G.W.M. Visser, E.L. Diemer (Radiochem. Radioanal. Letters **51** [1982] 135/40). – [49] L. Vasaros, Yu.V. Norseev, V.I. Kuzin et al. (Radiochim. Acta **26** [1979] 171/6). – [50] G.A. Krestov (Radiokhimiya **4** [1962] 690/6; Soviet Radiochem. **4** [1962] 612/7).

[51] C. Aaij, W.R.J.M. Tschroots, L. Lindner, T.E.W. Feltkamp (Intern. J. Appl. Radiat. Isotop. **26** [1975] 25/30). – [52] K. Berei, L. Vasaros, Yu.V. Norseyev, V.A. Khalkin (Radiochem. Radioanal. Letters **26** [1976] 177/84). – [53] G.-J. Meyer, K. Roessler, G. Stoecklin (Radiochim. Acta **24** [1977] 81/5). – [54] G. Samson, H.W. Aten (Radiochim. Acta **9** [1969] 53/4). – [55] G.W.M. Visser, E.L. Diemer, F.M. Kaspersen (Rec. Trav. Chim. **99** [1980] 92/103).

[56] A.H.W. Aten, T. Doorgeest, U. Hollstein, H.P. Moeken (Analyst [London] **77** [1952] 774/8). – [57] G.-J. Beyer, R. Dreyer, H. Odrich, F. Roesch (Radiochem. Radioanal. Letters **47** [1981] 63/6). – [58] L. Lindner, G.A. Brinkman, T.H.G.A. Suer et al. (IAEA-SM-171-63-Vol. 1 [1973] 303/16; Radiopharm. Labelled Compounds Proc. Symp., Copenhagen 1973, Vol. 1, pp. 303/16; C.A. **81** [1974] No. 162521). – [59] V.M. Vakhtel, G.V. Vinel, Ts. Vylov, I.I. Gromova et al. (Radiokhimiya **18** [1976] 886/93; Soviet Radiochem. **18** [1976] 752/8). – [60] M. Lefort, G. Simonoff, X. Tarrago (Compt. Rend. **248** [1959] 216/8).

[61] R.D. Neirinckx, J.A. Smit (Anal. Chim. Acta **63** [1973] 201/4). – [62] M.W. Parrott, W.M. Garrison, P.M. Durbin, M. Johnston et al. (UCRL-3065 [1955] 1/8; N.S.A. **9** [1955] No. 7063). – [63] G.W. Barton, A. Ghiorso, I. Perlman (Phys. Rev. [2] **82** [1951] 13/9). – [64] M. Lefort, G. Simonov, X. Tarrago (Bull. Soc. Chim. France **1960** 1726/7).

10.2 Astatine Compounds

Klara Berei, László Vasáros
Central Research Institute for Physics
Hungarian Academy of Sciences
Budapest, Hungary

General References:

E.H. Appelman, Astatine, MTP [Med. Tech. Publ. Co.] Intern. Rev. Sci. Inorg. Chem. Ser. One **3** [1972] 181/98; C.A. **76** [1972] No. 120925.

A.H.W. Aten Jr., The Chemistry of Astatine, Advan. Inorg. Chem. Radiochem. **6** [1964] 207/23.

K. Berei, L. Vasáros, Organic Chemistry of Astatine, in: S. Patai, Z. Rappoport, The Chemistry of Functional Groups, Suppl. D, Pt. 1, Wiley, New York 1983, pp. 405/40.

V.A. Khalkin[1)], E. Herrmann, Inorganic Chemistry of Astatine, Isotopenpraxis **11** [1975] 333/40; C.A. **84** [1976] No. 9632.

V.A. Khalkin, E. Herrmann, Yu.V. Norseev[2)], I. Dreyer, The Present Status of Astatine Chemistry, Chemiker-Ztg. **101** [1977] 470/81; C.A. **88** [1978] No. 28663.

A.J. Downs, C.J. Adams, The Chemistry of Chlorine, Bromine, Iodine and Astatine, Pergamon, Oxford 1975, pp. 1573/94.

A.K. Lavrukhina, A.A. Pozdnyakov, Analytical Chemistry of Technetium, Promethium, Astatine and Francium, Ann Arbor-Humphrey, Ann Arbor 1970, pp. 227/60; C.A. **75** [1971] No. 71094.

V.D. Nefedov, Yu.V. Norseev, M.A. Toropova, V.A. Khalkin, Astatine, Usp. Khim. **37** [1968] 193/215; Russ. Chem. Rev. **37** [1968] 87/98; C.A. **68** [1968] No. 100679.

J. Sedlet, Astatine: Atomic Number 85 in: I.M. Kolthoff, P.J. Elving, E.B. Shandel, Treatise on Analytical Chemistry, Part II, Sect. A, Vol. 6, Interscience, London 1964, pp. 487/501.

Introduction

Even though investigations on astatine chemistry can be traced back more than 40 years, much of the chemical behavior of this element is still uncertain. So far as inorganic compounds are concerned, the major part of the available information stems from extraction and coprecipitation experiments though neither type is easy to interpret. The use of iodine as a carrier for tracer amounts of astatine can also lead to erroneous conclusions. Besides the basic resemblance of the two halogens, significant differences in their oxidation potentials and other properties also exist. The recent introduction of chromatographic techniques, which allow investigations without using a carrier, seems to be more promising though the data provided by these latter studies and the conclusions derived are often preliminary and should, therefore, be considered with caution. A further point is that research in this area has expanded dynamically over the past few years and this expansion may in itself necessitate significant reevaluation of some of the earlier results and interpretations.

Some peculiarities related to the fact that astatine is available only in the form of its radioactive isotopes should always be kept in mind:

a) Consideration of compounds containing two or more astatine atoms is, at present, purely theoretical since the probability of more than one astatine atom being built into

---

1) Transliteration according to Chemical Abstracts, other used transliterations: W.A. Chalkin, V.A. Halkin.

2) Transliteration according to Chemical Abstracts, other used transliterations: J.V. Norszejev, Yu.V. Norseyev.

the same molecule is negligible due to its 10^{-11} to 10^{-15} M concentration in typical chemical investigations.

b) Disproportionation reactions are not applicable to astatine species for the same reason.

c) The excessive number of physicochemical properties calculated by extrapolation given in the following sections reflects the difficulties in establishing tracer amounts.

d) The relatively short half-lives of astatine isotopes mean that the time required for synthesizing and analyzing the compounds of astatine is of major importance.

10.2.1 Inorganic Compounds of Astatine

10.2.1.1 Astatine and Hydrogen

10.2.1.1.1 HAt

Though the formation of a volatile hydride in acidic solutions was indicated as long ago as in the earliest reports on astatine chemistry [1, 2], its real existence was first proved by Appelman et al. [3] in the sixties. Using a time of flight mass spectrometer with an ultimate sensitivity of about 40 atoms/mL, the mass lines of astatine species could be detected directly (see Fig. 10-23, p. 225, and Fig. 10-26, p. 234) instead of measuring the nuclear radiation of the astatine isotopes[1]. The formation of $H^{211}At^+$ – explained by the reaction of astatine with water or organic traces usually present in the apparatus – was always observed if the astatine sample was ionized in the absence of other macro components. A more recent study has also reported the relative amount of $H^{211}At^+$, viz. 3.2% of the total induced ^{211}At activity, forming in the plasma ion source of a mass separator in the absence of additives [4].

A molecular beam of HAt has been prepared [5 to 8] and utilized for studying hydrogen transfer reactions [9 to 12] by Grover et al. Instead of the commonly used longest-lived astatine isotopes, ^{217}At (a member of the artificial 4n+1 radioactive decay chain) has been employed since its short half-life of 0.032 s is especially convenient for investigating fast chemical processes by crossed-beam experiments. First, its long-lived precursor ^{225}Ac ($T_{1/2}$=10 days) is prepared and incorporated into lanthanum stearate powder [5, 6]. This emanates $H^{217}At$ presumably as a consequence of organic hydrogen abstraction by the recoil astatine atoms forming in the following chain of nuclear transformations:

$$^{225}Ac \xrightarrow[10\,d]{\alpha} {}^{221}Fr \xrightarrow[4.8\,min]{\alpha} {}^{217}At \xrightarrow[0.032\,s]{\alpha} {}^{213}Bi, \text{ etc.}$$

Fig. 10-19 shows a schematic diagram of the equipment serving as the HAt source for the beam experiments. Intensities of 2×10^4 molecules/s have been achieved at a beam purity of 0.9999 [11].

$$Cl + HAt \rightarrow HCl + At \quad (1)$$

$$Br + HAt \rightarrow HBr + At \quad (2)$$

The mechanism of the hydrogen transfer reactions (1) and (2) has been studied by crossing $H^{217}At$ beams with those of atomic chlorine and bromine, and the total cross sections of these processes as functions of impact parameters have been established [10,

[1] The only other known case is the direct measurement of the atomic absorption spectrum of astatine (see Section 5.11.1, p. 120).

References on pp. 231/2

Fig. 10-19

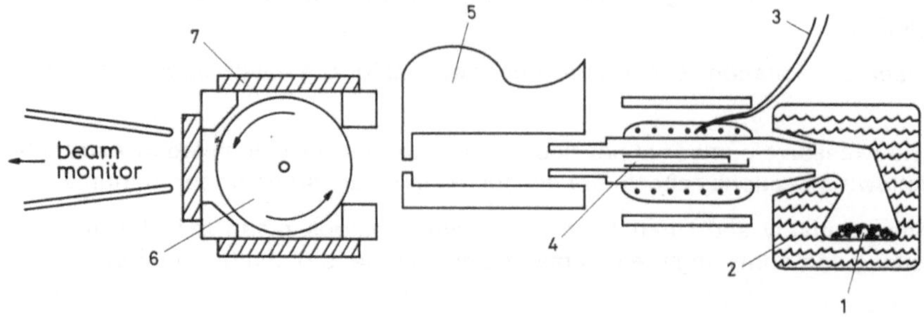

Schematic diagram of apparatus for producing $H^{217}At$ beam [7]: 1) Pyrex vessel with stearate powder. — 2) Water bath. — 3) Thermocouples. — 4) Activated Pt foil insert (scrubber) permitting passage of HAt only. — 5) Collimator cooled with liquid N_2. — 6) Rotating cylinder coated with Te film to distinguish, between At and HAt. — 7) Radioactivity detectors.

12]. An attempt was made to assess the contribution from astatine abstraction reactions:

$$Cl + HAt \rightarrow H + AtCl \quad (3)$$

$$Br + HAt \rightarrow H + AtBr \quad (4)$$

Reactions (3) and (4) were sought with good sensitivity but not observed at the center-of-mass collision energy of 25 kJ/mol used in the experiments [10].

The collision energy dependence of the process (2) has also provided information on the mechanism of this reaction which has turned out to be markedly different from the other known halogen atom — hydrogen halide hydrogen transfer processes. These latter are essentially direct reactions with a linear $X \cdots H \cdots Y$ transition state [13, 14]. As a means of explaining the unexpectedly low translational energy along with the broad angular distribution of the products in reaction (2), a "trapping model" has been proposed, i.e., the three atoms are temporarily trapped in each other's vicinity with an H atom rapidly moving between the two halogen atoms. Due to this mechanism, a considerable part of the reactant's translational energy is eventually converted into the excitation energy of the product HBr molecule [12].

The HAt bond energy (D) could roughly be estimated by investigating the gas phase reactions between HAt and several possible reactants, e.g., O_2, SO_3, NO_2, etc., as well as from the kinematic analysis of reactions (1) and (2). It was found to be in the range 52 to 64 kcal/mol [12] which is consistent with the value obtained by the same authors from pyrolysis experiments with HAt [15] and with the values of the bond energies established by different extrapolation techniques [16 to 19]. The values of bond energy together with boiling and melting temperature (t_b and t_m), heat of vaporization (ΔH_{vap}), ionization potential (IP), internuclear distance (r_e), bond refraction (R), and ground state vibrational frequency (ω_e) of hydrogen astatide are given in Table 10/6. Besides purely theoretical calculations [16 to 18, 21] or graphic extrapolations [20] from corresponding values of other hydrogen halides, some data are obtained by making use of an experimental parameter Z′ for astatine derived from the gas chromatographic behavior of volatile astatine compounds [19]. The bond refraction given in Table 10/6 represents the mean of two values, one extrapolated from bond refraction data for hydrogen halides, the other from the same data for the hydrides of astatine neighbors in the sixth row of the Periodic Table: Pb, Bi, and Po, according to the assumed amphoteric character of astatine [22].

Table 10/6
Some Properties of HAt (symbols see text).

t_b in °C	−3		[17]
	−16.4		[19]
	−1.5		[20]
t_m in °C	−26.2		[19]
ΔH_{vap} in kJ/mol . . .	21.3	(5.1 kcal/mol)[a)]	[19]
IP in eV	9.4		[19]
r_e in Å	1.68		[16]
	1.71		[18]
D in kJ/mol	243 ± 16	(58 ± 4 kcal/mol)[a)]	[15]
	245	(2.539 eV)[a)]	[16]
	243	(58 kcal/mol)[a)]	[17]
	231	(55.2 kcal/mol)[a)]	[18]
	205	(49 kcal/mol)[a)]	[19]
R in cm^3/mol	16.96		[22]
ω_e in cm^{-1}...	1840		[18]
	2079		[21]

a) Originally quoted values are given in parentheses.

The anharmonicity of vibrational frequency and the internuclear potential-function parameter for HAt have also been calculated [18].

10.2.1.1.2 Astatides

Only a few metal astatides have been identified. These were mainly in early studies, by the coprecipitation technique, using iodine as the carrier. However, extensive calculations have been carried out to assess physicochemical properties for predominantly unknown astatides of mono- and divalent cations.

In the first experiments with element 85, Segrè et al. observed that it did not coprecipitate with silver iodide from dilute nitric acid solutions whereas it was carried quantitatively on silver, mercury, bismuth and antimony sulfides [1, 23]. This phenomenon, which was at first attributed to the nonhalogen nature of astatine, could later be explained by the fact that in dilute nitric acid solutions astatine exists in higher valency states (as At^0 and partly At^+) and not as At^-. Further discrepancies observed in the early experiments can chiefly be attributed to iodine being the best though still not a perfect carrier for astatine due to the differences in their oxidation potentials and certain other chemical properties. Thus, for example, in contrast to iodine, At^0 is reduced only slowly and incompletely to At^- by SO_2 in acid solutions. This might have been the reason for its incomplete coprecipitation with thallium iodide experienced by Johnson et al. [24]. Another complicating factor is that At^0 can also be carried on iodide precipitates or on metallic silver produced in the course of reduction processes [25]. However, it is bound only on the surface and can be removed by organic solvents, e.g., acetone, as was later proved by Appelman [26].

References on pp. 231/2

Astatides of silver, thallium [24, 25, 27], palladium[1)] [24, 26], and lead[1)] [26] have been reported to form by coprecipitation of At^- with insoluble iodides of these metals. Zn or SO_2 in acidic solutions [24 to 27], As^{III} at pH>4, and $[Fe(CN)_6]^{2-}$ at pH≧2 [26, 28] were used as reducing agents. Aten [27] observed that even if astatine coprecipitated quantitatively with AgI during its formation, the precipitate was losing astatine activity with time. This phenomenon was apparently connected with the rapid recrystallization of AgI. In contrast, astatine losses from PdI_2 were negligible due to the extremely slow recrystallization, i.e., highly stable crystal lattice of the latter.

The values of the heat capacity (C), heat of formation (ΔH_f), and entropy (S) have been estimated for AgAt by Ozhigov [17] from corresponding quantities established for other silver halides, using the Mendeleev's rule of extrapolation, as follows:

$$C = 55.6\ J \cdot mol^{-1} \cdot K^{-1}\ (13.3\ cal \cdot mol^{-1} \cdot K^{-1})^{2)}$$

$$\Delta H_f = -45.2\ kJ/mol\ (-10.8\ kcal/mol)^{2)}$$

$$S = 133.1\ J \cdot mol^{-1} \cdot K^{-1}\ (31.8\ cal \cdot mol^{-1} \cdot K^{-1})^{2)}$$

While little is known about the properties of astatides actually obtained in chemical experiments, a number of characteristic parameters have been estimated by extrapolation for alkali and alkaline earth astatides.

Taking the Zachariasen value for the At ion radius, 2.27 Å, Ladd and Lee [29] have assumed from radius ratio considerations that CsAt would be likely to have a caesium chloride type structure, LiAt a zinc blende lattice, and the remaining alkali astatides would be isostructural with sodium chloride crystals. Equilibrium interionic distances (r_0) for these hypothetical crystalline compounds have been calculated and are given in Table 10/7. Another set of r_0 values has been estimated by Krestov and Krestova [30] using the At^- radius of 2.35 Å, extrapolated from the Goldschmidt system. The same method has been used to calculate r_0 values for gaseous alkali astatides taking $r_{At^-} = 1.97$ Å [31, 32].

Lattice energies, $U(r_0)$, for these systems have been calculated: (a) using simple extrapolation from corresponding values for other alkali halides [17]; (b) from equations taking into account the Coulomb interactions and the repulsive forces as well as the experimentally observed parallel changes in r_0 and in Madelung factors [32, 33]; (c) also considering the van der Waals interactions as well as the zero point energy [29]. These data together with the thence derived lattice entropy (ΔS) values [32] are also summarized in Table 10/7.

The standard heat of formation (ΔH_f°) for crystalline alkali astatides has been obtained (a) by simple extrapolation [29] from tabulated data for corresponding alkali halides [34] and (b) by calculation using the heat of formation for the corresponding cation and the At^- anion as well as extrapolated values of $U(r_0)$ [32]. This last type of extrapolation has also been used to estimate the ΔH_f° values for gas phase and aqueous solutions given in Table 10/8. It has also been used for establishing the standard Gibbs energy (ΔG°) and standard entropy (S°) values of these systems, listed in Table 10/9, p. 216. S° values for LiAt, NaAt, KAt, and RbAt: 99.4, 107.5, 116.0, and 123.0 $J \cdot mol^{-1} \cdot K^{-1}$, respectively, obtained more recently by extrapolation based on the principal quantum numbers of the elements, along with some other thermodynamic parameters [35], do not differ significantly from those given in Table 10/9, p. 216.

Melting and boiling temperature [30, 36], the heat capacity (C_p, C_v) [17, 30, 37], and the heat and entropy of vaporization [36] for alkali astatides are summarized in Table 10/10.

1) Only mixed $Me^{II}IAt$ can form due to the tracer amounts of astatine (see "Introduction", p. 210).

2) Originally quoted values are given in parentheses.

Table 10/7
Interionic Distances and Lattice Parameters Calculated for Alkali Astatides.

compound	r_0(cryst.) in Å [29]	r_0(cryst.) in Å [30]	r_0(gas) in Å [32]	$U(r_0)$ in kJ/mol (kcal/mol)[a] [17]	$U(r_0)$ in kJ/mol (kcal/mol)[a] [29]	$U(r_0)$ in kJ/mol (kcal/mol)[a] [32]	ΔS in $J \cdot mol^{-1} \cdot K^{-1}$ ($cal \cdot mol^{-1} \cdot K^{-1}$)[a] [32]
LiAt	2.93			−710 (−170)	−720 (−172)		
NaAt	3.31	3.33	2.74	−665 (−159)	−657 (−157)	−667.8 (−159.6)	220.1 (52.6)
KAt	3.61	3.68	3.09	−628 (−150)	−615 (−147)	−613.0 (−146.5)	215.5 (51.5)
RbAt	3.75	3.84	3.25	−586 (−144)	−594 (−142)	−590.4 (−141.1)	213.0 (50.9)
CsAt	4.04	4.00	3.44	−569 (−136)	−586 (−140)	−569.9 (−136.2)	210.9 (50.4)
FrAt	4.14	4.13	3.49		−573 (−137)	−554.4 (−132.5)	209.6 (50.1)

[a] Originally quoted values are given in parentheses.

Table 10/8
Heat of Formation Values Calculated for Alkali Astatides.

compound	ΔH_f° in kJ/mol (kcal/mol)[a] solid [17]	solid [29]	solid [32]	gas [32]	aqueous solution [32]
LiAt		−247 (−59)			
NaAt	−230 (−55)	−264 (−63)	−247.7 (−59.2)	−48.5 (−11.6)	−248.1 (−59.3)
KAt		−305 (−73)	−288.7 (−69.0)	−106.7 (−25.5)	−250.9 (−62.0)
RbAt		−305 (−73)	−286.2 (−68.4)	−110.1 (−26.3)	−254.8 (−60.9)
CsAt		−318 (−76)	−300.0 (−71.7)	−127.6 (−30.5)	−256.1 (−61.2)
FrAt		−326 (−78)	−298.3 (−71.3)	−133.1 (−31.8)	−255.2 (−61.0)

[a] Originally quoted values are given in parentheses.

References on pp. 231/2

Table 10/9
Gibbs Energy and Entropy Values Calculated for Alkali Astatides [32].

compound	$\Delta G°$ in kJ/mol (kcal/mol)[a)] solid	gas	aqueous solution	$S°$ in $J \cdot mol^{-1} \cdot K^{-1}$ ($cal \cdot mol^{-1} \cdot K^{-1}$)[a)] solid	gas	aqueous solution
NaAt	−245.2 (−58.6)	−91.2 (−21.8)	−270.3 (−64.6)	103.0 (24.7) 100 (24) [17]	254.85 (60.91)	185.8 (44.4)
KAt	−285.8 (−68.3)	−148.5 (−35.5)	−290.4 (−69.4)	115.1 (27.5)	264.22 (63.15)	228.0 (54.5)
RbAt	−285.3 (−68.2)	−153.1 (−36.6)	−290.4 (−69.4)	126.8 (30.3)	275.35 (65.81)	249.8 (59.7)
CsAt	−297.1 (−71.0)	−169.0 (−40.4)	−290.4 (−69.4)	134.3 (32.1)	282.29 (67.47)	258.6 (61.8)
FrAt	−295.0 (−70.5)	−173.6 (−41.5)	−288.7 (−69.0)	142.3 (34.0)	289.03 (69.08)	264.4 (63.2)

a) Originally quoted values are given in parentheses.

Table 10/10
Some Thermochemical Data Calculated for Alkali Astatides.

compound	T_m in K		T_b in K	C_p(300 K) in $J \cdot mol^{-1} \cdot K^{-1}$ ($cal \cdot mol^{-1} \cdot K^{-1}$)[a)]		C_v(300 K)	ΔH_{vap} in kJ/mol (cal/mol)[a)]	ΔS_{vap} in $J \cdot mol^{-1} \cdot K^{-1}$ ($cal \cdot mol^{-1} \cdot K^{-1}$)[a)]
	[30]	[36]	[36]	[17]	[30]	[37]	[36]	[36]
NaAt	920	908	1565	52.47 (12.54)	52.89 (12.64)	48.91 (11.69)	150.180 (35894)	95.93 (22.93)
KAt	910	904	1561		53.51 (12.79)	49.41 (11.81)	149.574 (35749)	95.81 (22.90)
RbAt	885	894	1551		53.93 (12.89)	49.66 (11.87)	148.214 (35424)	95.56 (22.84)
CsAt	880	887	1543		54.01 (12.91)	49.70 (11.88)	147.222 (35187)	95.40 (22.80)
FrAt	845	877	1534		54.22 (12.96)	49.75 (11.89)	145.983 (34891)	95.19 (22.75)

a) Originally quoted values are given in parentheses.

The temperature dependence of C_v and of S is given in [37]. Some volumetric, thermal and calorimetric coefficients for these compounds as functions of temperature, pressure and interionic distance are reported in [38].

Heat of solvation (ΔH_{solv}) values for alkali metal astatides in some nonaqueous solvents, calculated using an empirical method based on the Coulomb interactions in the molecules,

Table 10/11
Calculated Heat of Solvation Values for Alkali Astatides in Some Nonaqueous Solvents [39].

compound	solvent CH_3OH	C_2H_5OH	HCOOH	$HCONH_2$	$HCON(CH_3)_2$
	ΔH_{solv} in kJ/mol				
LiAt	821.5	818.5	783.8	809.1	843.7
NaAt	703.4	705.0	688.7	703.1	729.8
KAt	620.1	620.5	618.9	620.5	649.8
RbAt	594.6	594.6	597.0	586.9	623.4
CsAt	568.5	564.8	588.2	571.5	597.0
FrAt	552.7	548.6	561.0	559.0	582.0

Table 10/12
Interionic Distance and Lattice Parameters Calculated for Astatides of Alkaline Earth Metals [32].

alkaline earth metal	r_0 (cryst) in Å	$U(r_0)$ in kJ/mol (kcal/mol) a)	ΔS in $J \cdot mol^{-1} \cdot K^{-1}$ ($cal \cdot mol^{-1} \cdot K^{-1}$) a)
Mg	2.49	−2113 (−505)	354.0 (84.6)
Ca	2.80	−1962 (−469)	346.4 (82.8)
Sr	2.95	−1866 (−446)	341.4 (81.6)
Ba	3.13	−1795 (−429)	337.2 (80.6)
Ra	3.18	−1761 (−421)	335.1 (80.1)

a) Originally quoted values are given in parentheses.

are given in Table 10/11. The same method has been utilized to estimate some other thermodynamic parameters for hypothetical complex molecules containing two or more atoms of astatine [39].

Lattice characteristics and some thermodynamic parameters have also been extrapolated for the astatides of alkaline earth metals [32] and are given in Tables 10/12 and 10/13. Somewhat higher S° values: 150.5, 160.0, and 166.0 $J \cdot mol^{-1} \cdot K^{-1}$ have been obtained for beryllium, magnesium, and calcium astatides, respectively, by extrapolation according to the principal quantum numbers of the elements [35]. The significance of these data, similarly to those for the complex molecules mentioned above, seems to be purely theoretical since, e.g., $Me^{II}At_2$ type compounds are unlikely to be prepared under the given conditions (see "Introduction", p. 210).

References on pp. 231/2

Table 10/13
Some Thermodynamic Data Calculated for Astatides of Alkaline Earth Metals [32].

alkaline earth metal	ΔH_f^o in kJ/mol (kcal/mol) [a)] solid	aqueous solution	ΔG^o in kJ/mol (kcal/mol) [a)] solid	aqueous solution	S^o in $J \cdot mol^{-1} \cdot K^{-1}$ ($cal \cdot mol^{-1} \cdot K^{-1}$) [a)] solid	aqueous solution
Mg	−142 (−34)	−478.6 (−114.4)	−138 (−33)	−472.4 (−112.9)	145.6 (34.8)	133.1 (31.8)
Ca	−406 (−97)	−559.8 (−133.8)	−406 (−97)	−569.4 (−136.1)	159.4 (38.1)	195.8 (46.8)
Sr	−456 (−109)	−562.3 (−134.4)	−456 (−109)	−577.0 (−137.9)	175.3 (41.9)	224.7 (53.7)
Ba	−523 (−125)	−555.2 (−132.7)	−523 (−125)	−577.8 (−138.1)	184.1 (44.0)	263.6 (63.0)
Ra	−519 (−124)	−543 (−130)	−519 (−124)	−577 (−138)	192.0 (45.9)	305 (73)

a) Originally quoted values are given in parentheses.

Table 10/14
Vibration Frequency Values for Astatides of Alkali and Alkaline Earth Metals.

metal	ω_e in cm^{-1} [40]	metal	ω_e in cm^{-1} [41]
		Be	118
Na	260±10	Mg	94
K	181±6	Ca	80
Rb	120±5	Sr	75
Cs	101±4	Ba	71
Fr	88±4	Ra	69

The vibration frequency for the astatides of alkali [40] and alkaline earth metals as well as the force constant values for the latter type of molecules [41] have been calculated. The ω_e values are given in Table 10/14.

10.2.1.2 Astatine and Oxygen

10.2.1.2.1 Compounds of Astatine (I) Cation

The exact chemical form of astatine present in acidic solutions containing mild oxidizing agents has been one of the most discussed questions of inorganic astatine chemistry. Deposition of astatine on a gold cathode at −1.20 V electrode potential from sulfuric acid solutions containing $Cr_2O_7^{2-}$ ions and its decreasing extractability with CCl_4 from dilute nitric acid in the presence of Br_2 or $Fe(NO_3)_3$ was reported by Johnson et al. [24]. The exact valency of this species (+1 or +3) was uncertain for a long period and it is often designated in the literature as At(X).

Later evidence for the existence of monovalent astatine cation in dilute acid solutions containing $Cr_2O_7^{2-}$ has been gathered from the following experimental results:

(a) It coprecipitates more or less completely with insoluble salts of monovalent cations such as thallium and silver dichromates, silver iodate, tricaesium dodecatungstophosphate, etc. [42, 43].

(b) It deposits on Pt coated with a thin oxide film, as a result of a surface ion exchange process, competing with Tl^+, Hg_2^{2+} and BiO^+ cations [44 to 48].

(c) It adsorbs on monofunctional cation exchange resins [46 to 51] with an acid concentration dependence characteristic for +1 cations [46 to 48] as demonstrated in **Fig. 10-20a** and 10-**20b** for distribution of astatine and Tl^+, respectively, between Dowex 50 and acidic solutions.

(d) Cl^- ions compete with all the above mentioned processes forming chloro complexes of monovalent astatine [45, 46, 48 to 50, 52].

Fig. 10-20a

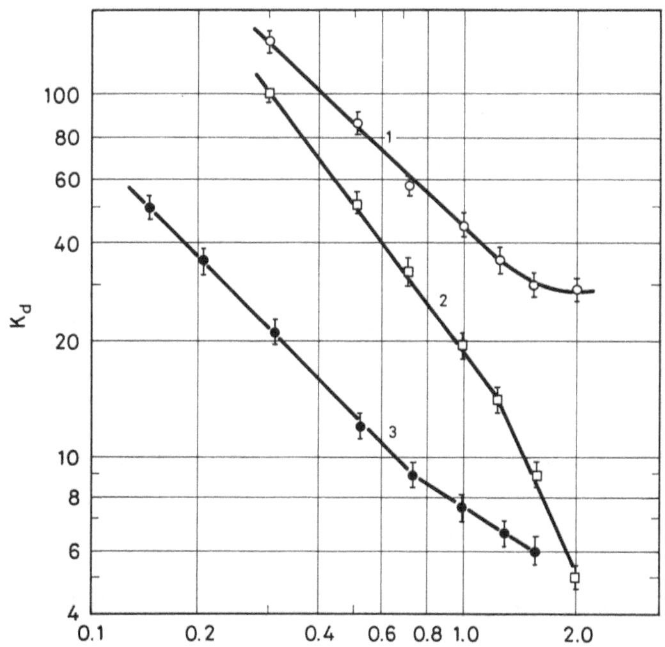

Distribution (K_d in mL/g) of AtI between Dowex 50 and 1) $HClO_4$; 2) HNO_3; 3) H_2SO_4 solutions containing 0.005 M $H_2Cr_2O_7$ at 20 °C [48].

Fig. 10-20b

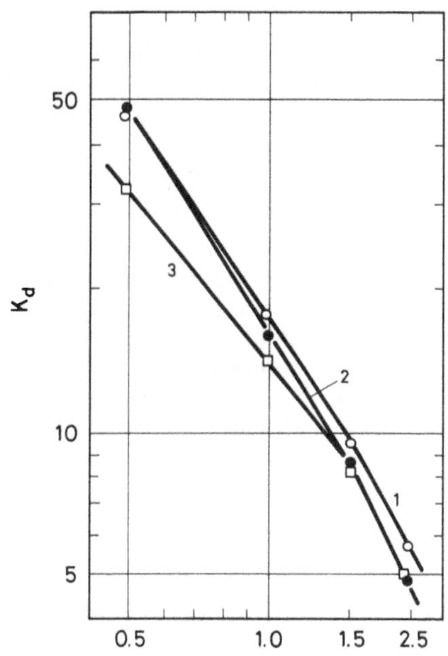

Distribution (K_d in mL/g) of Tl^+ between Dowex 50 and 1) $HClO_4$; 2) HNO_3; 3) H_2SO_4, at 20 °C [48].

References on pp. 231/2

Fig. 10-21

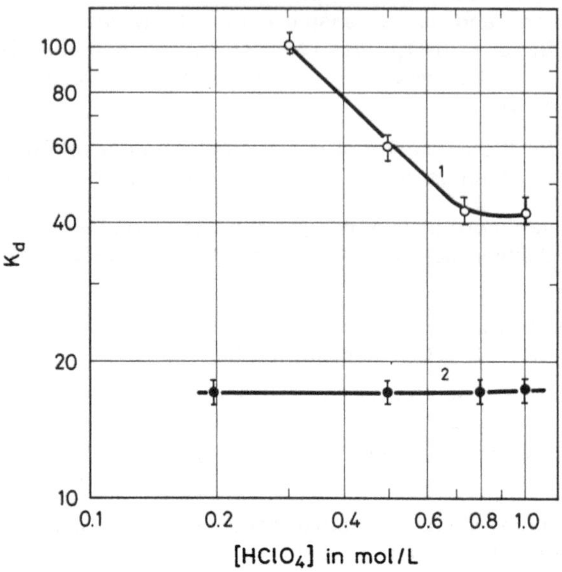

Distribution (K_d in mL/g) of 1) astatine cation, and 2) Tl^+, between Dowex 50 and $HClO_4$/$NaClO_4$ mixtures of constant ionic strength at 20 °C as a function of acidity [48].

Some significant differences in the behavior of astatine cation compared with other monovalent ions have, however, been observed. Its migration in an electric field is much slower than that of Ag^+ [53], Tl^+ or At^- [54]. The adsorption processes show, in some cases, different pH and temperature dependence from those of metal cations as shown in **Fig. 10-21** for the acid dependence of astatine cation and Tl^+ sorption on Dowex 50 cation exchange resin.

A complex cation, usually designated as $(At\ominus)^+$, has been suggested by Khalkin et al. [55, 56] as a means of explaining these peculiarities. It can easily be prepared by adding an equal volume of dilute HNO_3 containing a small amount of $K_2Cr_2O_7$ to the aqueous solution of At^0 while heating to 100 °C for 10 min [53]. As the most probable structure, a relatively stable aqua complex or protonated hypoastatous acid, $[H_2OAt]^+$, has been assumed, similarly to protonated hypoiodous acid reported to exist in aqueous solutions [57]. The equilibrium constant (K_{dp}) for the deprotonation reaction (5) is estimated, by extrapolation from corresponding data for lighter halogens, to be $<10^{-3}$ [56]. This indicates $[H_2OAt]^+$ to be a fairly weak acid:

$$[H_2OAt]^+ + H_2O \xrightleftharpoons{K_{dp}} HOAt + H_3O^+ \qquad (5)$$

A symmetric diaqua complex 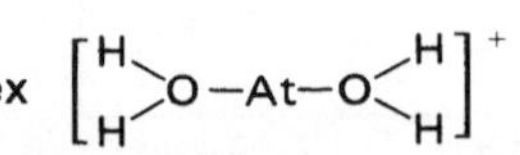 as a possible structure of complex astatine cation has also been proposed [55, 56].

Recent electromigration studies have provided a strong argument for the $[H_2OAt]^+$ structure of the astatine cation in aqueous solutions and have allowed a preliminary value ($K_{dp} = 0.032 \pm 0.05$) to be established experimentally [58] from the dependence of the astatine cation

Table 10/15
Astatine Migration Rates Measured by Free Solution Electrophoresis [58].
(Electrolyte: [0.4-n] M $HClO_4$ + n M $NaClO_4$ + 10^{-4} M $K_2Cr_2O_7$ at 25 °C.)

pH	1.68	1.56	1.42	1.21	1.09	0.93	0.81	0.71	0.63
$\mu \times 10^4$ in $cm^2 \cdot V^{-1} \cdot s^{-1}$	1.17	1.34	1.67	2.38	2.41	2.43	2.45	2.64	2.67

migration rate (μ) on the electrolyte acidity (see Table 10/15). It is actually equal to the deprotonation constant determined for hypoiodous acid [57] and shows $[H_2OAt]^+$ to be a stronger acid than previously expected.

Neumann's earlier finding that in acidic media astatine reacts with phenol and benzene, similarly to HOI or HOBr, [59] is well in line with the higher value of K_{dp}. Successful electrophilic astatination of aromatic compounds in heterogeneous systems using strong acidic media reported recently by Vasáros et al. [60] seems to validate reaction (5) and thus the assumed structure of the complex astatine cation.

Oxygen containing astatine species: $H_2O^{211}At^+$, $HO^{211}At^+$ and $^{211}AtO^+$, have been observed by Golovkov et al. [4] to form in small amounts (0.1 to 0.3% of the total ^{211}At activity) in the plasma ion source of a mass separator, apparently as a result of ^{211}At reacting with the water traces present.

The heat of formation for gaseous AtO has been estimated by extrapolation [61] to be $\Delta H_f^\circ = -123.4$ kJ/mol (-29.5 kcal/mol)[1)].

10.2.1.2.2 Compounds of Astatine(III)

Formation of AtO^+ instead of a monovalent astatine cation under mild oxidation conditions has often been assumed as a possibility (see "General References", p. 210) but was first identified from its electromigration behavior by Dreyer et al. [53, 62]. Thus, for example, in $HClO_4$ solution containing $Na_2S_2O_8$, At^0 is oxidized into a new species migrating to the cathode in an electric field much more rapidly than does the monovalent astatine cation (Table 10/16). Astatosyl ion AtO^+ or protonated astatous acid $H_2AtO_2^+$ have been proposed as the possible structures for this species. This assumption is supported by the finding that if heated in an alkaline solution the cation transforms into an anion, supposedly AtO_2^-, which again has a different migration rate from that of At^- (Table 10/16) obtained from the monovalent astatine cation under the same conditions. Furthermore, the halogen complexes of trivalent astatine $AtOX_2$ (see Section 10.2.1.3.4, p. 229) hydrolyze in alkaline solutions yielding the same anion AtO_2^- [63].

Formation of AtO_2^- is always observed in aqueous solutions and has been explained by oxidation of At^0 by the radiolysis products of water due to high local radiation doses caused by α-particle emission of astatine isotopes [64].

A strong tendency of trivalent astatine to form complex compounds, preferentially with H_2O, Cl^-, NO_3^-, SO_4^{2-} and dibutyl ether as ligands, has been suggested by Visser [65] on the basis of extraction experiments carried out with a variety of organic solvents. Visser also assumed that a compound of trivalent astatine, $HAtO_2$[AtO(OH)], forms complex bonds with nitrogen and oxygen atoms of proteins. It can be produced by oxidation of At^0 with H_2O_2 at pH=7 and is easily decomposed by reducing agents or in solutions with pH>8. Direct identification of the above form of trivalent astatine has not been reported.

---
1) Originally quoted value is given in parentheses.

10.2.1.2.3 Astatate

Several earlier studies reported the formation of astatate, oxo anion of At^{V} by the oxidation of astatine present in lower oxidation states with hypochlorite in alkaline solutions or with periodate, persulfate, and Ce^{IV} in acidic solutions [24, 26, 28]. It was identified by paper electrophoresis [66] showing a migration rate 1.6 times slower than that of the iodate ion, as well as by its being carried on some hydroxide [24] and iodate [24, 28] precipitates, assumedly in the form of insoluble astatates. The latter interpretation became doubtful when monovalent astatine cation was also shown to coprecipitate with $AgIO_3$ and $Ba(IO_3)_2$, which finding has been explained by the formation of insoluble $AtIO_3$ [42] (see Section 10.2.1.2.1, p. 218). Other discrepancies in the behavior of astatine oxidized by persulfate or Ce^{IV} as compared with that of iodate have also been observed: (a) it is extractable with isopropyl ether from strong acids H_2SO_4, $HClO_4$, and HNO_3 at certain acid concentrations whereas iodate is not, even if present in tracer amounts [56, 67]; (b) it completely fails to follow $^{123}IO_3^-$ when the latter is eluted from Aminex A 27 anion exchange resin [51].

More recently Dreyer et al. [53] have questioned whether astatine can, in fact, be in the At^{V} state if obtained by oxidation with Ce^{IV} or persulfate. Using free solution electrophoresis as a means of identifying the products, they were able to show that oxidation of At^0 with persulfate in $HClO_4$ solutions leads to the formation of At^{III} cation AtO^+ (see Section 10.2.1.2.2, p. 221) characterized by the migration rates of this cation and of the anion AtO_2^- obtained from the former by heating in alkaline solution.

According to the same authors [68], AtO_3^- ion can be obtained by oxidation of At^0 with KOCl in KOH solution, and with Chloramine-B, XeF_2, or KIO_4 in neutral aqueous solutions. Astatate can also be formed by acidifying a solution of AtO_4^- (see Section 10.2.1.2.4). The reaction is complete in 5 to 10 min at 90 °C for $pH < 1$.

The AtO_3^- was identified by paper chromatography, paper electrophoresis [68] and by free solution electrophoresis [53, 64]. The migration rate obtained by the last of these methods is given in Table 10/16.

Table 10/16
Migration Rates of Astatine and Some Iodine Ions Measured by Free Solution Electrophoresis (potential gradient: 30.5 V/cm, t = 25 °C).

ion	$\mu \times 10^4$ in $cm^2 \cdot V^{-1} \cdot s^{-1}$	electrolyte	ion	$\mu \times 10^4$ in $cm^2 \cdot V^{-1} \cdot s^{-1}$	Ref.
At^+	1.9	I			[53]
AtO^+	3.3	II			[53]
At^-	6.0	III	I^-	8.0	[64]
AtO_2^-	4.6	III or IV			[53]
AtO_3^-	2.3	III	IO_3^-	4.0	[64]
AtO_4^-	3.4	III	IO_4^-	4.9	[64]

I = 0.02 M $HClO_4$ + 0.001 M $K_2Cr_2O_7$; II = 0.02 M $HClO_4$ + 0.002 M $Na_2S_2O_8$; III = 0.04 M KNO_3; IV = 0.02 M $(NH_4)_2CO_3$ + 0.015 M NH_4OH.

10.2.1.2.4 Perastatate

After Appelman's unsuccessful attempts to produce AtO_4^- using an iodate-periodate couple for oxidation [28], it was long considered as being nonexistent despite Samson's brief remark on perastatate formation by the same method [69].

Khalkin et al. [70] reported obtaining AtO_4^- by adding solid XeF_2 to the hot alkaline (pH = 10) solution of At^-, in the same way that formation of perbromate had been achieved shortly before [71]. It was identified by (a) paper electrophoresis where it behaved similarly to IO_4^-, remaining at the starting point, and by (b) coprecipitation at pH = 6 with potassium and caesium metaperiodates [70]. The coprecipitation was shown to obey the Berthelot-Nernst homogeneous distribution law – indicating isomorphous replacement of periodate ions in the crystalline lattice by AtO_4^-.

Nagy et al. used anodic oxidation among other methods to generate the perastatate ion in aqueous solutions and identified it by paper chromatography [72]. Lately, oxidation of At^0 by KIO_4 in neutral or alkaline solutions while heating to 90 °C for 15 to 20 min has also been applied [63, 68, 73]. Separation techniques using paper as the support medium show that AtO_4^- like IO_4^- always remains at the starting point [68, 70, 72, 73], apparently due to some interactions with the cellulose. Such techniques are, therefore, not suitable for its separation from the At^0 form. The migration rate of AtO_4^- determined by free solution electrophoresis [53, 64] is faster than that of AtO_3^- (Table 10/16); this feature can be explained by the differences in solvation of these anions. Ion mobilities of perastatate and other astatine anions fit well into the series of these values obtained for other halogens [74] as is demonstrated in **Fig. 10-22**.

AtO_4^- is not stable in acidic solutions, similarly to IO_4^- [75]. Its transfer to AtO_3^- is complete in 5 to 10 min if the solution is heated to 90 °C at pH < 1 [68].

Fig. 10-22

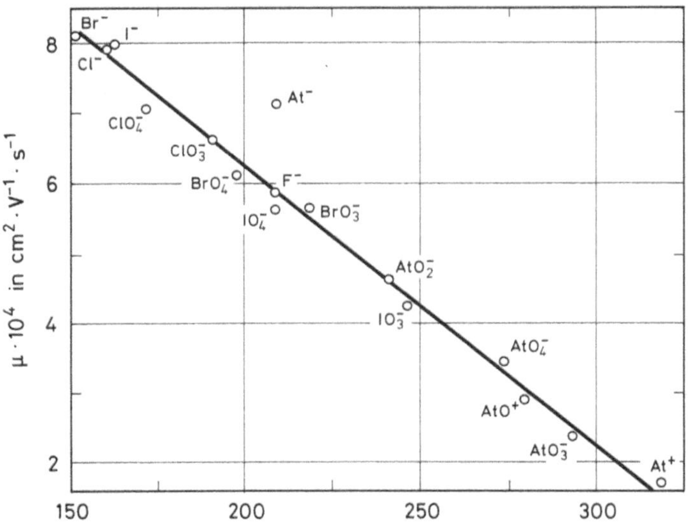

Relationship between migration rates and masses of hydrated ions [74].

References on pp. 231/2

10.2.1.3 Astatine and Halogens

Only dihalide molecules AtX and some trihalide anions of astatine AtX_2^-, $AtXY^-$ have been established so far.

10.2.1.3.1 Dihalides

Astatine forms interhalogen compounds with chlorine, bromine, and iodine. AtF could not be identified so far, maybe due to the extreme reactivity of this compound. However, the ground state vibration frequencies for the fluoride and the other halides of "ekaiodine" were estimated by Clark [76] years before the first isolation of the new element. Later these data were calculated in slightly different ways from the corresponding quantities established for halogen molecules. This made use of different extrapolated values for ω_e (At_2) [77] and also exploited the empirical relationship between electronegativity, reduced mass and ω_e values [78]. The ground state vibration frequencies for astatohalide molecules are given in Table 10/17.

$AtCl^+$, $AtBr^+$ and AtI^+ have been produced and directly identified by Appelman et al. [3]; these authors evaporated astatine together with chlorine, bromine or iodine into the ion source of a time of flight mass spectrometer. The mass spectra of $AtCl^+$ and $AtBr^+$ show doublet peaks according to the stable isotopes of chlorine and bromine as shown in **Fig. 10-23** for the $^{211}At^{35}Cl$-$^{211}At^{37}Cl$ doublet. Similar results have been obtained more recently by introducing halogen vapors together with astatine into a molybdenum discharge chamber serving as the plasma ion source of a mass separator. The relative amounts of interhalogen compounds forming under these conditions could be determined by measuring the α-radioactivity of ^{211}At in different fractions. $^{211}At^{35}Cl^+$ and $^{211}At^{37}Cl^+$ are produced at 0.37 and 0.13% with respect to the total amount of ^{211}At; $^{211}At^{79}Br^+$ and $^{211}At^{81}Br^+$ at 2.3 and 3.0%, respectively, $^{211}AtI^+$ at 1.7% [4].

The formation of AtCl, AtBr and AtI under thermochromatographic conditions has been reported by Merinis et al. [79] as a result of astatine reacting at elevated temperature, respectively, with chlorine, bromine, and iodine, present in the He carrier gas. Using a quartz column with a temperature gradient of 500 → 20 °C, the AtCl, AtBr, and AtI are deposited on the wall at 100, 70, and 35 °C, respectively[1]. Meyer and Rössler [80, 81] applied even simpler conditions to obtain interhalogen compounds of astatine: Cl_2, Br_2, or I_2 were frozen at liquid N_2 temperature in a capillary trap containing At^0, then warmed up to room

Table 10/17
Ground State Vibration Frequencies of Astatohalide Molecules.

molecule	ω_e in cm^{-1}			
	[76]	[77]	[77]	[78]
At_2	139	139.4	172.9	
AtF	509.5	523.0	546.7	553.1
AtCl	330.5	327.3	349.3	348.6
AtBr	228.0	224.6	244.0	
AtI	173.5	175.7	193.3	176.4

[1]) Deposition of At_2 is also reported which is doubtful under the given conditions (see "Introduction" p. 210).

Fig. 10-23

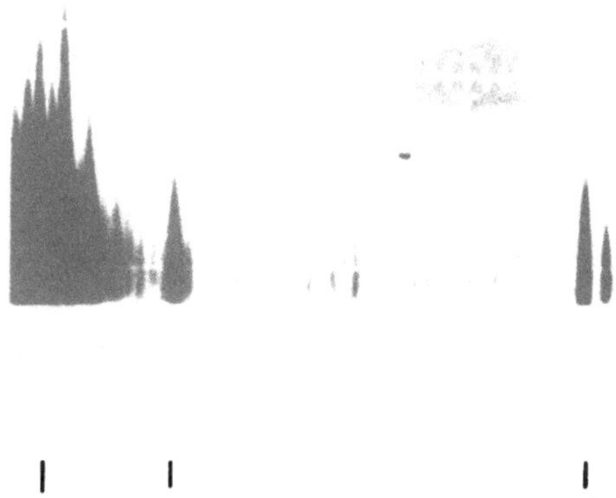

Mass spectrum of At^+ and astatochloride doublet [3].

temperature. About 90% of astatine transforms to AtX (X = Cl, Br, or I) in 5 min. The excess halogen is subsequently distilled off at −78 °C (Cl_2 and Br_2) or at room temperature (I_2) and AtX is removed from the wall of the trap with organic solvents.

AtBr and AtI can also be obtained in aqueous solutions [26, 82]. If the system contains iodine, the zero valency state astatine is present as AtI. In view of this, the best way to avoid ambiguity of the actual chemical form of At^0 in aqueous solutions is to use iodine as a "non-isotopic" carrier.

Coprecipitation of AtI with I_2 from organic media was studied by Aten et al. [27, 83]. Perchloric acid solution of astatine with a small amount of iodine present was first extracted with chloroform which was subsequently added to different amounts of solid iodine. On heating this mixture to about 100 °C, the iodine was dissolved and then precipitated again by cooling to 0 °C. The distribution of AtI between solution and precipitate follows the logarithmic law of Doerner and Hoskins with an initial distribution coeffient $\lambda = 4$. The validity of the Doerner-Hoskins distribution law and the high value of λ indicate a real coprecipitation of AtI with I_2 rather than an adsorption phenomenon. They also prove that astatine is present in the aqueous solution as well as being extracted into chloroform as AtI.

Boiling and melting temperature, heat of vaporization and ionization potential values for astatohalides calculated on the basis of the semiempirical value Z′ for astatine [19] (see Section 10.2.1.1.1, p. 211) are given in Table 10/18, p. 226. t_b for $AtCl_3$ was estimated in the same way to be 360 °C [19]. The boiling point of AtI has lately been reported as having been established by gas chromatography (GLC) to be 486 ± 2 K [84] though a fundamental mistake[1] implied in the interpretation of these results makes the above value dubious.

Certain properties of AtBr and AtI and their reactions were studied by Appelman [26, 82] who exploited solvent extraction with CCl_4 from dilute $HClO_4$ solutions. The distribution coefficients of these compounds in the series of some other interhalogen and halogen molecules are given in Table 10/19, p. 226. The relatively low values of D for astatohalides reflect their highly polar nature.

$$AtI + H_2O \rightleftharpoons AtOH + H^+ + I^- \qquad (6)$$

[1] A GLC peak for At_2 is also reported which is doubtful under the given conditions (see "Introduction", p. 210).

References on pp. 231/2

Table 10/18
Some Physicochemical Properties of Astatohalides [19].

AtX	t_b in °C	t_m in °C	ΔH_{vap} in kJ/mol (kcal/mol) [a)]	IP in eV
AtF	290			
AtCl	170	50	35.23 (8.42)	9.9
AtBr	140	100	39.87 (9.53)	9.5
AtI	200	156	45.94 (10.98)	8.8

a) Originally quoted values are given in parentheses.

Table 10/19
Distribution Constants for Some Interhalogen and Halogen Molecules.

XY	$D=[XY]_{CCl_4}/[XY]_{aq}$	t in °C	Ref.
AtBr	0.04	21	[82]
ICl	0.34	25	[85]
IBr	4.3	21	[82]
AtI	5.5	21	[82]
Br_2	27	25	[86]
I_2	85	25	[86]

An upper limit for the equilibrium constant (K) for the hydrolysis reaction (6) of AtI has been estimated to be 10^{-11} [26, 82], much lower than the corresponding quantity for IBr ($K=1.5\times10^{-7}$) in accordance with the expected higher stability of the heavier interhalogen compound towards hydrolysis.

Reactions of AtBr and AtI with halide ions in aqueous solutions leading to the formation of $AtICl^-$, $AtCl_2^-$, $AtIBr^-$, $AtBr_2^-$, and AtI_2^- have also been investigated by distribution experiments [26, 82].

$$AtI + Cl^- \rightleftharpoons AtICl^- \quad (7)$$

$$AtICl + Cl^- \rightleftharpoons AtCl_2^- + I^- \quad (8)$$

$$AtI + Br^- \rightleftharpoons AtIBr^- \quad (9)$$

$$AtIBr^- + Br^- \rightleftharpoons AtBr_2^- + I^- \quad (10)$$

$$AtI + I^- \rightleftharpoons AtI_2^- \quad (11)$$

The equilibrium coefficients of reactions (7) to (11), i.e., the formation constants for the trihalide anions mentioned above along with those for lighter halogens, are listed in Table 10/20. Astatine is complexed more strongly than iodine in corresponding compounds; this feature fits into the general tendency that the polyhalides become more stable with heavier constituent atoms.

Table 10/20
Formation Constants for Trihalide Anions.

reaction	K_f in L/mol	t in °C	method	Ref.
$Cl_2 + Cl^- \rightleftharpoons Cl_3^-$	0.191	25	spectrophotometry	[88]
$Br_2 + Br^- \rightleftharpoons Br_3^-$	16.53	25	distribution	[89]
$I_2 + I^- \rightleftharpoons I_3^-$	824	20	spectrophotometry	[87]
$Br_2 + Cl^- \rightleftharpoons Br_2Cl^-$	1.47	25	spectrophotometry	[87]
$I_2 + Cl^- \rightleftharpoons I_2Cl^-$	1.66	25	spectrophotometry	[90]
$ICl + Cl^- \rightleftharpoons ICl_2^-$	166	25	distribution	[85]
$IBr + Cl^- \rightleftharpoons IBrCl^-$	45.5	25	distribution	[85]
$AtI + Cl^- \rightleftharpoons AtICl^-$	9	21	distribution	[26, 82]
$AtICl^- + Cl^- \rightleftharpoons AtCl_2^- + I^-$	2×10^{-4}	21	distribution	[26]
$I_2 + Br^- \rightleftharpoons I_2Br^-$	13	22	spectrophotometry	[91]
$IBr + Br^- \rightleftharpoons IBr_2^-$	440	21	distribution	[26, 82]
$AtBr + Br^- \rightleftharpoons AtBr_2^-$	320	21	distribution	[26, 82]
	250	25	ion migration	[92]
$AtI + Br^- \rightleftharpoons AtIBr^-$	120	21	distribution	[26, 82]
$AtIBr^- + Br^- \rightleftharpoons AtBr_2^- + I^-$	1.1×10^{-3}	21	distribution	[26, 82]
$AtI + I^- \rightleftharpoons AtI_2^-$	2000	21	distribution	[26, 82]
	1600	25	ion migration	[92]

10.2.1.3.2 Trihalide Anions

Polyhalide ions are regarded as having more or less polarized covalent bonds with increasing ionic character for heavier halogens [87]. It is not contradictory, therefore, that trihalide anions of astatine, at least those existing in aqueous solutions, are considered to be the halogen compounds of At^+, especially since they form easily in solutions of complex monovalent astatine cation (see Section 10.2.1.2.1, p. 218).

Neumann found that in hydrochloric solutions containing HNO_3 or Cl_2 as the oxidizing agent, astatine is present as a chloro complex which can be extracted with isopropyl ether but not with nonpolar organic solvents such as benzene or CCl_4 [59]. His assumption of positive astatine species forming a complex with chloride ions has been confirmed by extraction and ion exchange studies [49, 52]. While originally both $AtCl_2^-$ and $AtCl_4^-$ were regarded as possible structures, by analogy with ICl_2^- and ICl_4^-, later the $AtCl_2^-$ was established as being much more probable. However, the existence of $AtCl_4^-$ containing trivalent astatine anion has not been completely excluded up to this day.

Khalkin et al. have observed that adsorption of astatine cation on Pt coated with an oxide film [45] and on cation exchange resin [46, 48, 50] from acidic media decreases with increasing HCl concentration (**Fig.** 10-**24**, curve 1). The phenomenon has been explained by complex formation of astatine cation with chloride ions. Moreover, the most probable number of ligands as well as the stability constants (β) for the complex compounds could be determined from the quantitative evaluation of the relationship between the distribution of astatine and HCl concentration ($K_d = f[HCl]$). The formation of AtCl and $AtCl_2^-$ could be concluded from the results; the β_1 and β_2 values given in Table 10/21, p. 230, show $AtCl_2^-$ to be much more stable [48, 50].

$AtCl_2^-$ can be obtained if HCl is added to the $HClO_4$ or HNO_3 solution of astatine containing $H_2Cr_2O_7$ which had previously been kept at about 100 °C for 30 min [48]. More recently,

References on pp. 231/2

Fig. 10-24

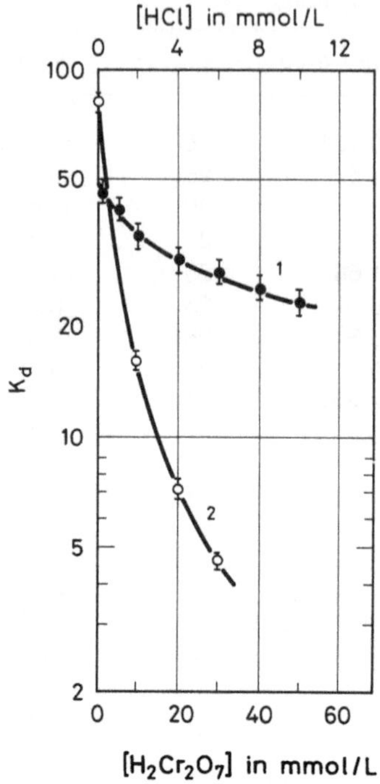

Distribution of astatine cation between Dowex 50 and acidic solutions at 20 °C [48], 1) $HClO_4+HCl$; 2) $HClO_4+H_2Cr_2O_7$.

direct synthesis of a chloro complex in the free solution electrophoresis procedure has also been described if astatine is dissolved in an NaCl-H_2SO_4 electrolyte [63, 92, 93].

$AtBr_2^-$ and AtI_2^- have been reported to form under analogous conditions using the acid or the sodium salt of the corresponding anion (Br^-, I^-) [63, 92, 93].

Lately, synthesis of $AtCl_2^-$, $AtBr_2^-$ and AtI_2^- from At^0 in neutral and acidic media as well as from At^- in acidic medium has also been reported. In the last of these cases, the formation of the complex is assumed to be preceded by oxidation of At^- to At^0 by means of oxygen present in the solution [93].

A rough estimate of the formation constants (K_f) for $AtBr_2^-$ and AtI_2^- could be made from the dependence of ion mobilities on halide ion concentration [92]. These values do not differ significantly from those obtained by Appelman in distribution experiments (see Table 10/15).

Using equimolar mixtures of Cl^- and Br^- ions in the electrolyte, astatine has been found to be present as $AtBr_2^-$ thereby indicating a higher stability for this complex than for $AtCl_2^-$ [63]. From the pH dependence of $AtCl_2^-$ and $AtBr_2^-$, it could be concluded that these species undergo hydrolysis and subsequent reduction at pH>8, probably according to reactions (12) to (14), transforming into At^- ions:

$$AtX_2^- + OH^- \rightleftharpoons AtOHX^- + X^- \quad (12)$$

$$AtOHX^- + OH^- \rightleftharpoons At(OH)_2^- + X^- \quad (13)$$

$$At(OH)_2^- \rightleftharpoons At^- + {}^1/_2O_2 + H_2O \quad (14)$$

where X = Cl or Br.

AtI_2^- is, however, stable even at pH = 12 [92].

Astatine in AtI_2^- has been found to be a much weaker electrophilic agent than iodine in I_3^-. Whereas KI_3 reacts with phenol and imidazole at room temperature giving the corresponding iodine derivatives, $KAtI_2$ fails to take part in astatination even at 80 °C [65].

Brinkman et al. have prepared AtI_2^- in the form of its solid salt $CsAtI_2$; this was done by exploiting its coprecipitation with CsI_3. AtI is extracted from chloroform by an aqueous solution of CsI to which a suitable amount of solid I_2 is then added. If this mixture is heated to 73 °C to dissolve I_2 and is subsequently cooled in ice water, brown needles of CsI_3 containing astatine are produced. According to the general rule, thermal decomposition of a polyhalide should leave the lightest halogen behind to form a monohalide. Indeed, if the $CsI_3(CsAtI_2)$ crystals are heated to 250 °C, essentially all of the AtI follows the I_2 vapor: less than 1% of astatine activity is retained in CsI after the decomposition [94].

10.2.1.3.3 Pseudohalide Anions

Dreyer et al. [95] have recently reported the formation of $At(SCN)_2^-$ in the process of free solution electrophoresis if acidic solution of monovalent astatine cation is added to the electrolyte of the composition: $HClO_4 + NH_4SCN$. The rate of migration for this ion does not differ essentially from that found for the halogen complex anions of monovalent astatine. From ion migration studies, the rough value for the complex ion formation constant, $K_f = 400 \pm 30$ L/mol has been established according to the reaction:

$$AtSCN + SCN^- \rightleftharpoons At(SCN)_2^- \quad (15)$$

The complex is stable in acidic and neutral media; at pH = 8 it transforms completely to At^- ion presumably due to reactions (16) and (17):

$$At(SCN)_2^- + 2OH^- \rightleftharpoons At(OH)_2^- + 2SCN^- \quad (16)$$

$$At(OH)_2^- \rightleftharpoons At^- + {}^1/_2O_2 + H_2O \quad (17)$$

Preliminary results on the possible formation of $At(CN)_2^-$ under similar conditions, using KCN as one of the components of the electrolyte, have also been reported [95].

10.2.1.3.4 Halogen Complex Anions of Trivalent Astatine

$AtOCl_2^-$, $AtOBr_2^-$, and $AtOI_2^-$ have been prepared using similar methods to those utilized to obtain trihalide anions of monovalent astatine. In this case, the corresponding acids HX (X = Cl, Br, I) were added to the solutions of At^{III} (see Section 10.2.1.2.2, p. 221) or were used as electrolytes in free solution electrophoresis of At^{III} solutions. Since the ion mobilities of $AtOX_2^-$ ions do not differ from those of AtX_2^-, their identification could be performed only in an indirect way. The hydrolysis of the two types of halogen complexes in dilute $(NH_4)_2CO_3 + NH_4OH$ solutions leads to two different products, viz. AtO_2^- and At^- [63]. These can easily be distinguished by their different migration rates (Table 10/16, p. 222).

10.2.1.4 Complexes of Astatine Cation with Nitrate, Sulfate, and Dichromate Ions

Khalkin et al. [48, 55, 56] have suggested that the astatine cation $(At\ominus)^+$ (see Section 10.2.1.2.1, p. 218) builds weak complexes with nitrate, sulfate, bisulfate, and dichromate

Fig. 10-25

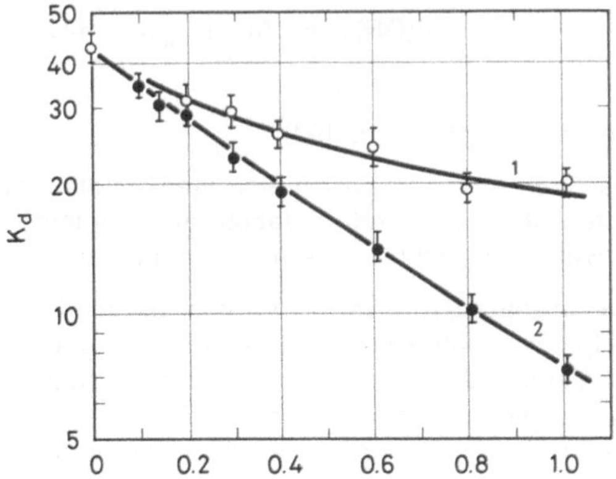

$[HNO_3]$; $[H_2SO_4]$ in mol/L

Distribution of astatine cation between Dowex 50 and acidic solutions at 20 °C [48]: 1) $HClO_4+HNO_3$; 2) $HClO_4+H_2SO_4$.

Table 10/21
Stability Constants for Complexes of Astatine(I) Cation at 20 °C.

ligand	solution	concentration region for ligand ions	stability constant β_1 in L/mol	β_2 in L^2/mol^2	Ref.
Cl^-	0.3 M $HClO_4$+n M HCl	$2\times10^{-3}\leqq n\leqq10^{-2}$	1.3×10^3	2.6×10^5	[48]
Cl^-	0.5 M $HClO_4$+n M HCl	$2\times10^{-3}\leqq n\leqq6\times10^{-3}$	1.6×10^3	2.5×10^5	[48]
Cl^-	0.5 M HNO_3+n M HCl	$2\times10^{-3}\leqq n\leqq6\times10^{-3}$	0.7×10^3	2.5×10^5	[50]
NO_3^-	(1-n) M $HClO_4$+n M HNO_3	$0.1\leqq n\leqq1$	1.35		[48]
SO_4^{2-}	(1-n) M $HClO_4$+n M H_2SO_4	$0.1\leqq n\leqq1$	1.75	3.1	[48]
$Cr_2O_7^{2-}$	1 M $HClO_4$+n M $H_2Cr_2O_7$	$10^{-3}\leqq n\leqq5\times10^{-2}$	22		[48]

ions in acidic solutions. The conclusion was drawn from ion exchange studies which show decreasing adsorption of astatine on Dowex 50 cation exchange resin with increasing ratio of HNO_3, H_2SO_4, and $H_2Cr_2O_7$ in their mixtures with $HClO_4$ at constant ionic strength [48] (see **Fig.** 10-**25**, curves 1 and 2 for HNO_3 and H_2SO_4, respectively, and also Fig. 10-24, p. 228, curve 2 for $H_2Cr_2O_7$). The stability constants (β) for these complexes have been established from the astatine distribution dependence on the concentration of corresponding acids and are given in Table 10/21.

10.2.1.5 Dipyridine Complexes of Astatine Salts

By analogy with the well-known pyridine addition compounds of monovalent iodine [96] and bromine [97], Schats and Aten have synthesized dipyridine complexes of astatine perchlorate and nitrate salts using iodine as a carrier [98].

$[At(C_5H_5N)_2]ClO_4$. Astatine dissolved in chloroform together with a small amount of iodine is added to the chloroform solution of $[Ag(C_5H_5N)_2]ClO_4$ followed by addition of solid iodine in stoichiometric amount. The mixture has to be stirred continuously until the reaction is complete as indicated by the color change from yellow to red. Silver iodide is separated from the liquid by centrifuging and filtering, and $[I(C_5H_5N)_2]ClO_4$ carrying astatine, presumably in the same chemical form, is precipitated by gradual addition of ether.

The At:I ratio in the solid compound was 2 to 3 times higher than in the solution before the formation of the complex. This is considered as proof that astatine forms the complex compound preferentially in agreement with the enhanced ability of astatine to form cationic species. Repeated recrystallizations, by dissolving in chloroform+pyridine and adding ether again, result in a slow decrease of this ratio without, however, falling to the original value.

$[At(C_5H_5N)_2]NO_3$. Since the pyridine complex of silver nitrate cannot be isolated, the chloroform solution of astatine and iodine carrier is added to $AgNO_3$ dissolved in a chloroform/pyridine mixture. From here, the synthesis is performed as in the case of perchlorate but the final chloroform solution must be added slowly to the ether accompanied by vigorous stirring. In this case, too, the At:I ratio has been found to be higher in the solid compound than in the solution before the formation of the complex.

The complex nitrate is less stable than perchlorate. The solid salt decomposes rapidly with concurrent formation of iodine [98].

10.2.1.6 Complexes of Astatine Cation with Thiourea and Its Derivatives

Preliminary results on possible formation of stable complexes of astatine cation with thiourea and with its trimethyl, tetramethyl, and pyrrolidine derivatives in aqueous solutions containing ClO_4^- or halide ions have been reported recently by Dreyer et al. The complex cations of assumed $[At\ominus X_2]^+$ structure (where X=thiourea or its derivative) have been identified by their migration rates using free solution electrophoresis [99].

References:

[1] D.R. Corson, K.R. MacKenzie, E. Segrè (Phys. Rev. [2] **58** [1940] 672/8). – [2] F.A. Paneth (Nature **149** [1942] 565/8, 567). – [3] E.H. Appelman, E.N. Sloth, M.N. Studier (Inorg. Chem. **5** [1966] 766/9). – [4] N.A. Golovkov, I.I. Gromova, M. Janicki, Yu.V. Norseyev [Norseev], V.G. Sandukovsky, L. Vasáros (Radiochem. Radioanal. Letters **44** [1980] 67/78). – [5] J.R. Grover, E. Lebowitz, E. Baker (J. Inorg. Nucl. Chem. **31** [1969] 3705/20).

[6] J.R. Grover, F.M. Kiely, E. Lebowitz, E. Baker (Rev. Sci. Instr. **42** [1971] 293/302). – [7] J.R. Grover, H.V. Lilenfeld (Rev. Sci. Instr. **43** [1972] 690/2). – [8] J.R. Grover, H.V. Lilenfeld (Nucl. Instr. Methods **105** [1972] 189/96). – [9] J.R. Grover, C.R. Iden (J. Chem. Phys. **61** [1974] 2157/9). – [10] J.R. Grover, C.R. Iden, H.V. Lilenfeld (J. Chem. Phys. **64** [1976] 4657/71).

[11] J.R. Grover, C.R. Iden, H.V. Lilenfeld, F.M. Kiely, E. Lebowitz (Rev. Sci. Instr. **47** [1976] 1098/108). – [12] J.R. Grover, D.E. Malloy, J.B.A. Mitchell (J. Chem. Phys. **76** [1982] 362/77). – [13] A.M.G. Ding, L.J. Kirsch, D.S. Perry, J.C. Polanyi, J.L. Schreiber (Faraday Discussions Chem. Soc. No. 55 [1973] 252/76). – [14] C.A. Parr, J.C. Polanyi, W.H. Wong (J. Chem. Phys. **58** [1973] 5/20). – [15] J.R. Grover, D.E. Malloy, J.B.A. Mitchell (Abstr. 7th Intern. Symp. Mol. Beams, Riva del Garda, Italy, 1979, pp. 90/2).

[16] S.P. Tandon, K. Tandon (Indian J. Phys. **38** [1964] 460/74; C.A. **62** [1965] 5909). – [17] E.P. Ozhigov (Zh. Obshch. Khim. **34** [1964] 3519/23; J. Gen. Chem. [USSR] **34** [1964] 3566/71). – [18] E.R. Lippincott, M.O. Dayhoff (Spectrochim. Acta **16** [1960] 807/34). – [19] Yu.V. Norseev, V.D. Nefedov (Issled. Khim. Tekhnol. Primen. Radioakt. Veshchestv **1977**

3/8; C.A. **94** [1981] No. 3539). – [20] K. Bächmann, P. Hoffmann (Radiochim. Acta **15** [1971] 153/63).

[21] S.P. Tandon, M.P. Bhutra, P.C. Mehta (Z. Physik. Chem. **245** [1970] 230/5). – [22] P.M. Christopher, J.M. Fitzgerald Jr. (Australian J. Chem. **18** [1965] 1705/9; C.A. **64** [1966] 11877). – [23] E. Segrè, K.R. MacKenzie, D.R. Corson (Phys. Rev. [2] **57** [1940] 1087, No. 166). – [24] G.L. Johnson, R.F. Leininger, E. Segrè (J. Chem. Phys. **17** [1949] 1/10). – [25] A.H.W. Aten Jr., T. Doorgeest, U. Hollstein, H.P. Moeken (Analyst [London] **77** [1952] 774/7).

[26] E.H. Appelman (UCRL-9025 [1960] 1/113; C.A. **1960** 16998). – [27] A.H.W. Aten Jr. (in: O.R. Frisch, F.A. Paneth, F. Laves, P. Rosbaud, Beiträge zur Physik und Chemie des 20. Jahrhunderts, Vieweg, Braunschweig 1959, pp. 121/3). – [28] E.H. Appelman (J. Am. Chem. Soc. **83** [1961] 805/7). – [29] M.F.C. Ladd, W.H. Lee (J. Inorg. Nucl. Chem. **20** [1961] 163/5). – [30] G.A. Krestov, N.V. Krestova (Radiokhimiya **10** [1968] 344/50; Soviet Radiochem. **10** [1968] 327/31).

[31] K.S. Krasnov (Zh. Strukt. Khim. **1** [1960] 209/16; J. Struct. Chem. [USSR] **1** [1960] 191/7). – [32] G.A. Krestov (Radiokhimiya **4** [1962] 690/6; Soviet Radiochem. **4** [1962] 612/7). – [33] A.F. Kapustinskii, K.B. Yatsimirskii (Zh. Obshch. Khim. **26** [1956] 941/8; J. Gen. Chem. [USSR] **26** [1956] 1069/76). – [34] O. Kubaschewski, E.L. Evans (Metallurgical Thermochemistry, 3rd Ed., Pergamon, Oxford 1958, pp. 1/426). – [35] L.I. Ivanova (Radiokhimiya **26** [1984] 206/10; Soviet Radiochem. **26** [1984] 191/5).

[36] L.I. Ivanova (Izv. Vysshikh Uchebn. Zavedenii Khim. Khim. Tekhnol. **19** [1976] 539/42; C.A. **85** [1976] No. 83861). – [37] G.A. Krestov, V.A. Lapin (Radiokhimiya **7** [1965] 311/5; Soviet Radiochem. **7** [1965] 309/13). – [38] G.A. Krestov, N.V. Krestova (Radiokhimiya **18** [1976] 387/92; Soviet Radiochem. **18** [1976] 341/6). – [39] D.S. Kaganyuk (Radiokhimiya **25** [1983] 814/6; Soviet Radiochem. **25** [1983] 778/80). – [40] K.S. Krasnov, A.I. Maksimov (Opt. Spektrosk. **8** [1960] 403/6; Opt. Spectrosc. [USSR] **8** [1960] 208/9; C.A. **56** [1962] 15060).

[41] V.I. Baikov (Opt. Spektrosk. **27** [1969] 923/9; Opt. Spectrosc. [USSR] **27** [1969] 502/5; C.A. **72** [1970] No. 72794). – [42] Fu-Chun Wang, N.G. Krilov, Yu.V. Norseev, Tao-Nan Ch'ao, V.A. Khalkin (in: V.M. Vdovenko, Soosazhdenie i Adsorptisya Radioaktivnikh Elementov, Nauka, Moscow 1965, pp. 80/8; AEC-TR-6716 [1967] 1/229; N.S.A. **21** [1967] No. 19591; C.A. **63** [1965] 7695). – [43] Yu.V. Norseev, V.A. Khalkin, Tao-Nan Ch'ao (Izv. Sibirsk. Otd. Akad. Nauk SSSR **1965** No. 3, pp. 21/7; C.A. **64** [1966] 16676). – [44] Yu.V. Norseev (Diss. Dubna-Leningrad 1965). – [45] Yu.V. Norseev, Tao-Nan Ch'ao, V.A. Khalkin (Radiokhimiya **8** [1966] 497/504; Soviet Radiochem. **8** [1966] 461/6).

[46] Yu.V. Norseev, V.A. Khalkin (Chem. Zvesti **21** [1967] 602/10; C.A. **68** [1968] No. 8848). – [47] Fu-Chun Wang, Yu.V. Norseev, V.A. Khalkin, Tao-Nan Ch'ao (Radiokhimiya **5** [1963] 351/5; Soviet Radiochem. **5** [1963] 318/22). – [48] Do-Kim Tuong, I.V. Dudova, V.A. Khalkin (Radiokhimiya **15** [1973] 548/53; Soviet Radiochem. **15** [1973] 552/6). – [49] Fu-Chun Wang, Meng-Hua Kang, V.A. Khalkin (Radiokhimiya **4** [1962] 94/8; Soviet Radiochem. **4** [1962] 81/5). – [50] Yu.V. Norseyev [Norseev], V.A. Khalkin (J. Inorg. Nucl. Chem. **30** [1968] 3239/43).

[51] K. Rössler, W. Tornau, G. Stöcklin (J. Radioanal. Chem. **21** [1974] 199/209). – [52] B.N. Belyaev, Yung-Yu Wang, E.N. Sinotova, L. Németh, V.A. Khalkin (Radiokhimiya **2** [1960] 603/13; Soviet Radiochem. [Jerusalem] **2** No. 5/6 [1960] 92/102). – [53] I. Dreyer, R. Dreyer, V.A. Chalkin [Khalkin] (Radiochem. Radioanal. Letters **36** [1978] 389/98). – [54] Do-Kim Tuong, I.V. Dudova, V.A. Khalkin (Radiokhimiya **14** [1972] 766/7; Soviet Radiochem. **14** [1972] 790/2).

[55] W.A. Chalkin [V.A. Khalkin], E. Herrmann (Isotopenpraxis **11** [1975] 333/40). – [56] W.A. Chalkin [V.A. Khalkin], E. Herrmann, J.W. Norseev, I. Dreyer (Chemiker-Ztg. **101** [1977] 470/81; C.A. **88** [1978] No. 28663). – [57] R.P. Bell, E. Gelles (J. Chem. Soc. **1951** 2734/40). – [58] M. Milanov, V. Doberenz, V.A. Khalkin, A. Marinov (J. Radioanal. Nucl. Chem. Articles **83** [1984] 291/9). – [59] H.M. Neumann (J. Inorg. Nucl. Chem. **4** [1957] 349/53). – [60] L. Vasáros, Yu.V. Norseyev [Norseev], D.D. Nhan, V.A. Khalkin (Radiochem. Radioanal. Letters **54** [1982] 239/47).

[61] B. Eichler, V.P. Domanov (JINR-P12-7928 [1974]; C.A. **83** [1975] No. 67358). – [62] R. Dreyer, F. Rösch, G.-J. Beyer (Wiss. Fortschr. **32** [1982] 251/5; C.A. **97** [1982] No. 151718). – [63] I. Dreyer, R. Dreyer, V.A. Chalkin [Khalkin], M. Milanov (Radiochem. Radioanal. Letters **40** [1979] 145/54). – [64] I. Dreyer, R. Dreyer, V.A. Chalkin [Khalkin] (Radiochem. Radioanal. Letters **35** [1978] 257/62). – [65] G.W.M. Visser, E.L. Diemer (Radiochim. Acta **33** [1983] 145/51).

[66] G.A. Nagy, V.A. Halkin [Khalkin], J.V. Norszejev [Yu. V. Norseev] (Magy. Kem. Folyoirat **73** [1967] 191/4; C.A. **67** [1967] No. 50005). – [67] Yung-Yu Wang, V.A. Khalkin (Radiokhimiya **3** [1961] 662/6; Soviet Radiochem. **3** No. 5/6 [1961] 154/8). – [68] I. Dreyer, R. Dreyer, Yu.V. Norseev, V.A. Chalkin [Khalkin] (Radiochem. Radioanal. Letters **33** [1978] 291/300). – [69] G. Samson (Diss. Univ. Amsterdam 1971, p. 10). – [70] V.A. Khalkin, Yu.V. Norseev, V.D. Nefedov, M.A. Toropova, V.I. Kuzin (Dokl. Akad. Nauk SSSR **195** [1970] 623/5; Dokl. Chem. Proc. Acad. Sci. USSR **193/195** [1970] 855/7).

[71] E.H. Appelman (J. Am. Chem. Soc. **90** [1968] 1900/1). – [72] G.A. Nagy, P. Gróz, V.A. Halkin [Khalkin], Do-Kim Tuong, J.V. Norszejev [Yu.V. Norseev] (KFKI [Kozp. Fiz. Kut. Int.] Közl. **18** [1970] 173/6; C.A. **75** [1971] No. 14412). – [73] E. Kroó, I. Dudova, V.A. Halkin [Khalkin] (Magy. Kem. Folyoirat **82** [1976] 1/10; C.A. **84** [1976] No. 127352). – [74] R. Dreyer, I. Dreyer, A. Kolaczkowski, S. Riedel (Isotopenpraxis **16** [1980] 117/9). – [75] I. Dreyer, R. Dreyer, Yu.V. Norseev, V.A. Chalkin [Khalkin] (Radiochem. Radioanal. Letters **33** [1978] 281/9).

[76] C.H.D. Clark (Trans. Faraday Soc. **33** [1937] 1398/401). – [77] K. Majumdar, Y.P. Varshni (Indian J. Phys. **28** [1954] 209/15). – [78] P.L. Goodfriend (Can. J. Phys. **45** [1967] 3425/7). – [79] J. Merinis, Y. Legoux, G. Bouissières (Radiochem. Radioanal. Letters **11** [1972] 59/64). – [80] G.-J. Meyer (JUEL-1418 [1977] 1/107, 18; C.A. **88** [1978] No. 71113).

[81] G.-J. Meyer, K. Rössler (Radiochem. Radioanal. Letters **25** [1976] 377/89). – [82] E.H. Appelman (J. Phys. Chem. **65** [1961] 325/31). – [83] A.H.W. Aten Jr., J.G. van Raaphorst, G. Nooteboom, G. Blasse (J. Inorg. Nucl. Chem. **15** [1960] 198/9). – [84] K. Otozai, N. Takahashi (Radiochim. Acta **31** [1982] 201/3). – [85] J.H. Faull Jr. (J. Am. Chem. Soc. **56** [1934] 522/6).

[86] A. Seidell (Solubilities of Inorganic and Metal Organic Compounds, 4th Ed., Vol. 1, Van Nostrand, New York 1958, pp. 1/1487, 446, 1268). – [87] A.I. Popov (in: V. Gutmann, Halogen Chemistry, Vol. 1, Academic, London 1967, pp. 225/64). – [88] G. Zimmerman, F.C. Strong (J. Am. Chem. Soc. **79** [1957] 2063/6). – [89] D.B. Scaife, H.J.V. Tyrrell (J. Chem. Soc. **1958** 386/92). – [90] D.L. Cason, H.M. Neumann (J. Am. Chem. Soc. **83** [1961] 1822/8).

[91] D. Meyerstein, A. Treinin (Trans. Faraday Soc. **59** [1963] 1114/20). – [92] R. Dreyer, I. Dreyer, F. Rösch, G.-J. Beyer (Radiochem. Radioanal. Letters **54** [1982] 165/75). – [93] R. Dreyer, I. Dreyer, F. Rösch (Z. Chem. [Leipzig] **22** [1982] 54/6; C.A. **96** [1982] No. 150056). – [94] G.A. Brinkman, J.Th. Veenboer, A.H.W. Aten Jr. (Radiochim. Acta **2** [1963] 48). – [95] R. Dreyer, I. Dreyer, M. Pfeiffer, F. Rösch (Radiochem. Radioanal. Letters **55** [1982] 207/13).

[96] Gmelin Handbuch "Jod" 1933, pp. 459/60. – [97] H. Carlson (Chem. Ber. **68** [1935] 2209/11). – [98] J.J.C. Schats, A.H.W. Aten Jr. (J. Inorg. Nucl. Chem. **15** [1960] 197). – [99] R. Dreyer, I. Dreyer, F. Rösch, S. Fischer (Z. Chem. [Leipzig] **23** [1983] 346/7; C.A. **100** [1984] No. 13379).

10.2.2 Organic Compounds of Astatine

The interactions of At^0 with organic impurities in aqueous solutions were recognized at an early stage but systematic studies on the organic chemistry of astatine did not start until the end of the sixties. The last decade, however, saw swift progress in this field. The rapidly growing interest in synthesizing organic astatine compounds and in studying their properties was mainly related to the potential application of the α-emitting ^{211}At isotope in tumor therapy [1 to 3].

While a mixture of the three longest-lived astatine isotopes: ^{209}At, ^{210}At, and ^{211}At is often used for the synthesis of simple organic compounds, pure ^{211}At is required for biological studies. This is not only due to its most favorable nuclear characteristics but also because ^{210}At decays into the radiotoxic ^{210}Po ($T_{1/2}$=138 d) thereby creating a health hazard.

10.2.2.1 Aliphatic and Nonaromatic Alicyclic Compounds

10.2.2.1.1 Alkyl Astatides

CH_3At^+ was first detected by its mass line (see **Fig.** 10-**26**) in a time of flight mass spectrometer [4], apparently forming as a result of astatine reacting with the organic impurities within the ion source.

CH_3At is one of the products of the recoil astatine atoms, originating from the $^{211}Rn(\varepsilon)^{211}At$ nuclear transformation, reacting with liquid and solid hydrocarbons: n-C_5H_{12}, n-C_6H_{14},

Fig. 10-26

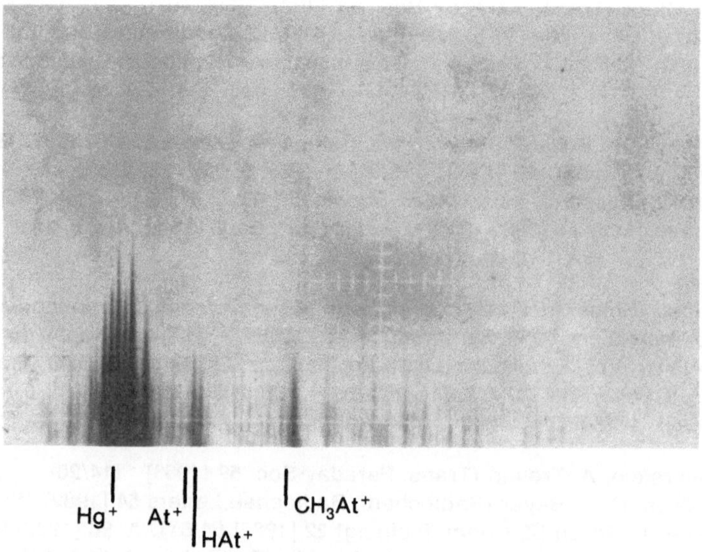

Mass spectrum of CH_3At^+ and other At species [4].

c-C_5H_{10}, c-C_6H_{12}, and C_6H_6, as found by Kuzin et al. [5]. The higher homologs, C_2H_5At, isomers of C_3H_7At, C_4H_9At, and $C_6H_{13}At$, as well as c-C_5H_9At and c-$C_6H_{11}At$, were also obtained as a result of recoil ^{211}At reactions with the saturated hydrocarbons mentioned above. Their separation and identification were done by gas chromatography (GLC). The relative amounts of the individual products were determined by measuring the radioactivity of the subsequent gas chromatographic fractions. These relative amounts vary for the different parent substances depending on the experimental conditions but are, as a rule, the highest for the astatinated derivatives of the original compounds [5].

Normal alkyl astatides n-$C_nH_{2n+1}At$ (n=2 to 6) were prepared by Samson and Aten [6] via homogeneous halogen exchange, by addition of I_2 containing AtI to the corresponding n-alkyl iodides at room temperature, according to the reaction:

$$RI + At^- \rightarrow RAt + I^- \tag{18}$$

The presence of ethyl alcohol seemed to accelerate the exchange reaction. Gas chromatographic halogen exchange was also used to obtain the same compounds. In this case, At^- was adsorbed on the solid phase whereas the alkyl iodide flowed through the GLC column in the carrier gas stream. A short precolumn packed with kieselguhr was utilized

Fig. 10-27

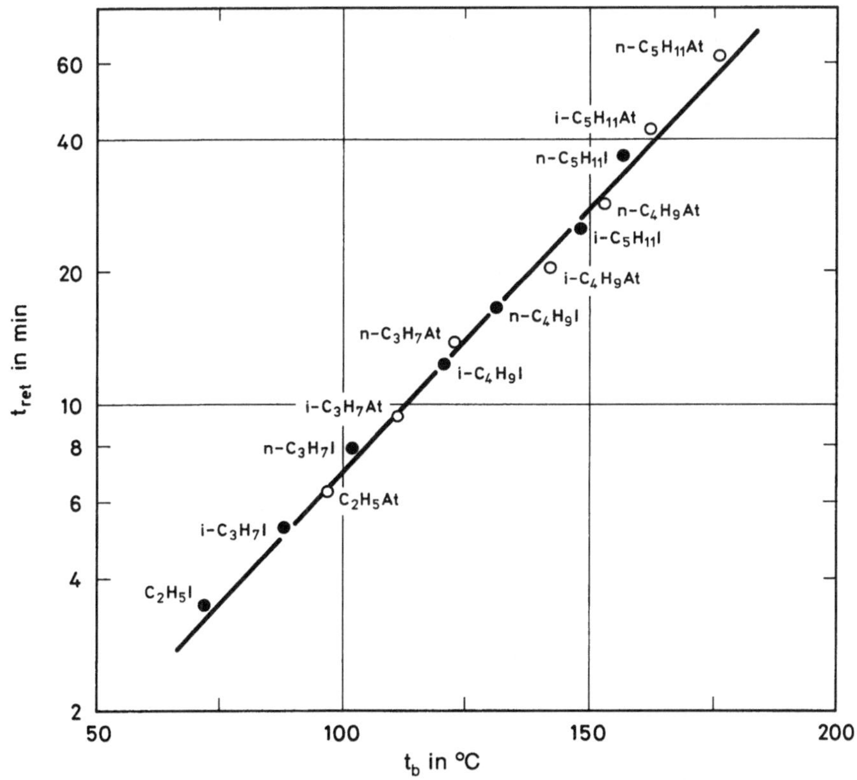

Dependence of GLC logarithmic retention time t_{ret} for alkyl iodides (●) and alkyl astatides (○), measured using dinonyl phthalate stationary phase, on boiling temperature of the compounds [8, 10].

References on pp. 257/9

and was connected to a longer analytical one to separate the products. The halogen exchange took place at 130 to 200 °C [6, 7].

Norseyev et al. have prepared n-alkyl astatides (n=2 to 5) and i-alkyl astatides i-$C_nH_{2n+1}At$ (n=3 to 5) in a simplified procedure of GLC halogen exchange using only one analytical column with At^- adsorbed at its inlet [8].

The boiling temperature for alkyl astatides obtained by halogen exchange has been determined by extrapolation from the t_b values of the corresponding lighter alkyl halides utilizing sequential GLC analysis [6 to 9]. The values are given in Table 10/22. A linear t_b dependence on the logarithmic GLC retention time has been observed for these astatine compounds similarly to the alkyl iodides [8], as shown in **Fig.** 10-**27**.

Studies dealing with the gas chromatographic behavior of alkyl astatides in relation to that of other alkyl halides [9] also provided a means of calculating an experimental parameter Z′ for astatine (see Section 10.2.1.1, p. 211). Using this parameter for extrapolation, a rough estimate of t_b (Tables 10/22 and 10/23), ΔH_{vap} and IP (Table 10/24), and of $D_{C\text{–}At}$ values (Table 10/25) for a number of simple organic astatine compounds was made [12]. Other extrapolation techniques based on the relationships between different physico-

Table 10/22
Boiling Temperature of Alkyl Astatides Established by GLC and by Extrapolation Methods.

compound	t_b in °C	method	Ref.
CH_3At	66±3	extrapolated	[11]
	65.8	extrapolated	[12]
	73±5	extrapolated	[13]
	72±2	extrapolated	[14]
	77±5	extrapolated	[15]
C_2H_5At	98±2	GLC	[6, 7]
	95.4	extrapolated	[12]
	103±5	extrapolated	[13]
i-C_3H_7At	112±2	GLC	[8]
	110.4	extrapolated	[12]
n-C_3H_7At	123±2	GLC	[6, 7]
	124.3	extrapolated	[12]
i-C_4H_9At	142±3	GLC	[8]
	140.4	extrapolated	[12]
n-C_4H_9At	152±3	GLC	[6, 7]
	152.5	extrapolated	[12]
i-$C_5H_{11}At$	163±3	GLC	[8]
	163	extrapolated	[12]
n-$C_5H_{11}At$	176±3	GLC	[6, 7]
	182	extrapolated	[12]
n-$C_6H_{13}At$	201±2	GLC	[6, 7]

Table 10/23
Extrapolated t_b Values for Some Aliphatic Astatine Compounds [12].

compound	t_b in °C	compound	t_b in °C	compound	t_b in °C	compound	t_b in °C
CH_2AtF	90	$CAtCl_3$	166	CH_2F-CHAtF	98	CHF_2-CHAtCl	133
CH_2AtCl	137	$CAtBrCl_2$	193	CH_2F-CHAtCl	143	CHF_2-CHAtBr	155
CH_2AtBr	168	$CAtCl_2I$	230	CH_2F-CHAtBr	165	CHF_2-CHAtI	185
CH_2AtI	208	$CAtBr_2Cl$	221	CH_2F-CHAtI	197	$CHCl_2$-CHAtCl	225
		$CAtClI_2$	295	CH_2Cl-CHAtF	143	$CHCl_2$-CHAtBr	250
CHAtClF	104	CAtBrClI	258	CH_2Cl-CHAtCl	190	$CHCl_2$-CHAtI	280
CHAtBrF	133			CH_2Cl-CHAtBr	214	$CHBr_2$-CHAtBr	300
CHAtFI	170	$CAtBr_3$	250	CH_2Cl-CHAtI	245	$CHBr_2$-CHAtI	330
$CHAtCl_2$	156	$CAtBr_2I$	286	CH_2Br-CHAtF	165	CHI_2-CHAtI	390
CHAtClBr	185	$CAtBrI_2$	325	CH_2Br-CHAtCl	214		
CHAtClI	224			CH_2Br-CHAtBr	237	1-At-2-CH_3-C_3H_6	140.6
$CHAtBr_2$	215	CH_2F-CH_2At	117	CH_2Br-CHAtI	268	2-At-2-CH_3-C_3H_6	123.5
CHAtBrI	254	CH_2Cl-CH_2At	158	CH_2I-CHAtF	197	1-At-2-CH_3-C_4H_8	166.7
$CHAtI_2$	292	CH_2Br-CH_2At	182	CH_2I-CHAtCl	245	CH_2OH-CH_2At	195.5
		CH_2I-CH_2At	212	CH_2I-CHAtBr	268	CH_2=CHAt	83
$CAtClF_2$	56			CH_2I-CHAtI	300	CH_2=CH-CH_2At	124.8
$CAtBrF_2$	83	CHF_2-CH_2At	100				
$CAtF_2I$	120	CHFCl-CH_2At	143				
$CAtCl_2F$	113	CHFBr-CH_2At	165				
$CAtBr_2F$	168	CHFI-CH_2At	197				
$CAtFI_2$	241	$CHCl_2$-CH_2At	190				
CAtBrClF	140	CHClBr-CH_2At	214				
CAtBrFI	205	CHClI-CH_2At	245				
		$CHBr_2$-CH_2At	237				
		CHBrI-CH_2At	268				
		CHI_2-CH_2At	300				

Table 10/24
Heat of Vaporization and Ionization Potential of Methyl and Ethyl Astatide.

compound	H_{vap} in kJ/mol (kcal/mol) [a] [12]	IP in eV [12]	IP in eV [15]
CH_3At	28.9 (6.9)	8.85	8.8
C_2H_5At	31.59 (7.55)	8.8	8.65

[a] Originally quoted values are given in parentheses.

chemical constants were used to establish the boiling temperature, ionization potential, and the carbon-astatine dissociation energy for methyl and ethyl astatide (Tables 10/22, 10/24, and 10/25, p. 238, respectively) [13 to 15], as well as to calculate the D_{C-At} values for a number of alkyl astatides [16]. More recently, Vasáros et al. experimentally established

References on pp. 257/9

Table 10/25
Dissociation Energy for Some Aliphatic Astatine Compounds.

compound	D_{C-At} in kJ/mol (kcal/mol) [a]		method	Ref.
CH_3At	139.3	(33.3)	extrapolated	[12]
	205	(49)	extrapolated	[15]
	176	(42)	extrapolated	[16]
C_2H_5At	167	(40)	extrapolated	[16]
n-C_3H_7At	163	(39)	extrapolated	[16]
	161.5 ± 10.5	(38.6 ± 2.5)	thermal decomposition	[17]
i-C_3H_7At	159	(38)	extrapolated	[16]
	151.9 ± 9.6	(36.3 ± 2.3)	thermal decomposition	[17]
n-C_4H_9At	163	(39)	extrapolated	[16]
CH_2AtCl	130.1	(31.1)	extrapolated	[12]
CH_2AtBr	124.7	(29.8)	extrapolated	[12]
CH_2AtI	118.0	(28.2)	extrapolated	[12]

[a] Originally quoted values are given in parentheses.

the dissociation energy of the carbon-astatine bond in n- and i-propyl astatide by measuring the kinetics of pyrolytic decomposition of these compounds [17]. The latter D_{C-At} values, also given in Table 10/25, are well in line with those obtained by extrapolation.

10.2.2.1.2 Astatoacetic Acid

$AtCH_2COOH$ was prepared by taking advantage of the fact that the halogen atoms of the haloacetic acids are readily replaced by heavier halogens in aqueous solutions [7, 18]. At^- in the presence of an iodide carrier was allowed to react in an aqueous solution of iodoacetic acid at 40 °C, according to the following equation:

$$At^- + ICH_2COOH \rightarrow AtCH_2COOH + I^- \tag{19}$$

The product was extracted with ethyl ether, after which the solvent was evaporated and the dry residue recrystallized from CCl_4.

$AtCH_2COOH$ was identified by ion exchange chromatography. The dissociation constant of this compound was established, by measuring its distribution between diisopropyl ether and aqueous solutions of varying acidity, to be $K_a = 1.5$ to 1.8×10^{-4} mol/L ($pK_a = 3.7$ to 3.8) at temperatures between 0 and 27 °C [7, 18].

10.2.2.2 Aromatic Compounds

10.2.2.2.1 Benzene Derivatives

10.2.2.2.1.1 Astatobenzene

Several synthesis routes successfully followed in order to obtain C_6H_5At are listed in Table 10/26. Halogen exchange reactions, At for I or Br, both in homogeneous and heteroge-

Table 10/26
Astatobenzene Syntheses.

method	substrate	assumed state of At	experimental conditions	radiochemical yield in %	Ref.
homogeneous electrophilic substitution	C_6H_6	At^+	CH_3COOH-$HClO_4$($H_2Cr_2O_7^{2-}$) 120 °C, 90 min	50	[19]
heterogeneous electrophilic substitution	C_6H_6	At^+	benzene-$HClO_4$ benzene-H_2SO_4 180 to 190 °C, 20 min	90	[20]
"electrophilic" halogen exchange (ipso-attack)	C_6H_5X X=Cl	AtCl	60 °C, 60 min	29 a)	[21,22]
	X=Br			45 a)	
	X=Cl	AtBr		19 a)	
	X=Br			30 a)	
homogeneous halogen exchange	C_6H_5I	At^- (I_2)	room temp., ionizing field	—	[7, 24]
	C_6H_5X X=Cl	At^-	prim-$C_4H_9NH_2$, 210 °C, 30 to 60 min	73	[25,26]
	X=Br			85	
	X=I			99	
heterogeneous halogen exchange	C_6H_5I	At^- (KI)	GLC, 130 to 200 °C, on-line	—	[7, 24]
	C_6H_5Br	At^- (NaI)	155 °C, 2 h	60	[27]
	C_6H_5Br	At^- (NaOH)	sealed ampule, 250 °C, 30 min	70	[16,28]
through diazonium intermediate	$C_6H_5NH_2$	At^- (KI)	normal conditions for diazotation and decomposition	80	[29,30]
	$C_6H_5NH_2$			—	[7, 24]
	$C_6H_5NHNH_2$			—	[7, 24]
through (C_6H_5)AtI (halogen exchange and thermal decomposition)	$(C_6H_5)_2I_2$	At^- (KI)	hot C_2H_5OH, crystallization, dec. at 175 °C in sealed tube	—	[7, 24]
recoil astatination	$(C_6H_5)_3Bi$		via $^{209}Bi(\alpha, 2n)^{211}At$	—	[7, 24]
	C_6H_6	$[At]^*$	via $^{211}Rn(\varepsilon)^{211}At$	44	[5, 31]
	C_6H_6	$[At]^*$	via $^{211}Rn(\varepsilon)^{211}At$	23	[32]
	C_6H_5X X=Cl	$[At]^*$	via $^{211}Rn(\varepsilon)^{211}At$	35	[32]
	X=Br	$[At]^*$	via $^{211}Rn(\varepsilon)^{211}At$	40	[32]
	X=I	$[At]^*$	via $^{211}Rn(\varepsilon)^{211}At$	45	[32]
	X=Cl		with amines	60	[33]

a) Values brought in question by more recent investigations [23] (see text).

neous systems as well as electrophilic substitution, especially if performed in heterogeneous systems using strong acids in the aqueous phase, are the most promising techniques to produce astatobenzene in high yields. It is identified by GLC.

References on pp. 257/9

The electrophilic substitution reaction of astatine cation to produce astatobenzene in a homogeneous mixture of benzene and acetic acid, containing $H_2Cr_2O_7$ as an oxidizing agent, was first used by Vasáros et al. The mixture was kept in sealed ampules at 100 to 120 °C for 30 to 60 min enabling astatobenzene to be obtained with ~50% radiochemical yield[1] [19]. More efficient electrophilic astatination — with radiochemical yields of about 90% — has been achieved by the same authors in heterogeneous systems if astatine is present in the aqueous phase containing H_2SO_4 or $HClO_4$, even in the absence of the $H_2Cr_2O_7$ oxidizing agent [20]. This finding is interpreted on the basis of the complex structure of monovalent astatine cation in aqueous solutions (see Section 10.2.1.2.1, p. 218) by the fact that in the presence of strong acids the equilibrium in the following reaction is shifted to the right:

$$HOAt + H_3O^+ \rightleftharpoons [H_2OAt]^+ + H_2O \qquad (20)$$

It is assumed that the heterolytic fission of hypoastatous acid $[H_2OAt]^+$, resulting in the formation of the most reactive At^+, is the rate determining step of astatination, immediately preceeding the substitution in the aromatic ring. From kinetic studies, the activation energy for the electrophilic astatination of benzene is 134 ± 8 kJ/mol (32 ± 2 kcal/mol)[2] [20].

The unexpectedly observed halogen replacement in monohalobenzenes by AtCl and AtBr yielding astatobenzene was explained by Meyer et al. by the "ipso-attack" of the highly polarized interhalogen molecule at the electronegative site of the substrate molecule. This results in replacement of the halogen atom competing with that of the hydrogen in ortho position [21, 22]. However, in a more recent study Cavallero has found that the yields for the halogen as well as for the hydrogen replacement were not reproducible, probably due to the ill-defined chemical state of the astatine under the given conditions [23].

Kolaczkovski and Khalkin obtained C_6H_5At via a heterogeneous halogen exchange reaction between At^- adsorbed on solid NaI and C_6H_5Br at the boiling temperature of the latter [27]. This technique was further developed by Norseyev et al. At^-, adsorbed on NaOH evaporated to dryness, reacted with C_6H_5Br in sealed ampules permitting the temperature to be increased to 250 °C thus resulting in higher yields of about 70% [16, 28].

Reactions of At^- in homogeneous mixtures of halobenzenes C_6H_5X (X = Cl, Br, I) with butylamine, di-, and triethylamine at 210 °C lead to the formation of C_6H_5At with 75 to 99% radiochemical yields as shown by Nhan et al. [25, 26]. A two-step process according to reactions (21) and (22) is assumed, the latter being the rate determining step:

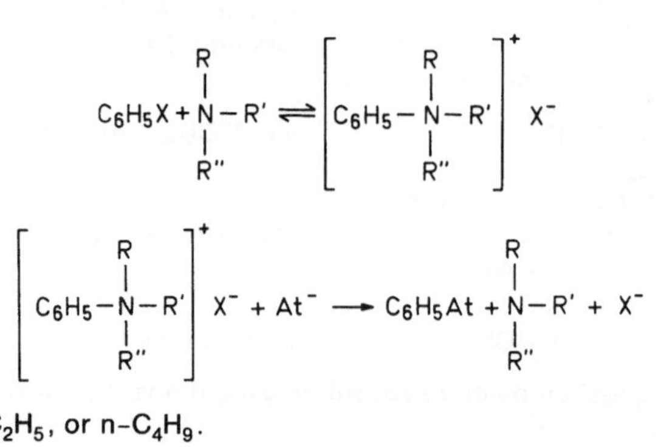

where R, R′, R″ = H, C_2H_5, or n-C_4H_9.

[1] The radiochemical yields given hereafter represent the ratio of a certain product's activity to the total astatine activity in percent, corrected for the radioactive decay.
[2] Originally quoted value is given in parentheses.

This assumption could be confirmed by the kinetic investigations showing that the activation energy (E_{act}) of the halogen exchange remains the same in the case of different leaving groups (Cl, Br, or I). However, it is significantly influenced by the nature of the amine, probably due to steric effects, as follows:

	E_{act} in kJ/mol (kcal/mol) [a]	
$At^- + C_6H_5Br + (C_2H_5)_3N$	111.7	(26.7)
$At^- + C_6H_5Br + (C_2H_5)_2NH$	21.8	(5.2)
$At^- + C_6H_5Br + C_4H_9NH_2$	17.2	(4.1)
$At^- + C_6H_5Cl + C_4H_9NH_2$	17.6	(4.2)
$At^- + C_6H_5I + C_4H_9NH_2$	17.2	(4.1)

[a] Originally quoted values are given in parentheses.

Accordingly, the highest yields could be obtained using butylamine [25].

Hot homolytic replacement reactions by recoil ^{211}At "in situ" is also used to produce astatobenzene [31 to 33]. Recoil astatine can be obtained, for example, via electron capture from ^{211}Rn, one of the spallation products of Th or U bombarded with 660 MeV protons. After separation from the other spallation products and subsequent purification [34], ^{211}Rn is introduced into thoroughly evacuated glass ampules filled with benzene or with the corresponding benzene derivative. The ampules are sealed and the ^{211}Rn ($T_{1/2}$=14.6 h) is allowed to decay for 14 to 20 h until the radioactive equilibrium with ^{211}At ($T_{1/2}$=7.2 h) is reached. The products are identified by GLC. The highest yield is achieved with chlorobenzene diluted with compounds of lower IP than that of astatine, thus promoting the neutralization of the originally multicharged recoil ^{211}At [33].

Most of the properties of astatobenzene have been determined by extrapolation from the other halobenzenes based on a comparison of their gas chromatographic behavior. An extensive study has been carried out by Vasáros et al. with a variety of stationary phases

Fig. 10-28

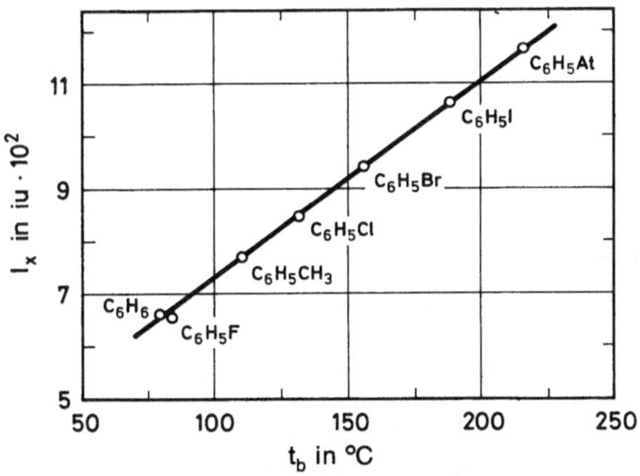

Retention index values (I_x), measured using squalane stationary phase, versus boiling temperature of benzene, toluene and halobenzenes [10] (iu=index units).

References on pp. 257/9

to establish the retention indices (I_x) for substituted benzene derivatives including those of astatine [35]. This method allows a more reliable estimation of physicochemical constants than the earlier extrapolations based on simple comparison of retention volumes under specific conditions. **Fig.** 10-**28** demonstrates the means of establishing the boiling temperature of astatobenzene based on the linear dependence of I_x values for monosubstituted benzene derivatives on their normal t_b values [10].

The dissociation energy of the C-At bond in astatobenzene has also been determined experimentally by measuring the kinetics of its thermal decomposition using a modified version of the toluene carrier gas technique [17, 36].

The properties of C_6H_5At established so far are listed in Table 10/22.

Table 10/27
Properties of Astatobenzene.

quantity	value	method	Ref.
t_b in °C	222 ± 3	GLC	[37]
	212 ± 2	GLC	[7, 24]
	211	extrapolated	[12]
	217	GLC	[16]
	219 ± 3	GLC	[21]
	216 ± 2	GLC (I_x)	[38, 39]
H_{vap} in kJ/mol	41.63 (9.95 kcal/mol) a)	extrapolated	[12]
	43.38 (10369 cal/mol) a)	GLC (I_x)	[38]
IP in eV	8.8	extrapolated	[12]
R_{C-At} in cm^3/mol	20.1 ± 0.1	GLC (I_x)	[39, 40]
μ_{C-At} in Debye	1.60	GLC	[21]
	1.66 ± 0.4	GLC (I_x)	[40]
	1.53	GLC (I_x)	[39]
D_{C-At} in kJ/mol	205 (49 kcal/mol) a)	extrapolated	[16]
	187.9 ± 21.3 (44.9 ± 5.1 kcal/mol) a)	thermal decomposition	[17]
	180.7 ± 9.1	thermal decomposition	[36]

a) Originally quoted values are given in parentheses.

10.2.2.2.1.2 Astatotoluenes

Decomposition of diazonium salts is utilized to produce the isomers of astatotoluene from the corresponding toluidines, with relatively low radiochemical yields (10 to 20%) [21, 41, 42]. The toluidines are dissolved in HCl or H_2SO_4 and converted into the corresponding diazonium salts by adding an aqueous solution of $NaNO_2$ at −5 °C. After destroying the excess of $NaNO_2$ with urea, At^- in Na_2SO_3 solution is added, the mixture is slowly heated to 50 to 80 °C and cooled again. The products are extracted with diethyl ether which is subsequently washed with NaOH solution, dried over $CaCl_2$ and analyzed by GLC [21, 41] or by thin layer chromatography (TLC) [42].

The low radiochemical yields are explained by the competing hydrolysis of the diazonium salts leading to phenol formation, due to the concentration of hydroxyl ions being much higher than that of At^- since the latter is present only in tracer amounts.

A reaction mechanism is proposed to interpret some peculiarities of the decomposition reaction and the isomer distribution of the products [21, 41]. This involves complex formation between the At^- and the diazonium ion followed by electron transfer leading to the release of nitrogen while the phenyl radical recombines with astatine, according to the following reaction:

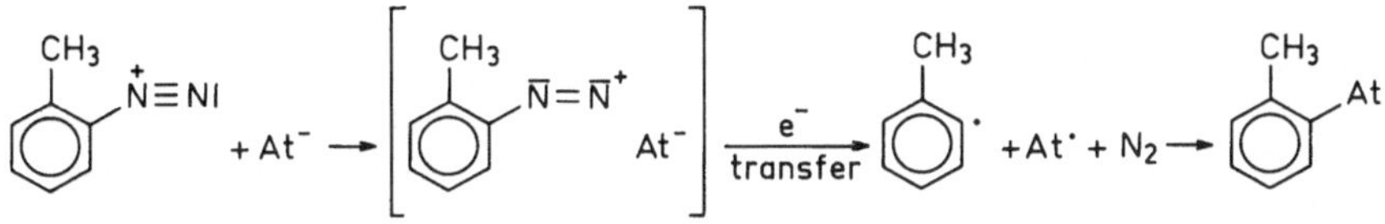

(23)

The physicochemical properties of astatotoluene isomers estimated by extrapolation from those of the corresponding lighter halotoluenes based on their GLC behavior [21, 38 to 40], calculated directly from the GLC retention volumes [38], and determined experimentally by measuring the kinetics of their thermal decomposition [36], are given in Table 10/28.

Table 10/28
Properties of Astatotoluenes.

compound	t_b in °C [21]	t_b in °C [38]	H_{vap} in kJ/mol (cal/mol) [a)] [38]		$R_{C\text{-}At}$ in cm^3/mol [40]	$\mu_{C\text{-}At}$ in Debye [39]	$D_{C\text{-}At}$ in kJ/mol [36]
ortho-$AtC_6H_4CH_3$	237±4	237±2	46.3	(11074)	20.3	—	180.7±9.8
meta-$AtC_6H_4CH_3$	240±4	237±2	46.6	(11134)	19.9	—	181.2±8.6
para-$AtC_6H_4CH_3$	237±4	236±2	46.7	(11172)	19.9	1.51	181.6±9.5

a) Originally quoted values are given in parentheses.

The isomers of $AtC_6H_4CF_3$ were recently prepared via homogeneous halogen exchange, At for Cl from the corresponding isomers of $ClC_6H_4CF_3$ in the presence of butylamine [43]. The radiochemical yields of 30, 45, and 36% for ortho-, meta-, and para-astatotrifluorotoluene, respectively, were significantly lower than the yields observed for astatobenzene formation under the same conditions (see Section 10.2.2.2.1.1, p. 238). The products were identified by GLC. The method was also used to estimate the ΔH_{vap} and T_b values for these compounds [43] while measuring the kinetics of their thermal decomposition was utilized for determining the dissociation energy of the C-At bond [36]. These values are given below:

	T_b in K [43]	ΔH_{vap} in kJ/mol (cal/mol) [a)] [43]		$D_{C\text{-}At}$ in kJ/mol [36]
ortho-$AtC_6H_4CF_3$	485±2	44.5	(10640)	176.6±8.8
meta-$AtC_6H_4CF_3$	478±1	43.4	(10370)	177.0±8.6
para-$AtC_6H_4CF_3$	481±1	42.6	(10170)	176.0±9.0

a) Originally quoted values are given in parentheses.

References on pp. 257/9

10.2.2.2.1.3 Astatohalobenzenes

The isomers of AtC_6H_4X (X=F, Cl, Br, I) are obtained in essentially the same ways as astatobenzene (see Section 10.2.2.2.1.1, p. 238) and belong to the most extensively studied group of organic astatine compounds [10].

Whereas the yields of electrophilic substitution by astatine cation in homogeneous systems of halobenzenes are negligible [19], astatination of fluorobenzene in its heterogeneous mixtures with strong inorganic acids occurs with 90% radiochemical yield. The procedure takes place at 190 °C in 30 min. The ortho:meta:para (25:5:70) distribution of the AtC_6H_4F isomers reflects the electrophilic character of the reacting astatine. This is less true for the distribution of the ortho:meta:para (30:20:50) AtC_6H_4Cl isomers produced with a total yield of 72% from C_6H_5Cl under the same conditions. The yield sharply decreases for the isomers of AtC_6H_4Br and AtC_6H_4I (8.1 and 2.5%, respectively). This finding has been explained by the decreasing capability of halobenzenes to undergo electrophilic substitution with the increasing atomic number of the halogen [20].

Hydrogen substitution in C_6H_5F, C_6H_5Cl, and C_6H_5Br by AtCl and AtBr is less efficient than the competing halogen replacement in these systems (see Section 10.2.2.2.1.1, p. 238) leading to total radiochemical yields of only a few percent [21, 22] with poor reproducibility [23].

The most convenient way to obtain a given isomer of an astatohalobenzene is to utilize the heterogeneous At for Br exchange reaction, starting from the corresponding bromohalobenzene isomer, under the same conditions as described in Section 10.2.2.2.1.1, p. 238, for producing C_6H_5At. Radiochemical yields of 60 to 70% can be achieved for meta- and para-isomers but only 40 to 50% for ortho-astatohalobenzenes [16, 28]. Homogeneous nucleophilic substitution of the bromine atom is also used for the preparation of astatohalobenzenes. At 210 °C, the reaction is complete in 1 h for mixtures of the meta- and para-BrC_6H_4X (where X=CH_3, F, Cl, NO_2) with 15 mol% butylamine [44]. No isomerization has been observed in the course of the halogen exchange reactions.

Direct synthesis of astatohalobenzenes can be carried out in recoil experiments (see Sections 10.2.2.1.1, p. 234, and 10.2.2.2.1.1, p. 238) in two ways: (a) via hydrogen replacement by recoil astatine in monohalobenzenes where a nearly statistical mixture of ortho-, meta-, and para-astatohalobenzene is produced with a total yield of 5 to 15% [32, 33]; (b) via halogen replacement by recoil astatine in dihalobenzenes as has been shown for AtC_6H_4F obtained from the corresponding ClC_6H_4F isomers, without noticeable isomerization of the products, with a yield of 14% [45].

Decomposition of diazonium salts has also been used to produce astatohalobenzenes under similar conditions as described for astatotoluenes (see Section 10.2.2.2.1.2, p. 242) starting from the corresponding haloaniline isomers. Here again relatively low radiochemical yields (10 to 26%) are obtained due to the competing reaction of OH^- ions, present in the aqueous solution in much higher concentration than At^-, leading to the by-product formation of phenols [21, 41, 46].

All astatohalobenzenes were identified by GLC and some of their physicochemical properties were estimated on the basis of their gas chromatographic behavior related to that of the corresponding lighter dihalobenzene isomers [10, 21, 38, 40]. The values of the C-At bond energies were determined experimentally by measuring the kinetics of the thermal decomposition using a modified version of the toluene carrier gas technique [36]. The established t_b, ΔH_{vap}, R_{C-At}, and D_{C-At} values for astatohalobenzenes are listed in Table 10/29.

Table 10/29
Properties of Astatohalobenzenes.

compound		t_b in °C	H_{vap} in kJ/mol (cal/mol) [a)]	R_{C-At} in cm^3/mol	D_{C-At} in kJ/mol
		[38]	[38]	[40]	[36]
AtC_6H_4F	ortho	213±2	44.6 (10652)	—	179.5±8.8
	meta	206±2	43.4 (10380)	20.3	179.9±9.0
	para	209±2	42.5 (10167)	—	179.9±8.2
		[38]	[10]	[40]	[36]
AtC_6H_4Cl	ortho	258±2	50.8	—	173.6±8.6
	meta	255±3	49.0	20.2	175.3±8.7
	para	253±2	47.5	—	179.5±9.1
		[21]			[36]
AtC_6H_4Br	ortho	303±3			176.1±7.9
	meta	304±3			177.0±8.6
	para	305±3			178.2±9.3
		[21]			
AtC_6H_4I	ortho	336±4			
	meta	337±4			
	para	337±4			

a) Originally quoted values are given in parentheses.

10.2.2.2.1.4 Astatophenols and Derivatives

A mixture of ortho- and para-AtC_6H_4OH is obtained via electrophilic astatination of the phenol by AtCl or AtBr with 20 to 30% radiochemical yields — leading predominantly to the para-isomer [21, 22].

The same isomers have been prepared in much higher yields (95%) by Visser et al. via astatination of the corresponding chloromercury derivatives [47, 48]. The latter compounds have been obtained by the well-known mercuration of phenol with $Hg(OAc)_2$ followed by the reaction with NaCl [49]. At^-, in NaOH solution containing sulfite, is added to the chloromercury derivative, followed by iodine carrier in $CHCl_3$ and by KI_3. After stirring the mixture for 30 min at room temperature, the HgI_2 precipitate is filtered or dissolved by excess KI.

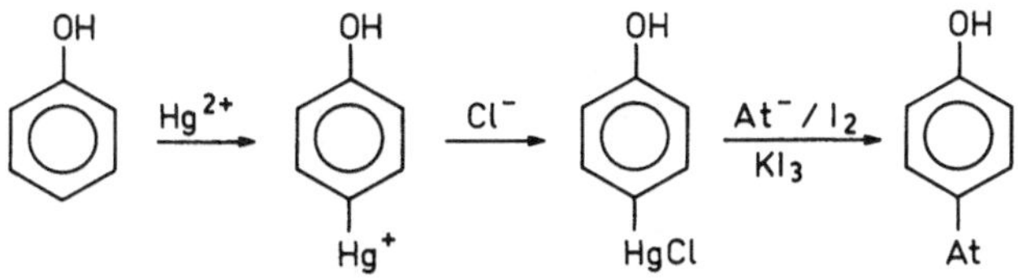

(24)

The astatinated products formed according to reaction (24) are extracted from the reaction mixture with CH_2Cl_2 and identified by TLC. Small amounts (1 to 5%) of astatoiodophenols have also been observed, presumably forming by astatination and subsequent iodination of dimercurated phenol.

References on pp. 257/9

The higher reactivity of astatine compared with that of iodine in the reaction with the chloromercury derivatives is reflected in the higher yields of the astatinated products. Furthermore, astatination, though with lower yields, is also possible without iodine carrier whereas ^{131}I in tracer amounts fails to react with some aromatic substances (e.g., tyrosine, aniline, nitrobenzene). If astatine is reacting in the absence of an iodine carrier, a strong indication of the radical mechanism for the astatination of chloromercury derivative has been found which is explained by the easy oxidation of At^- into At^0 at lower pH values [47].

meta-Astatophenol has been prepared via decomposition of the diazonium salt of meta-astatoaniline, according to the reaction:

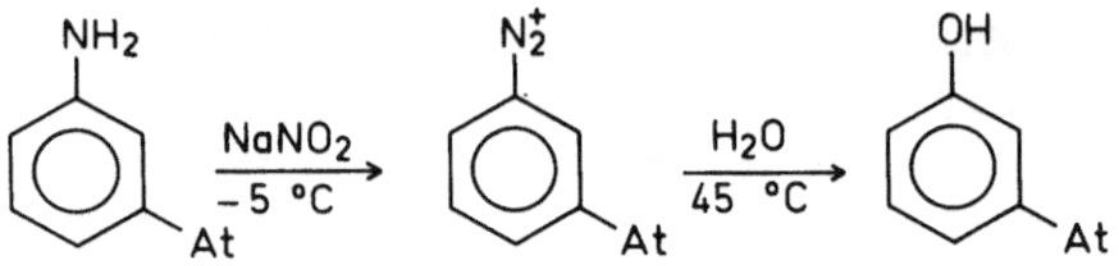

(25)

This synthesis only gives a 30% radiochemical yield, due to the side reactions between the diazonium salt and astatophenol which cannot be eliminated even by continuous extraction of the product with heptane [50].

High pressure liquid chromatography (HPLC) [21, 22] and TLC [47, 48, 50] have been used to identify astatophenols. Their dissociation constants (K_a) have been established from extraction experiments by measuring the distribution between aqueous borax buffer solutions and n-heptane as a function of acidity. The pK_a values are given in Table 10/30. Based on these values, the Hammett σ-constants and hence the field and resonance effects for these compounds have also been estimated [50], as follows:

	σ_{meta}	σ_{para}	F	R
astatophenols	+0.26	+0.18	+0.28	−0.08

para-$AtC_6H_4OCH_3$ has been prepared, using the reactions of At^- with the chloromercury [47] and with the Tl^{III}-di-trifluoroacetate derivative [51] of the anisole, with 70 to 90% radiochemical yields. The product was identified by TLC.

Table 10/30
pK_a Values for Astatophenols, Astatoanilines, and Astatobenzoic Acids at 0 °C [50].

compound		pK_a
AtC_6H_4OH	ortho	8.92±0.03
	meta	9.33±0.03
	para	9.53±0.03
$AtC_6H_4NH_2$	ortho	3.03±0.03
	meta	3.90±0.03
	para	4.04±0.02
AtC_5H_4COOH	ortho	2.71±0.02
	meta	3.77±0.02
	para	4.03±0.02

10.2.2.2.1.5 Astatoanilines and Derivatives

Besides the inefficient methods of recoil astatination of aniline [21, 32] or electrophilic substitution by AtCl and AtBr [21, 22], ortho- and para-$AtC_6H_4NH_2$ can be produced by the reaction of At^- with the corresponding arsanilic acids, according to reaction (26) with radiochemical yields of 10 to 15% as reported by Visser et al. [50].

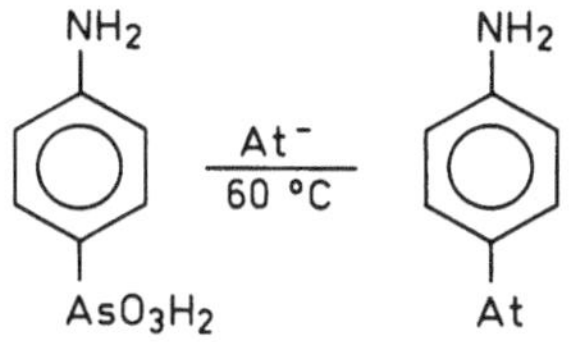
(26)

Much higher efficiency could be achieved by the same authors in synthesizing these compounds from the chloromercury derivatives [47, 50], in a similar way as described in Section 10.2.2.2.1.4, p. 245, for obtaining astatophenols. Mercuration in this case is performed under more drastic conditions (3 to 4 h stirring with $Hg(NO_3)_2$ or $Hg(ClO_4)_2$ in strong acid solutions at 60 °C) because of the lower reactivity of aniline compared with phenol. The astatoanilines are obtained with 80% radiochemical yield and extracted from the reaction mixture with n-heptane. The formation of astatoiodoanilines in small amounts (1 to 5%) has also been observed, apparently from the dimercurated intermediates (see Section 10.2.2.2.1.4, p. 245) [47].

meta-Astatoaniline has been produced by reducing meta-astatonitrobenzene (see Section 10.2.2.2.1.6) with $SnCl_2$ at 60 °C, with 90% radiochemical yield [50].

HPLC [21, 22, 32] and TLC [47, 48, 50] may be used to identify astatoanilines. The pK_a values established by the extraction experiments, using n-heptane and citrate buffer solutions, are given in Table 10/30; the Hammett σ-values, the field and resonance effects have been estimated [50] as follows:

	σ_{meta}	σ_{para}	F	R
astatoanilines	+0.24	+0.18	+0.25	−0.05

para-Astato-N,N-dimethylaniline has been prepared by astatination of its chloromercury derivative with 65% yield and identified by TLC [47].

10.2.2.2.1.6 Astatonitrobenzenes

Heterogeneous At for Br exchange was employed to obtain all three isomers of $AtC_6H_4NO_2$ from the corresponding bromo compounds with 70% radiochemical yields. The conditions of the halogen exchange were similar to those described for synthesizing astatobenzene (see Section 10.2.2.2.1.1, p. 238), except that in this case the temperature was kept at 50 to 60 °C to avoid thermal decomposition of the initial as well as of the product compounds. GLC, though leading to partial decomposition, and HPLC were used for identification [52].

meta-Astatonitrobenzene was also obtained from the chloromercury derivative applying $Hg(ClO_4)_2$ in strong acidic solution for the mercuration of nitrobenzene [53]. The astatonitrobenzene was obtained with 95% radiochemical yield, separated by extraction with CH_2Cl_2 and identified by TLC [47, 50]. This astatonitrobenzene isomer was used for tissue distribution studies in rats [54].

References on pp. 257/9

Extrapolation, making use of the GLC retention indices, was the means of establishing the boiling temperature of the $AtC_6H_4NO_2$ isomers [52]:

		t_b in °C
astatonitrobenzenes	ortho	303
	meta	297
	para	303

10.2.2.2.1.7 Astatobenzoic Acids and Derivatives

One of the first organic derivatives of astatine was para-AtC_6H_4COOH prepared through its diazonium intermediate, similarly to the procedure described for the astatotoluene isomers (see Section 10.2.2.2.1.2, p. 242), by Hughes et al. in the fifties [29, 30]. Later, the same procedure was used by a number of research groups to produce all three isomers of the astatobenzoic acid with yields up to 90%, utilizing TLC and column chromatography for their identification [42, 50, 55 to 58].

ortho-AtC_6H_4COOH has also been obtained by reaction of At^- with the Tl^{III}-di-trifluoroacetate derivative of the benzoic acid resulting in 70 to 90% radiochemical yields [51]. Shiue et al. employed the heterogeneous At for Br exchange (see Section 10.2.2.2.1.1, p. 238) at 200 to 250 °C to prepare meta-AtC_6H_4COOH, its methyl ether and the meta-astatohippuric acid(I) from their bromo analogs with radiochemical yields of about 60%. The time of the synthesis did not exceed 1.5 h. meta-AtC_6H_4COOH and meta-$AtC_6H_4COOCH_3$ were identified by GLC, meta-astatohippuric acid by TLC [59].

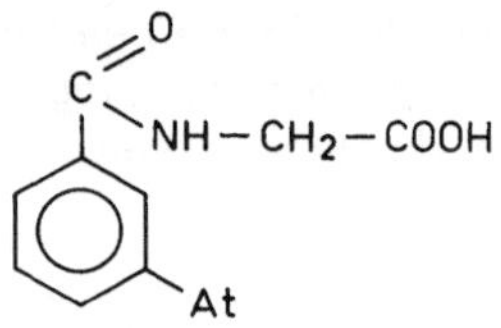

I

The pK_a values of astatobenzoic acids estimated from the extraction experiments using n-heptane and citrate buffer solutions (see Table 10/30, p. 246) were also used for identification purposes [50].

para-AtC_6H_4COOH often serves as an intermediate for labeling proteins with ^{211}At [55 to 58, 60 to 62], by means of an acylation reaction (see Section 10.2.2.6, p. 254).

10.2.2.2.2 Astatinated Aromatic Amino Acids

Building astatine into the molecules of the aromatic amino acids: phenylalanine, 4-methoxyphenylalanine, tyrosine and 3-iodotyrosine, can be carried out most effectively via astatination of their chloromercury derivatives, as has been shown by Visser et al. [47, 63].

To achieve mercuration of the phenylalanine and of the 4-methoxy-phenylalanine, the suspension of these substrates with $HgSO_4$ and H_2SO_4 has to be stirred for 5 h at 60 °C, because of the deactivation of hydrogen atoms in the benzene ring. 4-astato-phenylalanine (I) and 3-astato-4-methoxy-phenylalanine(II), obtained by subsequent reaction of the mercurated compounds with At^- in about 30 min (see Section 10.2.2.2.1.4, p. 245) with radiochemical yields of 85 and 70%, respectively, are identified by paper electrophoresis [47].

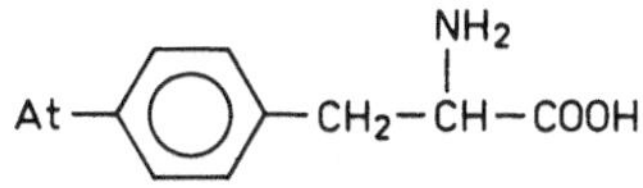

I

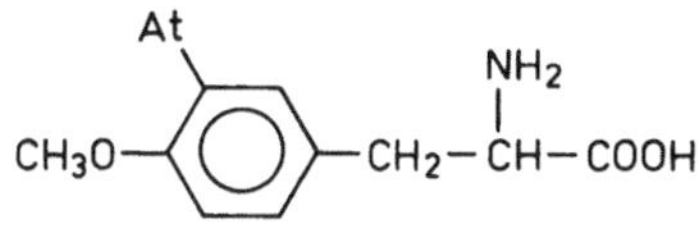

II

The tissue distribution of I studied in rats after intravenous administration indicated the loss of the ^{211}At label in vivo [54].

The mercuration of tyrosine and iodotyrosine proceeds smoothly at room temperature and after subsequent astatination (see Section 10.2.2.2.1.4, p. 245), leads to 3-astato-tyrosine (III) and 3-astato-5-iodotyrosine(IV):

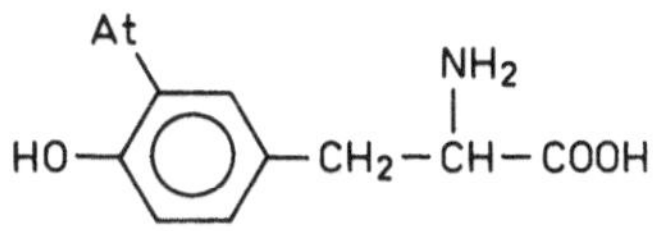

III

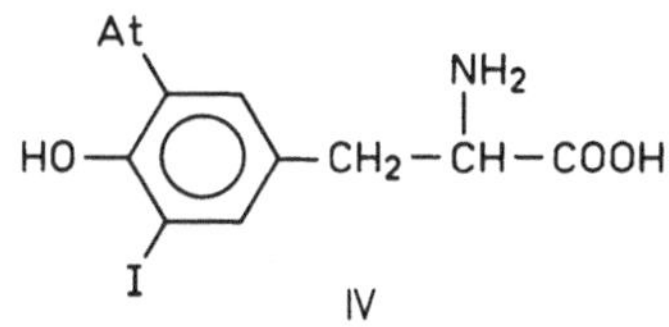

IV

Radiochemical yields of 60 to 80% were obtained [47]. Paper electrophoresis [47] and paper chromatography [63] were used for their identification.

The exact positions of the astatine atoms in the aminoacids mentioned above (compounds I to IV) have not been verified but are assumed on the basis of the electrophilic substitution pattern for mercuration [47].

Vaughan et al. have reported the formation of astatotyrosine by electrophilic astatination of L-tyrosine using H_2O_2 as the oxidizing agent. Only low yields (<15%) could be achieved and the product appeared to be very unstable [64]. This phenomenon has been explained by later investigations on the stability of astatinated amino acids in different media, carried out by Visser et al. [63]. These authors showed that compounds III and IV, while stable in acidic solutions, decomposed rapidly at pH$\geqq$8, especially in the presence of oxidizing agents [63, 65, 66].

The method used by Hughes et al. to produce astatoiodotyrosine was simply to allow astatine to react with tyrosine in the presence of N-iodosuccinimide [67, 68].

10.2.2.2.3 Salts of 6-Astato-2-methyl-1,4-naphthoquinol bis (dihydrogenphosphate)

Salts of 6-(astato-^{211}At)-2-methyl-1,4-naphthoquinol bis (dihydrogenphosphate) (6-^{211}At-MNDP, I) is a potential anti-tumor drug.

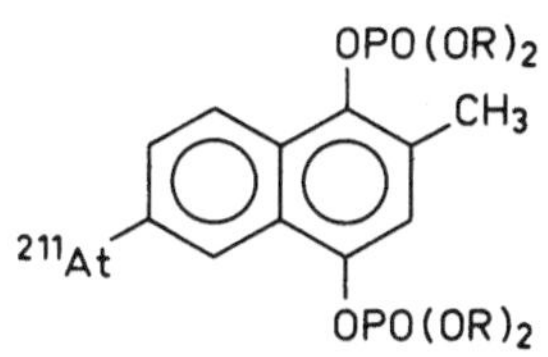

I

where R = Na, Li, or NH_4.

It has been prepared by Brown in two different ways: (a) by astatination of 6-chloromercury-2-methyl-1,4-naphthoquinone followed by reduction, phosphorilation and hydrolysis;

this five-step procedure takes about 7 h thereby considerably decreasing the original 60 to 70% yield of astatination [69]; (b) by heterogeneous halogen exchange in vacuum between At^- and 6-I-MNDP which takes only 10 minutes and gives a radiochemical yield of 40 to 60% [70]. In both cases the astatinated product is obtained together with its iodo analog as a carrier. This is an advantage from the viewpoint of biological studies. TLC was employed as a means of separating and identifying the product I [69, 70]. It is used in preliminary experiments on tumor therapy in mice [71] and for establishing its intra-cellular localization by α-particle track radiography [72, 73].

10.2.2.3 Astatosteroids

A mixture of 2-astatoestradiol (I), 4-astatoestradiol (II), and 2-astato-4-iodoestradiol (III) with radiochemical yields of 55, 19, and 18%, respectively, was obtained by Visser et al. [74].

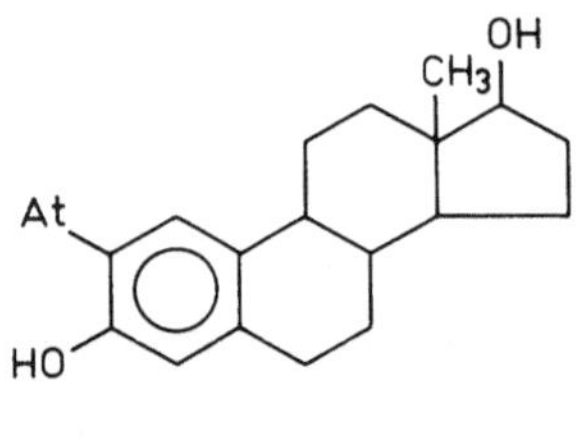

I

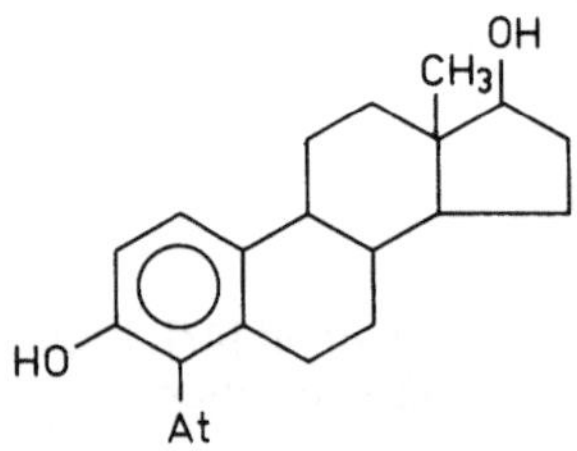

II

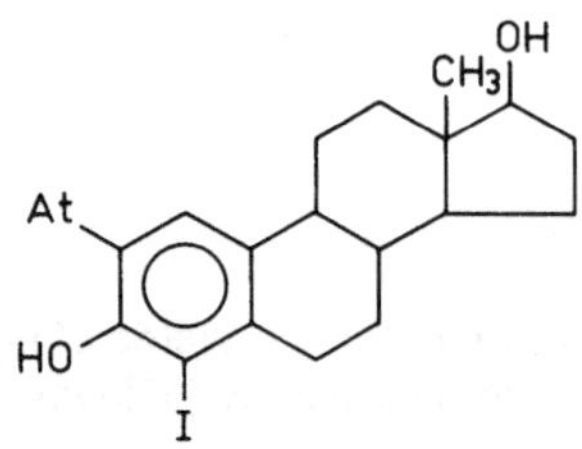

III

They mercurated estradiol by reaction with $Hg(OAc)_2$ in an ethyl alcohol-water solution for 16 h at room temperature and then allowed it to react with At^- in H_2SO_4 in the presence of KI_3 for 1 h. The products were separated and identified by TLC [74].

Astatoestradiols are less stable than the analogous iodine compounds. Their deastatination was measured in their mixtures with ethyl alcohol and aqueous buffer solutions. No decomposition could be observed at room temperature in acidic media up to pH = 7; however, at 50 °C and pH = 7, more than 75% of the astatine was lost from the molecule in 20 h. Higher pH and addition of H_2O_2 enhanced the decomposition [74].

Astatination of the corresponding chloromercury derivative has also been used to obtain 6-astatocholesterol(IV); the radiochemical yield was 95%; IV was identified by TLC.

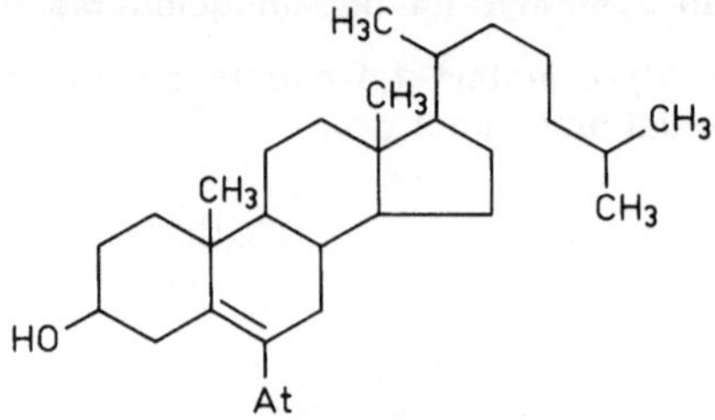

IV

No decomposition of this compound was observed by heating its ethyl alcohol-water solutions to 70 °C and by incubation at room temperature with H_2O_2 or with $NaHSO_3$ [74]. Organ distribution studies carried out on rats showed no specific accumulation of IV in tumor tissues, nor was there any evidence of its deastatination in vivo [54].

10.2.2.4 Heterocyclic Compounds

10.2.2.4.1 Astatoimidazoles

4-Astatoimidazole(I), 5-astato-4-methylimidazole(II), and 5-astatohistidine(III) were prepared by astatination of the corresponding chloromercury derivatives [75].

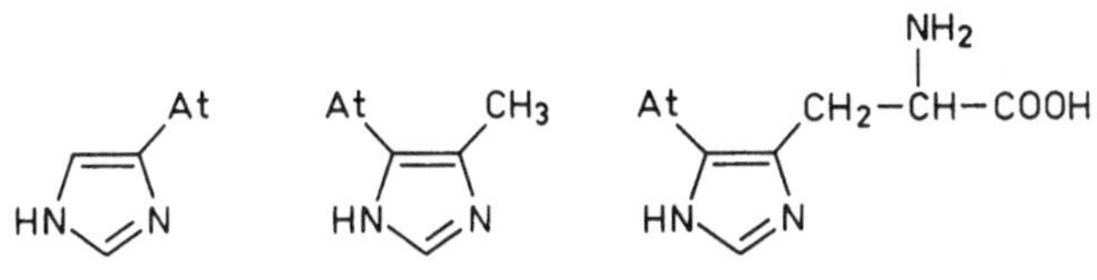

I II III

Mercuration occurred in 3 to 5 h by the reaction of imidazoles with (a) $HgSO_4$ in H_2SO_4 solution at 55 °C [75]; (b) with $Hg(OAc)_2$ in aqueous NaOAc solution (pH=5) at 50 °C [76]. Astatination took 30 min in H_2SO_4 solution at room temperature using At^- with an I_2 carrier (see Section 10.2.2.2.1.4, p. 245). Radiochemical yields of 50 to 80% were obtained; I and II were identified by TLC, III by paper electrophoresis. 2-Astato-4-iodoimidazole and 2-astato-5-iodo-4-methylimidazole also formed (with radiochemical yields of 5%) in the processes described above, presumably due to reactions of the iodine carrier [75].

The astatoimidazoles are stable in aqueous solutions at room temperature in the 0 to 14 pH range over a period of 15 h. If the temperature is increased to 80 °C, the subsequent addition of reducing (Na_2SO_3) or oxidizing (H_2O_2) agents leads to their rapid decomposition. Excess iodine also gives rise to the decomposition of III [75] in a similar way as reported for the analogous iodine compound, 5-iodohistidine [77].

10.2.2.4.2 Astatopyrimidines, Their Nucleosides and Nucleotides

Using 5-aminouracil Meyer et al. prepared 5-astatouracil (AtU, I) via diazonium salt decomposition, in a similar way to that described for obtaining astatohalobenzenes (see Section 10.2.2.2.1.3, p. 244).

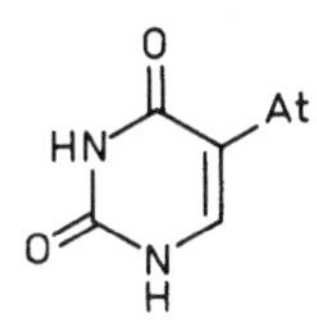

I

A radiochemical yield of about 30% [21, 78, 79] was obtained and the product I identified by HPLC utilizing the sequential analysis of uracil and the other halouracils [21, 78, 80], as shown in **Fig.** 10-**29**, p. 252.

Synthesis of AtU via astatination of the chloromercury derivative of uracil leads to a higher yield (~80%) and to a very pure product (>95%) because of the absence of side reactions; this is one of the advantages of using this procedure rather than the decomposition of the diazonium salt. Uracil can easily be mercurated by stirring it with $HgSO_4$ in H_2SO_4 at room temperature for 3 h; the chloromercury compound is then allowed to react with At^- according to the procedure described in Section 10.2.2.2.1.4, p. 245, for preparing astatophenols. The product I is isolated from the reaction mixture by extraction with benzene or n-butyl alcohol, and identified by TLC [47].

Fig. 10-29

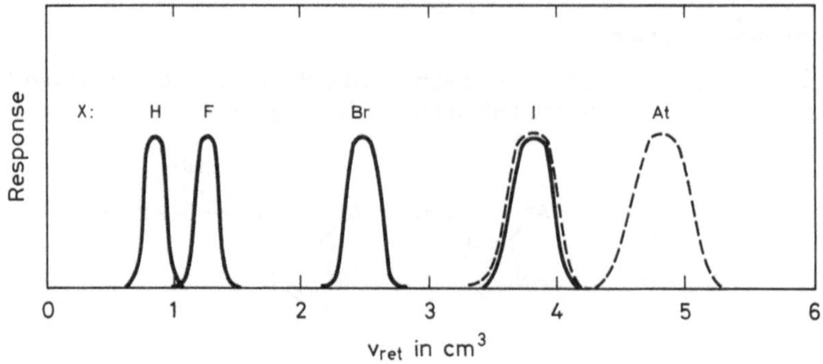

Sequential ion exchange HPLC analysis of uracil, 5-X-uracils (X=F, Br, I), and carrier free 5-(iodo-^{125}I)-uracil, and 5-(astato-^{211}At)-uracil using Aminex A25 at 25 °C, 5×10^6 Pa, with 2 M HCOOH eluent [78].

Attempts to produce AtU by allowing AtCl and AtBr to react with uracil and iodouracil have not been successful [21].

No decomposition of AtU has been observed in the 1 to 11.5 pH range at room temperature or in 1 to 7 pH range at 50 °C over a period of 20 h, similarly to the analogous iodine compound (IU). If it is heated to 50 °C at pH=11.5 the loss of 20 to 30% of astatine occurs (as well as of iodine in IU) after 20 h, presumably due to the direct attack of OH^- on the 5-position of the pyrimidine nucleus. Both halouracils are stable in the presence of $NaHSO_3$ or H_2O_2 [74].

From distribution studies utilizing benzene and aqueous borax buffers, $pK_a=8.97\pm0.01$ was established for AtU at 0 °C [50].

To produce 5-astatouridine (II) and its mono- and triphosphate (At-UMP and At-UTP), the corresponding chloromercury derivatives are formed first by reaction with $Hg(OAc)_2$ in an NaOAc buffer (pH=5) at 50 °C in 3 to 5 h [74]. Astatination can be accomplished in 1 h (see Section 10.2.2.2.1.4, p. 245) with 75% radiochemical yield.

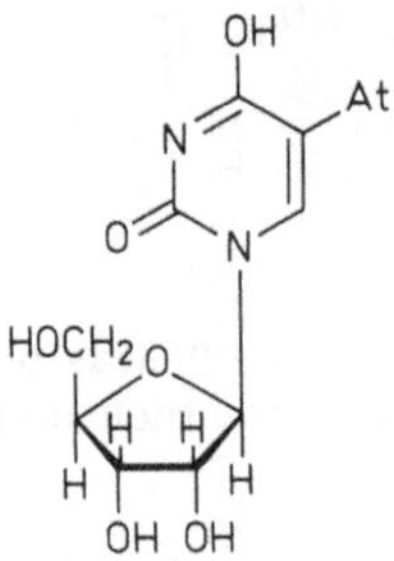

II

At-uridine and At-UMP were purified and isolated by TLC, At-UTP by paper electrophoresis and extraction. TLC was used for their identification.

The stability of At-uridine is very similar to that of AtU. The astatinated nucleotides are sensitive to hydrolysis. Fifty percent dephosphorilation of At-UTP and At-UMP can be observed in 20 h if their acidic solutions (pH=1) are kept at 50 °C [74].

5-Astato-2-deoxyuridine (At-UdR, III) is obtained from 2-deoxy-uridine with 85% radiochemical yield in the same way as described for preparing At-uridine [74].

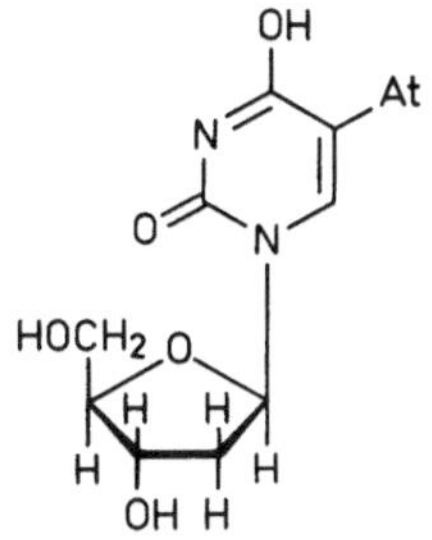

III

Diazotation of 5-amino-2-deoxyuridine with subsequent decomposition of diazonium salt according to reaction (23) (see Section 10.2.2.2.1.2, p. 242) leads to only a 2 to 3% yield of At-UdR whereas 20 to 25% of astatine is found in the form of AtU, apparently due to the hydrolysis of the N-glycosyl bond in the course of diazotation reaction [21, 79]. At-UdR is identified by TLC, paper electrophoresis [74] and by HPLC [21, 79, 80].

5-Astatocytosine(IV), 5-astatocytidine(V) as well as the monophosphate of the 5-astatocytidine (At-CMP) and of the 5-astato-2-deoxycytidine (At-dCMP) have been prepared via astatination of the chloromercury derivatives, similarly to 5-astato-uridine, with radiochemical yields of 70 to 90%; isolated by TLC and identified by TLC and by paper electrophoresis [74].

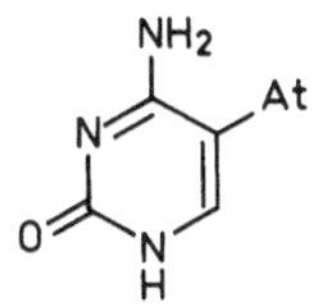

IV

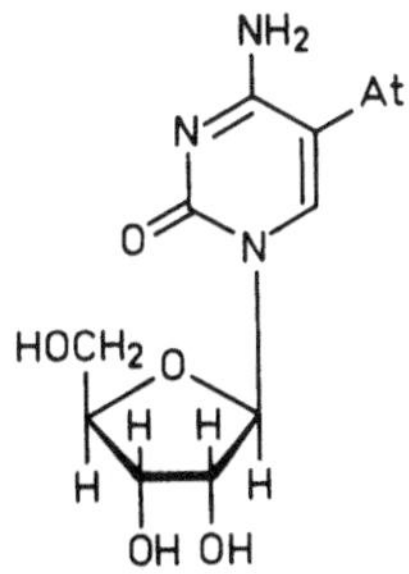

V

At-cytosine and At-cytidine are stable at room temperature over a wide pH range, similarly to AtU. Addition of reducing ($NaHSO_3$) or oxidizing (H_2O_2) agents, however, results in the decomposition of the former compounds similarly to their iodo analogs. In acidic solutions (pH = 1), dephosphorilation of At-CMP has been observed as well as a slow decomposition of the sugar-pyrimidine bond. These effects are more pronounced for At-dCMP which decomposes completely in 20 h at 50 °C in acidic solutions with formation of 5-astatocytosine and 5-astato-2-deoxycytidine [74].

Astatinated DNA and RNA obtained via the chloromercury derivatives (similarly to At-uridine), with radiochemical yields of >90%, have been isolated and purified by gel filtration. No deastatination of these products could be observed after incubation for 20 h at pH = 2 to 11.5 nor does this process take place at pH = 7 in the presence of small amounts of $NaHSO_3$ or H_2O_2 at room temperature. Heating to 50 °C causes a slow deastatination: 15 to 20% astatine loss in 20 h. Slight degradation of the nucleic acids experienced on heating does not involve breaking the C-At bond [74].

References on pp. 257/9

The potential importance of ^{211}At-labeled compounds in nuclear medicine has led to a number of tissue distribution studies with AtU, AtUdR [21, 79], At-UMP, At-cytidine and At-RNA [54]. The accumulation of AtU and of At-UdR in tumor tissues was found to be three times higher than that of the corresponding iodinated derivatives [79]. Significant deastatination of all these compounds in vivo was also indicated [21, 54].

10.2.2.5 Organic Compounds of Multivalent Astatine

In view of the more pronounced positive character of astatine compared with the other halogens, one of the early studies in this field was aimed at preparing organometallic compounds of astatine, viz. some aromatic derivatives of At^{3+} and At^{5+}. Norseyev et al. prepared $ArAtCl_2$ (I), Ar_2AtCl (II), and $ArAtO_2$ (III), where $Ar=C_6H_5$ or p-$CH_3C_6H_4$, along with the analogous iodine compounds due to macro amounts of iodine used as a carrier [81, 82].

$$Ar_2ICl + At^- \rightarrow Cl^- + Ar_2IAt \xrightarrow{170\,°C} ArI + ArAt \xrightarrow[0\,°C]{Cl_2} \underset{\text{I}}{ArAtCl_2} \quad (28)$$

To prepare $ArAtCl_2$ in accordance with equation (28), KI containing At^- is first added to a solution of Ar_2ICl. The crystalline $Ar_2I_2(Ar_2IAt)$ formed is centrifuged, washed and heated in sealed glass ampules at 170 to 190 °C for some minutes. The product of thermal decomposition, ArI(ArAt), is dissolved in $CHCl_3$ and chlorinated at 0 °C to obtain $ArICl_2(ArAtCl_2)$. The iodine compound is a yellow precipitate which can be recrystallized from $CHCl_3$ together with I.

This substance is used as the starting material for synthesizing $Ar_2ICl(Ar_2AtCl)$ by adding Ar_2Hg slowly to its hot $CHCl_3$ solution according to the reactions:

$$\underset{\text{I}}{ArAtCl_2} + Ar_2Hg \rightarrow \underset{\text{II}}{Ar_2AtCl} + ArHgCl \quad (29)$$

$$\underset{\text{I}}{ArAtCl_2} + ArHgCl \rightarrow \underset{\text{II}}{Ar_2AtCl} + HgCl_2 \quad (30)$$

After cooling, $HgCl_2$ precipitates leaving a mixture of I and II (together with their iodo analogs) in the $CHCl_3$ solution. Compound II can be extracted into the aqueous phase.

$ArIO_2(ArAtO_2)$ is formed, in accordance with reaction (31), if NaOH and CH_3COOH are added to the crystals of $ArICl_2(ArAtCl_2)$, the mixture is heated to its boiling point and Cl_2 gas is introduced to oxidize $I^{3+}(At^{3+})$ into $I^{5+}(At^{5+})$:

$$\underset{\text{I}}{ArAtCl_2} + OCl^- + 2OH^- \xrightarrow{70\text{ to }100\,°C} \underset{\text{III}}{ArAtO_2} + 3Cl^- + H_2O \quad (31)$$

$ArIO_2(ArAtO_2)$ precipitates as a white crystalline substance when the mixture is cooled; it can be recrystallized from a small amount of water.

The products I, II, and III were identified, along with the corresponding iodo analogs, by paper chromatography [81] and by TLC [82].

10.2.2.6 Bioorganic Systems Labeled with ^{211}At

In view of the potential application of the α-emitting ^{211}At isotope for tumor therapy and research, attempts have been made to build this isotope into a number of proteins and even lymphocytes (see Table 10/31).

The most reliable and unambiguous way of producing ^{211}At-labeled proteins is first to obtain para-AtC_6H_4COOH (p-At-BA), together with the analogous iodine compound, by

Table 10/31
Incorporation of ^{211}At into Bioorganic Systems.

system	method of incorporation	biological application	Ref.
bovine serum albumin	acylation via p-$AtC_6H_4SO_2Cl$	–	[30]
	acylation via p-At-BA	–	[29]
	acylation via p-At-BA	tissue distribution (mice)	[55]
	acylation via p-At-BA	immunocompetence studies	[56]
	electrophilic (H_2O_2)	tissue distribution (rats)	[54, 87]
γ-globulin	acylation via p-At-BA	–	[30]
	At + KI_3	–	[68]
	At + N-I-succinimide	disappearance from plasma (rabbits)	[67, 68]
ovalbumin	At + KI_3	–	[68]
fibrinogen	acylation via p-At-BA	–	[30]
streptokinase phytohemoagglutinin tuberculin (PPD)	electrolytic	antigen-suicide studies	[83]
keyhole limpet hemocyanine human γ-globulin tuberculin	electrolytic electrophilic (H_2O_2)	immunological studies (in vitro)	[85]
β-lactoglobulin hemoglobin (ox) cytochrome-C lysozyme	electrophilic (H_2O_2)	–	[87]
rabbit IgG immunoglobulin and the light chain fragment	electrophilic (H_2O_2)	–	[86]
rabbit antimouse thymocyte IgG	acylation via p-At-BA	cytotoxicity assay	[57]
concanavalin A	acylation via p-At-BA	cell survival studies	[58, 60]
human serum albumin		inhibition of cells	[60]
monoclonal antibody BK 19.45		proliferative	[60]
IgG class monoclonal antibody BK 19.9		capacity	[61, 62]
lymphocytes	electrolytic	selective immunosuppression (baboons)	[84]

diazotation of the para-aminobenzoic acid with subsequent decomposition of the diazonium salt (see Section 10.2.2.2.1.7, p. 248). These species are then bound to the protein via acylation of the amino groups of the latter [55 to 58, 60 to 62]. The labeling of proteins in this way was first reported by Hughes et al. [29, 30].

References on pp. 257/9

To prepare a mixed anhydride acylation agent, the halobenzoic acids are dissolved in tetrahydrofuran containing tert-butylamine, and, to this mixture, 1,4-dioxane containing i-butyl-chloro-carbonate is added. After being allowed to stand for 30 min at room temperature, the solvents are removed by vacuum pumping and the solution of the corresponding protein in a borate buffer (pH=9) is added. The acylation is completed in 1 h at 4 °C and the labeled protein is separated from the components of lower molecular weight by gel filtration using sterile phosphate buffered saline (PBS) as eluent. The overall labeling yields do not exceed 10 to 28% [55, 57, 58] and an approximate ratio of ^{211}At-labeled proteins to the unlabeled ones of 1:3000 could be achieved [58].

The ^{211}At-labeled proteins obtained via acylation (Table 10/31, p. 255) were applied by Zalutsky et al. in tissue distribution and immunocompetence studies [55, 56]. Vaughan et al. used such proteins for cytotoxicity assay [57] as well as for binding ^{211}At specifically to the tumor cells in order to investigate the effect of the α-radiation on cell survival [58] and on the inhibition of the proliferative capacity of cells [60 to 62].

Neirinckx et al. made use of the monovalent astatine cation originating from electrolysis for direct labeling of proteins [83] and lymphocytes [84]. Although the bioorganic systems thus obtained lost the ^{211}At label rapidly, it was possible to use them for antigen-suicide studies [83] and for selective immunosuppression experiments during liver transplants carried out on baboons [84].

It was shown by Aaij et al. that a low value of electrode potential (1 V) was essential to avoid denaturation of the electrolytically labeled ^{211}At proteins [85]. A more reliable method for electrophilic astatination of proteins, using H_2O_2 as an oxidizing agent in a solution of astatine containing KI as a carrier and neutralized by sodium phosphate buffer (pH=7.4), was also suggested by these authors. Labeling yields of 60% were obtained and the synthesis was performed in 30 to 60 min. Gel filtration served for purification of the labeled proteins from the nonbound astatine and from the fragment products. ^{211}At labeled keyhole limpet hemocyanine (KLH) was used in immunological studies [85].

Although electrophilic astatination of proteins has also been performed by other groups [64, 86, 87], controversial views have been expressed about the mechanism of labeling and about the true nature of the At-protein bond thus obtained. Vaughan has assumed that astatine is originally bound to the tyrosine residue of the protein but is readily released due to the unstable C-At bond. It then reacts, as At^0, nonspecifically with the other functional groups of the protein, eventually being trapped by the tertiary structure of the latter [86]. Visser et al. consider the electrophilic astatination of tyrosine residues in the proteins to be unlikely under the given conditions due to the decomposition of astatotyrosine observed in the presence of oxidizing agents at pH≧8 (see Section 10.2.2.2.2, p. 248) [63, 65]. These authors have suggested two different types of labeling mechanism and of the resulting At-protein bond [87]. Namely, astatination of proteins containing S-H groups results in relatively stable labeling with high radiochemical yield — even in the absence of an oxidizing agent — by formation of S-At bonds. In contrast, to label proteins containing no free S-H groups, the presence of H_2O_2 as the oxidizing agent is absolutely necessary. Even in this case, the astatination process is slower, the yield and the stability of the astatine label are much lower. Visser assumes formation of an At^{3+} species, $HAtO_2[AtO(OH)]$, in the presence of H_2O_2 at pH=7 (see Section 10.2.1.2.2, p. 221) which tends to form a complex bond with oxygen and nitrogen atoms of the proteins [88], similarly to some oxometallic species. These complex bonds are easily destroyed by reducing agents at pH>8 and are not stable enough to even survive interaction with the chromatographic gel used to purify the labeled proteins [87].

References:

[1] W.D. Bloomer, W.H. McLaughlin, R.D. Neirinckx, S.J. Adelstein, P.R. Gordon, T.J. Ruth, A.P. Wolf (Science **212** [1981] 340/1). — [2] W.D. Bloomer, S.J. Adelstein (Radioakt. Isotop. Klin. Forsch. **15** No. 1 [1982] 227/34; C.A. **99** [1983] No. 18894). — [3] W.D. Bloomer, W.H. McLaughlin, R.M. Lambrecht, R.W. Atcher, S. Mirzadeh, J.L. Madara, R.A. Milius, M.R. Zalutsky, S.J. Adelstein, A.P. Wolf (Intern. J. Radiat. Oncol. Biol. Phys. **10** [1984] 341/8). — [4] E.H. Appelman, E.N. Sloth, M.N. Studier (Inorg. Chem. **5** [1966] 766/9). — [5] V.I. Kuzin, V.D. Nefedov, Yu.V. Norseev, M.A. Toropova, E.S. Filatov, V.A. Khalkin (Khim. Vysokikh Energ. **6** [1972] 181/3; High Energy Chem. [USSR] **6** [1972] 161/3; C.A. **77** [1972] No. 27357).

[6] G. Samson, A.H.W. Aten Jr. (Radiochim. Acta **12** [1969] 55/6). — [7] G. Samson (Diss. Univ. Amsterdam 1971). — [8] M. Gesheva, A. Kolachkovsky [Kolaczkowskí], Yu.V. Norseyev [Norseev] (J. Chromatog. **60** [1971] 414/7). — [9] A. Kolachkovski [Kolaczkowskí], Yu.V. Norseyev [Norseev] (J. Chromatog. **84** [1973] 175/80). — [10] K. Berei, L. Vasáros (in: S. Patai, Z. Rappoport, The Chemistry of Functional Groups, Suppl. D, Pt. 1, Wiley, New York 1983, pp. 405/40).

[11] V.I. Kuzin (Diss. Leningrad State Univ. 1971). — [12] Yu.V. Norseev, V.D. Nefedov (Issled. Khim. Tekhnol. Primen. Radioakt. Veshchestv **1977** 3/8; C.A. **94** [1981] No. 3539). — [13] K. Bächmann, P. Hoffmann (Radiochim. Acta **15** [1971] 153/63). — [14] P. Hoffmann (Radiochim. Acta **17** [1972] 169/70). — [15] P. Hoffmann (Radiochim. Acta **19** [1973] 69/75).

[16] L. Vasáros, K. Berei, Ju.V. Norszejev [Yu.V. Norseev], V.A. Halkin [Khalkin] (Magy. Kem. Folyoirat **80** [1974] 487/90; C.A. **82** [1975] No. 42845). — [17] L. Vasharosh [Vasáros], Yu.V. Norseev, V.A. Khalkin (Dokl. Akad. Nauk SSSR **263** [1982] 119/24; Dokl. Phys. Chem. Proc. Acad. Sci. USSR **262/267** [1982] 161/4). — [18] G. Samson, A.H.W. Aten Jr. (Radiochim. Acta **9** [1968] 53/4). — [19] L. Vasharosh [Vasáros], Yu.V. Norseev, V.A. Khalkin (Dokl. Akad. Nauk SSSR **266** [1982] 120/2; Dokl. Chem. Proc. Acad. Sci. USSR **262/267** [1982] 297/9). — [20] L. Vasáros, Yu.V. Norseyev [Norseev], D.D. Nhan, V.A. Khalkin (Radiochem. Radioanal. Letters **54** [1982] 239/47).

[21] G.-J. Meyer (JUEL-1418 [1977] 1/107; C.A. **88** [1978] No. 71113). — [22] G.-J. Meyer, K. Rössler, G. Stöcklin (Radiochim. Acta **24** [1977] 81/5). — [23] A. Cavallero (Diss. Univ. Cathol. Louvain, Belg., 1981). — [24] G. Samson, A.H.W. Aten Jr. (Radiochim. Acta **13** [1970] 220/1). — [25] L. Vasáros, Yu.V. Norseyev [Norseev], D.D. Nhan, V.A. Khalkin (Radiochem. Radioanal. Letters **47** [1981] 313/22).

[26] L. Vasáros, Yu.V. Norseyev [Norseev], D.D. Nhan, V.A. Khalkin (Radiochem. Radioanal. Letters **47** [1981] 403/7). — [27] A. Kolachkovskii [Kolaczkowski], V.A. Khalkin (JINR-12-9473 [1976] 1/10; C.A. **88** [1978] No. 89231; INIS Atomindex **8** [1977] No. 337070). — [28] L. Vasáros, K. Berei, Yu.V. Norseyev [Norseev], V.A. Khalkin (Radiochem. Radioanal. Letters **27** [1976] 329/39). — [29] W.L. Hughes, J. Klinenberg (BNL-367 [1955] 42/3; N.S.A. **10** [1956] No. 3143). — [30] W.L. Hughes, E. Smith, J. Klinenberg (BNL-406 [1956] 44/5; N.S.A. **11** [1957] No. 12).

[31] V.D. Nefedov, M.A. Toropova, V.A. Khalkin, Yu.V. Norseev, V.I. Kuzin (Radiokhimiya **12** [1970] 194/5; Soviet Radiochem. **12** [1970] 176/7). — [32] L. Vasáros, Yu.V. Norseyev [Norseev], G.-J. Meyer, K. Berei, V.A. Khalkin (Radiochim. Acta **26** [1979] 171/6). — [33] L. Vasáros, Yu.V. Norseyev [Norseev], K. Berei, V.A. Khalkin (Radiochim. Acta **31** [1982] 75/8). — [34] A. Kolachkovskii [Kolaczkowski], Yu.V. Norseev (JINR-P6-6923 [1969] 1/17; C.A. **79** [1973] No. 83749; N.S.A. **28** [1973] No. 226). — [35] L. Vasharosh [Vasáros], Yu.V. Norseev, V.A. Khalkin (JINR-12-12188 [1979]; JINR-R-12-80-439 [1979] 1/8; C.A. **96** [1982] No. 198792).

[36] L. Vasáros, Yu.V. Norseyev [Norseev], D.D. Nhan, V.A. Khalkin, N.Q. Huan (J. Radioanal. Nucl. Chem. Letters **87** [1984] 31/40). – [37] V.I. Kuzin, V.D. Nefedov, Yu.V. Norseev, M.A. Toropova, V.A. Khalkin (Radiokhimiya **12** [1970] 414; Soviet Radiochem. **12** [1970] 385). – [38] L. Vasharosh [Vasáros], Yu.V. Norseev, V.A. Khalkin (JINR-P6-80-158 [1980]). – [39] L. Vasáros, K. Berei (Kem. Kozlemen in press). – [40] L. Vasharosh [Vasáros], Yu.V. Norseev, V.A. Khalkin (JINR-P12-81-511 [1981]).

[41] G.-J. Meyer, K. Rössler, G. Stöcklin (J. Am. Chem. Soc. **101** [1979] 3121/3). – [42] G.W.M. Visser, E.L. Diemer (Radiochem. Radioanal. Letters **51** [1982] 135/41). – [43] L. Vasáros, Yu.V. Norseyev [Norseev], D.D. Nhan, V.A. Khalkin (Radiochem. Radioanal. Letters **59** [1983] 347/54). – [44] L. Vasáros, Yu.V. Norseyev [Norseev], D.D. Nhan, V.A. Khalkin (Radiochem. Radioanal. Letters **50** [1982] 275/81). – [45] K. Berei, L. Vasáros, Yu.V. Norseyev [Norseev], V.A. Khalkin (Radiochem. Radioanal. Letters **26** [1976] 177/84).

[46] G.-J. Meyer, K. Rössler, G. Stöcklin (Radiochem. Radioanal. Letters **21** [1975] 247/59). – [47] G.W.M. Visser, E.L. Diemer, F.M. Kaspersen (J. Labelled Compounds Radiopharm. **17** [1980] 657/65). – [48] G.W.M. Visser (Diss. Univ. Amsterdam 1982). – [49] O. Dimroth (Chem. Ber. **31** [1898] 2154/6). – [50] G.W.M. Visser, E.L. Diemer, F.M. Kaspersen (Rec. Trav. Chim. **99** [1980] 93/6; C.A. **93** [1980] No. 94592).

[51] G.W.M. Visser, E.L. Diemer (Intern. J. Appl. Radiat. Isotop. **33** [1982] 389/90). – [52] L. Vasharosh [Vasáros], Yu.V. Norseev, V.I. Fominykh, V.A. Khalkin (Radiokhimiya **24** [1982] 95/9; Soviet Radiochem. **24** [1982] 84/7). – [53] W.J. Klapproth, F.H. Westheimer (J. Am. Chem. Soc. **72** [1950] 4461/5). – [54] G.W.M. Visser, E.L. Diemer, C.M. Vos, F.M. Kaspersen (Intern. J. Appl. Radiat. Isotop. **32** [1981] 913/7). – [55] M.R. Zalutsky, A.M. Friedman, F.C. Buckingham, W. Wung, F.P. Stuart, S.J. Simonian (J. Labelled Compounds Radiopharm. **13** [1977] 181/2).

[56] A.M. Friedman, M.R. Zalutsky, W. Wung, F. Buckingham, P.V. Harper Jr., G.H. Scherr, B. Wainer, R.L. Hunter, E.H. Appelman, R.M. Rothberg, F.W. Fitch, F.P. Stuart, S.J. Simonian (Intern. J. Nucl. Med. Biol. **4** [1977] 219/24). – [57] A.T.M. Vaughan (Intern. J. Appl. Radiat. Isotop. **30** [1979] 576/7). – [58] A.T.M. Vaughan, W. Bateman, J. Cowan (J. Radioanal. Chem. **64** [1981] 33/9). – [59] C.-Y. Shiue, G.-J. Meyer, T.J. Ruth, A.P. Wolf (J. Labelled Compounds Radiopharm. **18** [1981] 1039/46). – [60] A.T.M. Vaughan, W.J. Bateman, G. Brown, J. Cowan (Intern. J. Nucl. Med. Biol. **9** [1982] 167/71).

[61] A.T.M. Vaughan, W.J. Bateman, D.R. Fisher (Intern. J. Radiat. Oncol. Biol. Phys. **8** [1982] 1943/6). – [62] W.J. Bateman, A.T.M. Vaughan, G. Brown (Intern. J. Nucl. Med. Biol. **10** [1983] 241/4). – [63] G.W.M. Visser, E.L. Diemer, F.M. Kaspersen (Intern. J. Appl. Radiat. Isotop. **30** [1979] 749/52). – [64] A.T.M. Vaughan, J.H. Fremlin (Intern. J. Appl. Radiat. Isotop. **28** [1977] 595/8). – [65] G.W.M. Visser, F.M. Kaspersen (Intern. J. Nucl. Med. Biol. **7** [1980] 79).

[66] A.T.M. Vaughan (Intern. J. Nucl. Med. Biol. **7** [1980] 80). – [67] W.L. Hughes, D. Gitlin (BNL-314 [1954] 48; N.S.A. **9** [1955] No. 2099). – [68] W.L. Hughes, D. Gitlin (Fed. Proc. Fed. Am. Soc. Exptl. Biol. **14** [1955] 229). – [69] I. Brown (Intern. J. Appl. Radiat. Isotop. **33** [1982] 75/6). – [70] I. Brown (Radiochem. Radioanal. Letters **53** [1982] 343/49).

[71] I. Brown, R.N. Carpenter, J.S. Mitchell, S.F. Russell (Proc. 7th Intern. Congr. Radiat. Res., Amsterdam 1983, Ref. D4-05). – [72] J.S. Mitchell, I. Brown, R.N. Carpenter (Proc. 7th Intern. Congr. Radiat. Res., Amsterdam 1983, Ref. D4-20). – [73] J.S. Mitchell, I. Brown, R.N. Carpenter (Experientia **39** [1983] 337/9). – [74] G.W.M. Visser, E.L. Diemer, F.M. Kaspersen (J. Labelled Compounds Radiopharm. **18** [1981] 799/807). – [75] G.W.M. Viser, E.L. Diemer, F.M. Kaspersen (Intern. J. Appl. Radiat. Isotop. **31** [1980] 275/8).

[76] R.M.K. Dale, D.C. Livingston, D.C. Ward (Proc. Natl. Acad. Sci. U.S. **70** [1973] 2238/42). – [77] L. Schutte, E. Havinga (Rec. Trav. Chim. **86** [1967] 385/92). – [78] G.-J. Meyer, K. Rössler, G. Stöcklin (J. Labelled Compounds Radiopharm. **12** [1976] 449/58). – [79] K. Rössler, G.-J. Meyer, G. Stöcklin (J. Labelled Compounds Radiopharm. **13** [1977] 271). – [80] K. Rössler, W. Tornau, G. Stöcklin (J. Radioanal. Chem. **21** [1974] 199/209).

[81] V.D. Nefedov, Yu.V. Norseev, Kh. Savlevich, E.N. Sinotova, M.A. Toropova, V.A. Khalkin (Dokl. Akad. Nauk SSSR **144** [1962] 806/9; Proc. Acad. Sci. USSR Chem. Sect. **142/147** [1962] 507/10). – [82] Yu.V. Norseev, V.A. Khalkin (Chem. Zvesti **21** [1967] 602/10; C.A. **68** [1968] No. 8848). – [83] J.A. Smit, J.A. Myburg, R.D. Neirinckx (Clin. Exptl. Immunol. **14** [1973] 107/16; C.A. **79** [1973] No. 30327). – [84] R.D. Neirinckx, J.A. Myburg, J.A. Smit (Radiopharm. Labelled Compounds Proc. Symp., Vienna 1973, Vol. 2, pp. 171/81). – [85] C. Aaij, W.R.J.M. Tschroots, L. Lindner, T.E.W. Feltkamp (Intern. J. Appl. Radiat. Isotop. **26** [1975] 25/30).

[86] A.T.M. Vaughan, J.H. Fremlin (Intern. J. Nucl. Med. Biol. **5** [1978] 229/30). – [87] G.W.M. Visser, E.L. Diemer, F.M. Kaspersen (Intern. J. Appl. Radiat. Isotop. **32** [1981] 905/12). – [88] G.W.M. Visser, E.L. Diemer (Radiochim. Acta **33** [1983] 145/51).

10.3 Chemical Behavior of Astatine in Solution

S.H. Eberle
Universität und Kernforschungszentrum Karlsruhe
Institut für Radiochemie
Karlsruhe, Federal Republic of Germany

Introductory Remarks

From a theoretical point of view, it would be desirable to arrange the known facts of the chemical reactions of astatine according to the valency state of the element. But in many cases, the authors themselves felt unable to tell the valence state with certainty or the valence state given seems doubtful in the light of more recent findings. Therefore, this compilation is arranged according to the investigation methods and the composition of the solutions is given if available.

Whenever possible, the valence state is indicated as was done by the authors or how it has to be assumed from the chemical treatment applied. These assignments must not be regarded as being fully proven. The valence state is indicated by Roman numbers as usual. For an understanding of the results, it is sometimes useful to indicate the probable chemical form or the charge of the species under consideration. The notation used for this is:

astadide ion	At^-
zerovalent astatine	At^0
monovalent astatine	At^I, At^+, AtO^-, AtX_2^-
trivalent astatine	At^{III}, AtO^+
pentavalent astatine	At^V, AtO_3^-
heptavalent astatine	At^{VII}, AtO_4^-

The monovalent positively charged astatine species At^+ occurs in acid solution and is frequently assumed to be $At(OH_2)^+$ generated by protonation of AtOH. Analogously, At^I forms the anionic species AtO^- in alkaline solutions. AtX_2^-, with X=Cl, Br, I, is the species assumed to be formed by complexation of At^I in acid solution containing the respective halide ion.

10.3.1 Coprecipitation

The coprecipitation behavior of astatine depends more on the redox condition during the experiment than on the valence state of the astatine stock solution used. Without controlling the redox potential, a change of valency has been frequently observed, e.g., oxidation of At^- by dissolved oxygen [1].

With Slightly Soluble Compounds as Carriers

Corson, MacKenzie and Segrè [2], the discoverers of astatine, first investigated coprecipitation behavior with a solution of the bombarded Bi target in nitric acid. This solution most probably contained At^0. It is therefore understandable that they observed no coprecipitation with TlCl, $PbCl_2$ and AgI. They reported complete precipitation with Bi^{III}, Hg^{II}, Ag^I, and Sb^{III} as carrier cations and H_2S as a precipitation reagent. Later, Johnson et al. [3] have shown that HgS carries astatine to >90% in 0.5 to 10 M HCl.

Coprecipitation was extensively used by Appelman [4] in his experiments on the redox potentials of the astatine valence states. Some of his results, together with results of other authors, are given in Table 10/32. The general procedure was: mixing of the carrier with the astatine solution to give a final concentration of 0.02 to 0.04 M NaI or 0.02 to 0.05 M

$NaIO_3$; adding the precipitant solution (0.5 M $AgNO_3$, 0.25 M $TlNO_3$ or $TlClO_4$, 1 M $Pb(ClO_4)_2$, 0.6 M $Ba(ClO_4)_2$) to give a final concentration about twice the concentration of the precipitant and shaking for a few minutes. The precipitate was centrifuged, the supernatant withdrawn, and the precipitate washed one or more times with a washing solution (0.01 M $AgNO_3$+ 0.01 M $HClO_4$; 0.05 M $Pb(ClO_4)_2$, pH 2.3; 0.1 M $Ba(ClO_4)$, pH 2; 0.05 M $TlNO_3$+1% acetic acid).

Table 10/32
Coprecipitation of Astatine by Slightly Soluble Compounds.

compound precipitant	solution composition	probable valence state of At	% carried	Ref.
AgI	1 M H_2SO_4+Zn	At^-	100	[3]
	0.5 M HNO_3+SO_2+KI	At^-	100	[3, 4]
	0.001 M $HClO_4$, 0.02 M NaI, 0.04 M $AgNO_3$	At^-	90 to 100	[4]
	0.01 M HNO_3 +0.02 M NaI+0.04 M $AgNO_3$	At^0	0.1	[4]
	nitric acid+SO_2	At^-	>90	[5]
TlI	1 M H_2SO_4+Zn	At^-	100	[3]
	0.5 M HNO_3+SO_2+KI	At^-	100	[3, 4]
	0.01 M HNO_3 +0.02 M NaI+0.03 M $TlNO_3$	At^0	4	[4]
	0.00015 M H_2SO_4, 0.003 M SO_2, 0.02 M NaI, 0.04 M $TlClO_4$	At^-	91	[4]
HgS	0.5 to 10 M HCl+H_2S	At^-?	80	[3]
	6 M HCl	At^-?	100	[2]
$La(OH)_3$	0.5 M HNO_6+NaOH	At^0	5 to 8	[3]
	0.5 M 4 NO_3+$K_2S_2O_7$+NaCH	At^{III}?	97 to 99	[3]
	2 M NaOH, 0.1 M Na_2SnO_2	At^-	<5	[6]
$Fe(OH)_3$	0.5 M HNO_3+HaCH	At^0	40 to 50	[3]
	0.5 M HNO_3+$K_2S_2O_8$+NaOH	At^{III}	97 to 99	[3]
	2 M NaOH, 0.1 M Na_2SnO_2	At^-	13	[6]
	2 M NaOH, 0.1 M NaClO	At^I	95	[6]
$Pb(IO_3)_2$	pH 2 0.05 M IO_3^-, 0.01 M H_5IO_6	At^V	100	[8]
$AgIO_3$	0.5 M HNO_3, $H_2S_2O_8$ or +HOCl	At^{III} or At^V	>90	[3]

References on pp. 288/9

Beside different results due to uncontrolled redox conditions of [2] and [3] there is now no doubt that the astatide ion is completely coprecipitated with AgI, TlI and PbI_2. The following values are given as an example of the necessity of controlling redox conditions in some results of Bochvarova et al. [10]:

Coprecipitation of astatine with TlCl under different experimental conditions:

starting solution	5 mL 4 M HCl+0.035 M KI with At tracer added being At^-			
$SnCl_2$ added to reach 0.2 M	no	yes	no	yes
precipitant:				
0.035 M $TlNO_3$ in mL	5	5	—	—
Tl suspension 8 mg/mL in mL	—	—	5	5
At in solution above precipitate in %	45±3	~1	65±1	3±2
At in 10 mL of wash solution in %:				
2 M HCl	17±6	~1	9	1±0.5
2 M NaCl	~4	~1	—	—
first aceton	10±3	1.5±0.5	~4	4±2
second aceton	~3	1.5±0.5	~1	2.0±0.5
At in final precipitate in %	21±6	95±2	20±1	92±2

Bochvarova et al. [10] started with an alkaline At solution, but upon acidification without reducing reagent present only 20% of the activity appeared in the final TlI precipitate. It is assumed that oxidation by atmospheric oxygen occurred if no $SnCl_2$ was present. Aten et al. [9] have used PdI_2 carrier to determine astatine quantitatively. The hydroxides of Fe,La,Th and Bi have been shown to coprecipitate At^- and At^0 only to a small extent, presumably due to irregular sorption, but they carry 100% of the astatine after it is oxidized by $Na_2S_2O_8$ in HNO_3 or by NaOCl in NaOH [6]. According to the present knowledge [11] the astatine in the solutions used was At^{III} and At^I, respectively. According to [5] the astatine in an HNO_3 solution of irradiated Bi metal is not coprecipitated with $BiPO_4$.

After oxidation of astatine in weakly acid solution with $(NH_4)_2S_2O_8$; $(NH_4)_2Ce(NO_3)_6$ or IO_3^-/IO_4^- on heating or with HOCl in the cold a 90±10% coprecipitation with $AgIO_3$, $Pb(IO_3)_2$ and $Ba(IO_3)$ was observed by Appleman [3, 8]. He assumed At^V to be the carried At species but according to the findings of Dreyer et al. [11] at least with $(NH_4)_2SO_8$ one would expect that At^{III} has been formed.

Khalkin et al. [12] have identified At^{VII} by its cocrystallization with periodates. In recrystallization experiments in water the distribution coefficients were independent of the fraction of salt that has crystallized out: KIO_4 D=0.068±0.021, $CsIO_4$ D=0.31±0.08 (0 °C).

Brinkman et al. [13] have shown the cocrystallization of astatine with CsI_3. About 95% of the astatine was found in the iodine that sublimed off after thermal degradation of the compound at 250 °C, according to $CsI_3 \rightarrow CsI + I_2$.

With Tellurium

The carrying of astatine by a tellurium precipitate produced by SO_2 in HCl was first mentioned by Corson, MacKenzie, Segrè [2]. They also mentioned that astatine was not carried if tellurium is precipitated by sodium stannite from alkaline solution. These reactions seem to be rather selective for astatine and permit its separation from irradiated targets.

The reduction can be achieved by SO_2, $SnCl_2$ or SO_2+hydrazine [14] and is reported to result in complete coprecipitation of the astatine with a few milligrams of tellurium. It allows the quantitative determination of astatine. Garrison et al. [15] used the following procedure: the astatine containing tissue was oxidized by 9 M $HClO_4$ + 5 M HNO_3, evaporated to 10 to 15 mL, diluted with water to 3 M $HClO_4$ and spiked with 5 mg H_2TeO_3 in mL 12 M HCl. A stream of SO_2 is passed through the solution, precipitating the metallic tellurium. Lefort et al. [14] started with about 3 mL of 3 M HCl containing the astatine and 5 mg of Te, added 2 mL 10% $SnCl_2$ in HCl and obtained complete coprecipitation within 10 min in a water bath. A sequence of tellurium precipitation from acid and alkaline solution was used by Lefort et al. [14] and Belyaev et al. [6] for the isolation of astatine from proton irradiated thorium, bismuth and lead. The latter authors report 80 to 90% yield from 25 to 30 mL 11.5 M HCl containing 0.2 mg/mL Te with $SnCl_2$ as reducing agent.

With Elemental Iodine

Aten et al. [9] have investigated the cocrystallization of astatine in $CHCl_3$ solution with iodine, the crystallization was initiated by cooling the saturated solution. The results are in "reasonable" agreement with a logarithmic distribution law:

$$\log\left(\frac{\text{At in solution}}{\text{At in solution}+\text{At in precipitate}}\right)=\lambda\cdot\log\left(\frac{\text{I in solution}}{\text{I in solution}+\text{I in precipitate}}\right)$$

The average value of the distribution coefficient was $\lambda=4$ at the temperature of an ice/water mixture.

With Miscellaneous Carriers

Appleman [4] reports that astatine is coprecipitated up to 50% by TlI and PbI_2 from a mixed solution of At^0 iodine and iodine, but in most cases the radioactivity could be washed from the precipitate by acetone. Astatine is coprecipitated with the sulfur deposit occurring upon addition of 0.5 mL 0.5 M $Na_2S_2O_3$ solution to 1 mL of 0.4 to 9 M H_2SO_4 containing the radioactivity [1]. This was interpreted by the formation of S_7At^+ and would indicate that oxidation occurred during the experiment.

10.3.2 Adsorption and Deposition

Onto Metallic Surfaces from Aqueous Solution

Astatine is adsorbed from acid solution by plates or foils of certain metals. Corson, MacKenzie and Segrè [2] were the first to report the adsorption of astatine from its acid solution (0.25 M HNO_3) containing Bi and Hg onto copper. By heating cautiously, the mercury which is also deposited on the plate disappears and the astatine stays behind. The adsorption onto a silver foil from 3 M $HClO_4$ or 0.3 to 0.5 M HNO_3 was used for its quantitative determination by Garrison et al. [15] and Lefort et al. [14, 16], respectively. The reaction required one hour for completion (Table 10/33, p. 264) in an experiment where the astatine was

References on pp. 288/9

in an $HClO_4/HNO_3$ mixture then evaporated until $HClO_4$ fumes appeared and afterwards diluted to 3 M acid.

Table 10/33
Rate of Astatine Deposition onto Silver Foil from 3 M $HClO_4$ [15].

time in min	% deposited
0 to 10	73.5
10 to 20	20.0
20 to 30	8.0
30 to 60	2.0
sum	103.5

Onto Glass from Aqueous Solution

Appelman [4] has investigated the adsorption of astatine onto glass wool, using 0.3 g Pyrex wool and 1 mL of various solutions containing At^0. His results are collected in Table 10/34. The adsorption under these conditions, which are severe with respect to the surface/volume ratio, was in no case more than 75%. No adsorption occurred from 3 M HCl, 0.2 M NH_4OH and 0.1 M NaOH.

Table 10/34
Adsorption of At^0 onto Glass Wool from Aqueous Solution [4].

solution	time in h	% left in solution
a) HNO_3		
0.5 M	24	70
	44	25
0.15 M	6	93
	20	69
	48	52
0.01 M	12	62
	39	44
0.001 M	5	88
	20	84
	40	80
b) HCl		
3 M	22	100
	49	91
0.5 M	46	59
0.1 M	24	62
c) NH_4NO_3		
0.1 M	27	89
	19	79
	46	69
d) water		
	5	70
	49	70
e) $NH_3+NH_4NO_3$		
0.2 M+0.1 M	47	100
f) NaOH		
0.1 M	49	100
	46	95

Onto Solids from Organic Solution

Visser et al. [1] have carried out what they call interaction experiments consisting in extraction of astatine from different aqueous solutions and shaking the organic phase with different organic and inorganic solids. A large number of data as to how much astatine is adsorbed from the organic solvents (hexone, pentandione, cyclohexane, dibutyl ether) is given in the paper showing that up to 80% of the activity is held back by sulfur containing solids like cysteine, Na_2S, thioacetamide, sodium diethyl dithiocarbamate and by CaO. There is only a minor influence of the pretreatment before extraction besides the low adsorption after oxidation with peroxidisulfate.

Onto Platinum from Aqueous Solution

Norseev et al. [17] have made an extensive study of the adsorption of astatine onto platinum surfaces, that were previously boiled for 2 to 3 hours in a solution of 13 M HNO_3 + 0.01 M Cr^{VI}, washed with doubly distilled water and heated to yellow-white heat in an oxygen-gas flame. The researchers believe that an oxide film has formed on the platinum which acts as a cation exchanger.

The deposition occurred from a solution 1 to 10 M in nitric acid and 0.005 M in Cr^{VI}. It is assumed that the adsorbed species is At^{I} in cationic form. The adsorption decreased with increasing nitric acid concentration (**Fig.** 10-**30**). The deposit redissolved upon cathodic polarization of the platinum.

The rate of deposition is described satisfactorily by the following first order reaction equation:

$$\frac{dq}{dt} = k \cdot (m_0 - x) \cdot \frac{S}{V}$$

with m_0 denoting the initial amount of At in the solution, q the amount adsorbed, S the apparent surface in m^2 and V the volume of the solution in m^3. Data for k at 23 °C are collected in Table 10/35. Additional adsorption data at 90 °C may be found in the original article [17]. **Fig.** 10-**31**, p. 266, shows that the deposition increased and accelerated with temperature.

Fig. 10-30

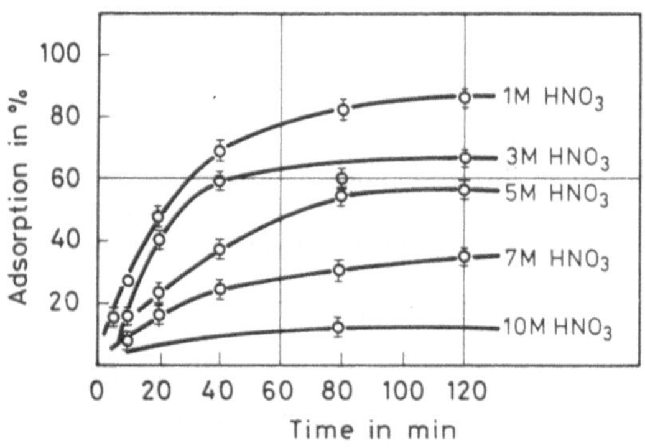

Dependency of At adsorption onto oxidized Pt surface on time and nitric acid concentration [18].

References on pp. 288/9

Table 10/35
Constant k of the Rate Equation for the Adsorption of Astatine onto Platinum at 23 °C [17].

HNO_3 in mol/L	$10^5 \cdot k$ in m/s
1	1.27 ± 0.17
3	1.0 ± 0.1
5	0.48 ± 0.08
7	0.321 ± 0.023

Fig. 10-31

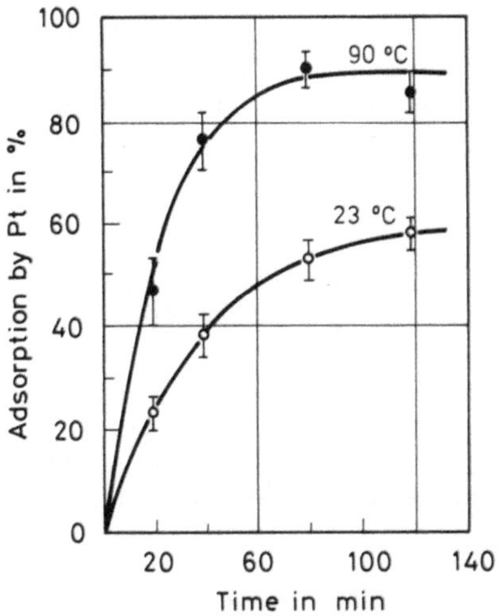

Adsorption of At^+ onto oxidized Pt surface from 5 M HNO_3 at two different temperatures [20].

The deposition was lower in the presence of nitrates of Tl, Hg^I and Bi^{III}. Hg^I prevented the deposition almost completely. Hydrochloric acid had a regular negative influence on the deposition (Table 10/36) which was consistent with the equation:

$$\log (q_0 \cdot q - 1) = \log B_2 + 2 \cdot \log (Cl^-)$$

The identification of the variables is: q_0 = ratio of At on Pt and in solution in absence of chloride, q = the same in presence of chloride, B_2 the stability constant of the complex $AtCl_2^-$. Norseev et al. [17] reported $\log B_2 = 5.87 \pm 0.05$ as calculated from these experiments. This is not too far from the result of other investigations by more classical methods, which gave $\log B_2 = 5.4$ [19].

Table 10/36
Deposition of Astatine on Platinum in the Presence of Hydrochloric Acid [17].

HCl in mol/L	At on Pt in %
0	53.7 ±3.1
2×10^{-3}	23.5 ±2.8
2×10^{-3}	12.5 ±2
5×10^{-3}	4.9 ±0.3
8×10^{-2}	2.1 ±0.12
1.3×10^{-2}	0.9 ±0.13
2×10^{-2}	0.52±0.02

1 M HNO_3 – 5×10^{-3} M Cr^{VI},
$S(Pt) = 4 \times 10^{-3}$ m^2,
$V = 10 \times 10^{-6}$ m^3,
$\tau = 2.4 \times 10^3$ s

Onto Caesium Phosphotungstate from Aqueous Solution

The cationic At^I as formed in H_2SO_4 containing chromate is readily adsorbed onto preformed precipitates of $Cs_3PW_{12}O_{40}$ from 3 M HNO_3+0.02 M $CsNO_3$ [19]. **Fig.** 10-**32** shows that the distribution coefficient decreases with increasing temperature. The following temperature dependency is reported $\log D = 14.15 - 0.04 \cdot T$ with T in K and D in cm^3/g (presumably).

Fig. 10-32

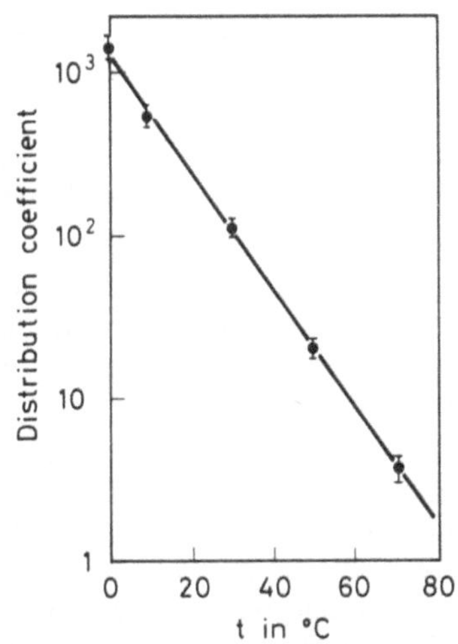

Distribution coefficient for At^+ between $Cs_3PW_{12}O_{40}$ and HNO_3+0.02 M $CsNO_3$ [20].

References on pp. 288/9

Onto Titanium Hydroxide from Aqueous Solution

The cationic At^{I} is adsorbed up to 90% onto hydrated titanium hydroxide from slightly acidic solutions 0.1 M in $NaNO_3$ and 0.001 M in $Cr_2O_7^{2-}$ [19]. The adsorption is strongly pH dependent (Fig. 10-6, p. 190) but different for At^+ and Tl^+. This was one of the important facts from which the conclusion that At^+ is not simply At^+ was drawn.

Onto Surfaces from the Gaseous State

Johnson et al. [3] first investigated the deposition of astatine from the gaseous state onto metal surfaces. Their experiments consisted in placing 0.3 × 2.5 cm strips of the metals in an evacuated glass cylinder and exposing them to "astatine vapor" for 16 h at room temperature and at 325 °C, respectively. How the "astatine vapor" was applied is not specified but was probably an astatine deposit on glass. Table 10/37 shows that Pt and Ag are rather good adsorbents. Au is active only at room temperature and Al, Ni, Cu are ineffective. The authors mention that an At deposit on glass at liquid nitrogen temperature was very volatile at room temperature. The evaporation followed an exponential law with a "half life" of about 1 hour.

Table 10/37
Collection of Astatine by Metallic Surfaces in Vacuum [3].

metal	% found room temperature	% found 325 °C [a)]
Al	0.3	0.2; 0.3
Ni	0.2	0.6; 0.7
Cu	0.5	0.6; 4.3
Pt	36.3; 20	33.5; 6.5; 16
Au	38.4; 69	0.1; 0.7; 4
Ag	24.3; 15	65; 87.5; 86

a) Replicate experiments in parentheses, the metal strips were present together.

Contrary to this, Appelman needed to heat the glass surfaces onto which he had collected the astatine for evaporating it [4]. He observed a colored layer of impurities at the region where the At had deposited, which might be responsible for this effect.

Several authors have investigated the deposition of astatine in a temperature gradient and using a stream of a carrier gas. Vakhtel et al. [21] used the apparatus shown in Fig. 10-13, p. 200, "filters" of Cu, Ag and Pt, respectively, of 20 to 30 cm^2 surface and O_2:He = 9:1 as a carrier gas (800 $cm^3 \cdot cm^{-2} \cdot min^{-1}$).

The data in Table 10/38 show that in this oxidative medium the astatine is collected very effectively by Cu and Pt at 160 and 240 °C, respectively, but not by Ag at 250 to 540 °C. This metal collects astatine better at lower temperatures (< 100 °C) and is very good under inert (vacuum or He carrier gas) conditions. This was used by Appelman [4] and Doberenz [22] for the distillative isolation of astatine from ^4He bombarded Bi targets. Appelman reported that the activity is strongly held by the silver and very little is washed away with hot water, methanol, and benzene.

Table 10/38
Deposition of Astatine from an O_2:He=9:1 Stream in a Temperature Gradient Apparatus [21].

temperature in °C	% adsorbed onto a 20 to 30 cm^2 foil filter	temperature in °C	% adsorbed onto a 20 to 30 cm^2 foil filter	temperature in °C	% adsorbed onto a 20 to 30 cm^2 foil filter
Cu		Ag		Pt	
160	90	250	20	240	100
240	26	270	19	310	38
280	16	300	26	420	14
360	9	400	20	550	0
440	0	450	13		
		480	3		

The temperature of deposition onto glass walls in a temperature gradient was determined by Merinis et al. [23] to be 16 °C with He carrier gas (200 $cm^3 \cdot cm^{-2} \cdot min^{-1}$) and by Eichler et al. [24] to be 210±10 °C with H_2 carrier gas (2800 $cm^3 \cdot cm^{-2} \cdot min^{-1}$). The gas velocity has a great influence.

10.3.3 Ion Exchange from Aqueous Solution

Resin Pretreatment

Astatine in its different valency states is known to undergo easily changes of the valency by impurities and this was found to be most important in ion exchange work. Therefore, a special pretreatment of the resin is required to get reproducible results.

Wang Fu-Chiung et al. [25] first washed the resin successively for several hours with 8 M HCl+Cl_2, 6 M HNO_3+0.1 M Ce^{IV} and H_2O at 60 to 70 °C. The static exchange capacity of the resin after this treatment was 4.2 mequiv/g. In elution experiments with 8 M HCl from a 2×50 mm column, 7±3% of the astatine, which should have been present as At^{I}, remained on the resin. Do Kim Tyung et al. [26] with the same aim pretreated their resin as follows: vigorous stirring for 5 to 6 hours with 3 M HCl through which chlorine was bubbled, eluting the resin in a glass column with 3 M HNO_3 containing 0.01 M $H_2Cr_2O_7$ at 0.1 $mL \cdot cm^{-2} \cdot min^{-1}$ for a week, washing the resin free from nitric acid with 1 M HCl and storing in this solution. The final capacity of the resin after this treatment was 3.7 mequiv/g. Prior to the determination of distribution coefficients, the resin was washed with the appropriate acid until the eluate was freed from chloride. Roessler et al. [27] also expressed the necessity for a careful conditioning of the resin to get reproducible results.

HCl and LiCl Solution

Fig. 10-**33**, p. 270, shows the cation exchange curves of chlorine oxidized astatine from HCl and LiCl solutions. Wang Fu-Chiung et al. [25] have pointed out that these curves are approximately the same as for gold and interpreted this by formation of chloride complexes. The maximum distribution coefficient (K_d in mL per gram of dry resin) was 63±3 at about 8 M HCl and 610±3 atm at about 6 M CsCl. It is mentioned that at HCl concentrations below 1 M the sorption is not reversible. **Fig.** 10-**34**, p. 270, shows elution curves for different HCl concentrations indicating sharper elution peaks with lower acid concentration, as might be expected. The recovery with hydrochloric acid below 1 M was low. For a preparation

References on pp. 288/9

Fig. 10-33

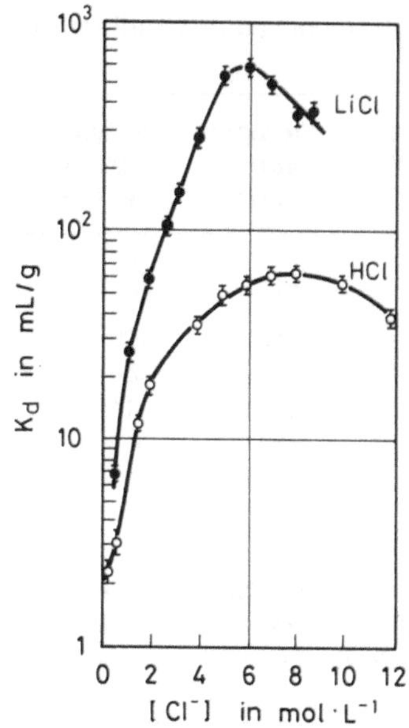

Sorption of chlorine oxidized astatine onto Dowex 50X8 [25].

Fig. 10-34

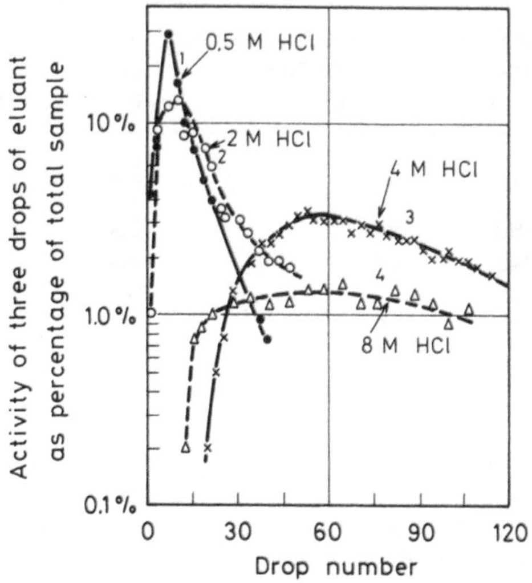

Elution of chlorine oxidized astatine from a Dowex 50X8 column 2×50 mm by HCl [25].

procedure, the authors recommended loading the astatine from 8 M HCl and desorbing it with chlorine water.

Do Kim Tyung et al. [36] have investigated the influence of HCl on the sorption of astatine from perchlorid acid and sulfuric acid solutions onto Dowex 50X8. The solutions contained 0.005 M $H_2Cr_2O_7$ to maintain constant redox conditions.

Do Kim Tyung et al. [26] mention that At^0 is adsorbed by Dowex 50X8 completely but was hardly eluted by perchloric acid containing no oxidant. They have intensively investigated the influence of HCl on the sorption of oxidized astatine from perchloric acid and sulfuric acid solutions onto Dowex 50X8. The solutions contained $HClO_4$ and 0.005 M $H_2Cr_2O_7$ to maintain a constant redox potential. The astatine is assumed to be in the At^I state. The distribution coefficient decreases with increasing HCl concentration (**Fig.** 10-**35**). A complex formation was deduced from this. Norseev et al. [28] have investigated the complex formation with chloride in nitric acid solution also using the sorption onto Dowex 50X8 and have calculated the stability constants. The data of both groups of researchers are given in Table 10/39 and show a rather good coincidence.

$HClO_4$ Solution

The distribution of At^0 between Dowex 50X8 and perchloric acid containing 0.005 M $H_2Cr_2O_7$ decreases with increasing acid concentration (Fig. 10-20a, p. 219). A charge of +1 for the astatine cation was deduced from the slope of the curves. The sorption decreases appreciably with a change in temperature (**Fig.** 10-**36**, p. 272). A plot of the distribution coefficient K_d against 1/T is linear and from this the sorption enthalpy was calculated as −3300 cal/mol [26]. This is substantially greater than the sorption enthalpy of simple monovalent cations.

Fig. 10-35

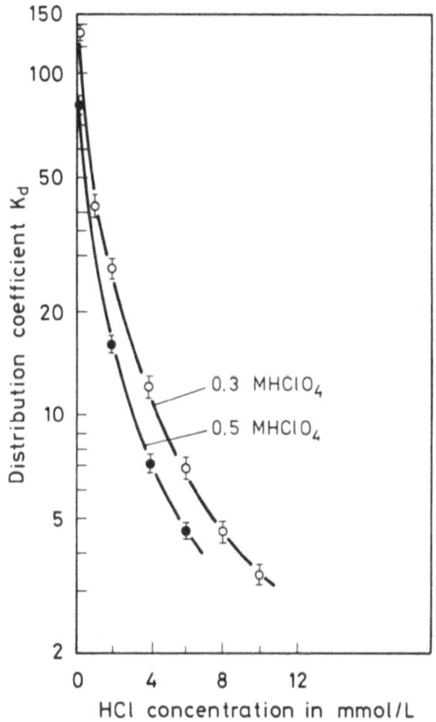

Sorption of astatine from solution $HClO_4$+HCl containing 0.005 M $H_2Cr_2O_7$ onto Dowex 50X8 [26].

Table 10/39
Stability Constants of Inorganic Complexes of At^I Measured by Cation Exchange Experiments at 20 °C.

anion	ionic strength	log β_1	log β_2	Ref.
Cl^-	0.5	3.20	5.40	[26]
	0.5	2.48	5.40	[28]
NO_3^-	1	0.13	–	[26]
SO_4^-	1	0.23	0.50	[26]
$Cr_2O_7^{2-}$	1	1.34	–	[26]

HNO_3 Solution

Data on the distribution coefficient of astatine between HNO_3 containing 0.005 M HNO_3 and Dowex 50X8 are given in Table 10/40, p. 272. The researchers [29] found that a plot

Fig. 10-36

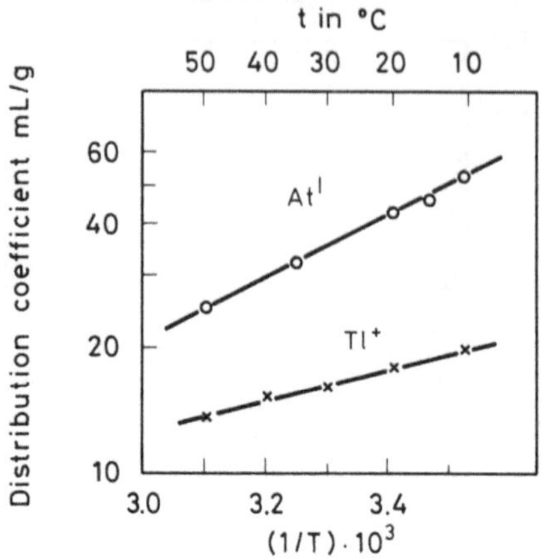

Temperature dependency of the sorption of At^I from 1 M $HClO_4$+0.005 M $H_2Cr_2O_7$ onto Dowex 50X8 [26].

of the distribution coefficient versus the pH is linear and follows the equation

$$\log\left(\frac{1}{K_d}\right) = Z \cdot \log(H^+) + K$$

They found $Z = 1.55 \pm 0.15$ for the oxidized astatine, $Z = 1.4$ for Cs and $Z = 2.2$ for Sr.

From this, they concluded that the astatine was adsorbed as a cation with charge +1. From other experiments with HNO_3 and Dowex 50X8, Do Kim Tyung et al. [26] identified a 1:1 complex of astatine with nitrate and calculated the stability constant (Table 10/39). Norseev et al. [28] report a distribution coefficient of $K_d = 80$ mL/g for Dowex 50X8 and 0.5 M HNO_3+0.005 M $H_2Cr_2O_7$.

H_2SO_4 Solution

Fig. 10-20a, p. 219, shows that the sorption of "oxidized" astatine decreases with increasing sulfuric acid concentration. Furthermore, the H_2SO_4 curve is well below the curves for $HClO_4$ and HNO_3. From this, the formation of the 1:1 and 1:2 complexes of At^I and HSO_4^- was deduced [26] and the stability constants calculated (Table 10/40).

Table 10/40
Distribution Coefficients of At^I Between HNO_3+0.005 M $H_2Cr_2O_7$ and Dowex 50X8, 200 to 500 mesh [29].

[HNO_3] in mol/L	K_d in mL/g
1.25	10.1 ± 0.6
0.9	17.9 ± 1.1
0.8	20.1
0.6	37.4 ± 2.7
0.9	44.4
0.4	52 ± 10

10.3.4 Chromatography

Paper Chromatography

Dreyer et al. [30] have investigated the transport of At^0, At^-, AtO_3^- and AtO_4^- on Whatman No. 1 FN11 paper (Table 10/41). They found a good resemblance between At^- and I^- but differences between AtO_3^- and IO_3^-. The perastatate ion did not move under the experimental conditions used as was also the case with the periodate ion.

Table 10/41
R_f values of Astatine and Iodine Ions on Whatman No. 1 Paper [30].

solvent system	dimethyl formamide/ 3 M NH_4OH 1:2 to 8:1	acetone/H_2O 4:1	n-butanol/ 3 M NH_4OH 5:1
At^0	0	0	0
At^-	0.95	0.9	0.3
I^-	0.95	0.9	0.3
AtO_3^-	0.95	0.25	0.95
IO_3^-	0	0.42	0
AtO_4^-	0	0	0
IO_4^-	0	0	0

Thin Layer Chromatography

Visser et al. [1] have made extraction experiments with astatine under a multitude of conditions and tested the extracts by thin layer chromatography on silica gel plates. The results are given in Table 10/42. A minor scatter of the R_f values was caused by different extraction solvents. If $S_2O_8^{2-}$ was added to the aqueous phase prior to the extraction, a second spot with $R_f = 0.65$ (dibutylether) was observed, presumably due to At^I.

Table 10/42
R_f Values of At^0 on Silica Gel Thin Layer Chromatography Plates [1].

solvent	R_f
CCl_4	0.3
CH_2Cl_2	0.6 to 0.67
hexone	0.59 to 0.63
dibutylether	0.48 to 0.56

High Pressure Liquid Chromatography

This technique was applied by Roessler et al. [27] for the separation of inorganic and organic compounds of astatine. The authors used Aminex A27 anion exchanger and Aminex A7 cation exchanger, both of them were pretreated with chlorine water. A typical example of a chromatogram is shown in **Fig. 10-37**, p. 274, indicating that At^- has the highest retention

Fig. 10-37

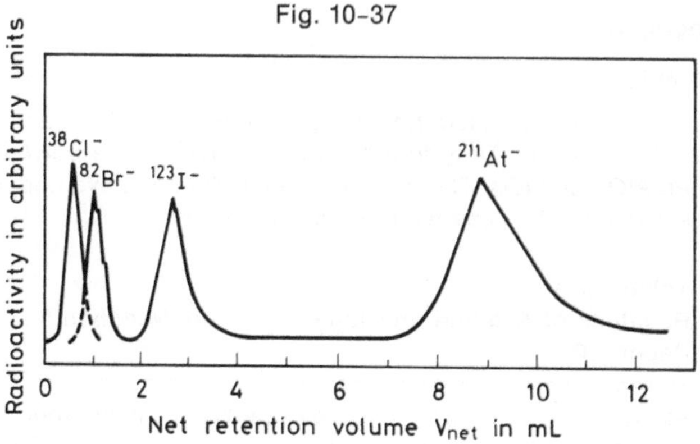

Separation of carrier-free radiohalides on an Aminex A27 anion exchanger column at 80 °C using 1 M $NaNO_3$+0.1 N Na_2SO_3 as eluent (column data: length 110 mm, diameter 1.8 mm, flow rate 0.4 to 1.0 mL/min) [27].

volume among the halogens. The retention volume decreases with increasing $NaNO_3$ concentration.

The AtO_3^- ion had a retention volume of about 0.6 mL at 23 °C with 0.1 M $NaNO_3$ as eluant on the column given in Fig. 10-37. The retention volumes of the At^+ cation was determined to be 11.0 mL (80 °C) using an Aminex A7 cation exchange column (110 mm length, 1.8 mm internal diameter, flow rate 0.2 to 0.8 mL/min) and 0.1 M HNO_3+0.005 M $Cr_2O_7^{2-}$. This fits well with the retention volumes of other univalent cations: Tl^+ 11.8 mL; Cs^+ 10.3 mL; Rb^+ 6.8 mL; K^+ 6.3 mL; Na^- 4.0 mL.

10.3.5 Extraction

Overview

The extraction of astatine into organic solvents was first investigated by Johnson et al. [3], who used carbon tetrachloride and benzene. They demonstrated the solubility of astatine deposited from the gaseous state, regarded as At^0, in carbon tetrachloride and benzene. A high distribution coefficient was obtained if the organic solution was shaken with 0.1 M HNO_3.

It is now generally accepted that At^0 is the only At valence state appreciably extractable by carbon tetrachloride and benzene [3, 8]. Diisopropylether extracts At^0, At^I, and At^{III} [31]. The latter valency state was not identified by the authors themselves but must be regarded as being present after the one hour oxidation with peroxidisulfate in HNO_3 [22]. The astatide ion, At^V and At^{VIII} are not extractable by inert organic solvents. The astatine in every valency state is back extracted or not extracted with alkaline solutions from every inert solvent. This may be understood as a disproportionation in the case of At^0 and the formation of anions in the case of At^I and At^{III}.

By Carbon Tetrachloride

Johnson et al. [3] investigated the behavior of astatine that had been deposited from the gaseous state onto glass and was dissolved with carbon tetrachloride. The organic solution should have contained At^0 and was repeatedly back extracted by 0.1 M HNO_3.

The results are compiled in Table 10/43 and show that At^0 is readily extractable by CCl_4 but with an increasing partition coefficient in successive back extractions. The authors assume this to be caused by impurities and concluded that the true distribution coefficient may equal the limit of the successive extractions. The same effect has been observed by Appelman [4], one example is presented in Fig. 10-1, p. 184. The At^0 is completely back extracted by sodium hydroxide solution. Johnson et al. [3] also conducted a number of experiments in order to test the extractability of other valence states than At^0. The results are summarized in Table 10/44. The experimental procedure was to shake the solution of At in carbon tetrachloride (or benzene) several times with 0.01 M HNO_3 to give the partition coefficient listed in the table and then to extract the remaining solvent layer once with an aqueous redox system. Both strong reducing agents and strong oxidizing agents (SO_2, Br_2, $K_2S_2O_8$, KIO_3) cause almost complete back extraction of the astatine. In accordance with this, Neumann [32] found that astatine is no longer extractable by CCl_4 after treatment by chlorine in 6 M HCl. The same author has investigated the behavior of the oxidized astatine (presumably At^I) after back extraction from diisopropylether with NaOH and reacidifying to 0.5 M H_2SO_4. It was found that pure CCl_4 does extract this astatine form only slightly but a solution of phenol in CCl_4 causes ≅80% extraction. The author assumes the formation of an organic astatine compound.

Table 10/43
Back Extraction of Astatine (At^0) from Organic Solvents by 0.01 M HNO_3 [3].

extraction number	partition coefficient with solvent CCl_4	C_6H_6
1	9.8	43*)
2	53	84*)
3	64	126*)
4	90	130*)
5	—	165*)
6	—	227

*) Numbers are mean values of two experiments.

From these results and a number of occasional observations of other researchers, it may be concluded that At^0 is the only valence state of astatine that is appreciably extracted by carbon tetrachloride. The extractability exists only with an acid or neutral aqueous layer but not under alkaline conditions where disproportionation may take place according to $At_2 \rightarrow HAt + HOAt$.

Starting with this assumption, Appelman has investigated the redox behavior [8] and the formation of interhalogen compounds of astatine at 21±0.5 °C [33]. Elemental iodine has only a slight effect on the extraction of astatine in the region 10^{-5} to 10^{-1} M. Halide ions generally decrease the extractability of astatine [4] both as the alkali salts and as the free acids. The influence of the iodine concentration on the partition coefficient D can be described by the equation

$$D = K_d/(1 + K_2[I^-])$$

References on pp. 288/9

Table 10/44
Partition Coefficients of Astatine Obtained on Sequential Back Extraction of an Organic Solution with 0.01 M HNO_3 and an Aqueous Redox System [3].

solvent	partition coefficient in 0.01 M HNO_3	aqueous redox system	partition coefficient with redox system
C_6H_6	91	0.25 M $FeSO_4$	89
CCl_6	90	saturated H_2AsO_3	49
CCl_4	50	1.1 M KI in 0.01 HNO_3	0.4
C_6H_6	200	saturated I_2 0.01 M HNO_3	64
C_6H_6	200	saturated SO_2 0.01 M HNO_3	3
CCl_4	85	same	0.8
CCl_4	85	0.25 M $Fe(NO_3)_3$, 0.01 M HNO_3	2
CCl_4	200	$Hg(NO_3)_2$ 0.5 M, 0.01 M HNO_3	0.01
CCl_4	85	Br_2 < 0.1 M, 0.01 M HNO_3	0.05
C_6H_6	200	cold 0.01 M $K_2S_2O_8$	5
C_6H_6	200	50 °C 0.1 $K_2S_2O_8$	0.1
C_6H_6	50	3 M HCl	0.66
C_6H_6	200	0.1 M KIO_3, 0.01 M HNO_3	16

$K_d = 5.5$ is the extraction coefficient of the astatine species AtI between the aqueous phase (pH 1 to pH 5) and CCl_4 and $K_2 = 2000$ is the formation constant of the AtI_2^- complex anion:

$$K_2 = [AtI_2^-]/([AtI^0] \cdot [I^-])$$

The dependency on the salt concentration may be represented approximately as

$$\log D = \log D^0 + 0.1\,[NaClO_4]$$

Appelman [4, 33] reports further a lot of extraction data for the systems I_2-IBr-Br^-, I_2-I^--Br^-, I_2-I^--Cl^- from which he has calculated the constants collected in Table 10/45.

Table 10/45
CCl_4 Extraction and Equilibrium Constants of Interhalogen Compounds of Astatine [4].

extraction coefficient at 21 ± 0.5 °C		equilibrium constants at 21 to 25 °C	
species	$K_d = [AtX]_{CCl_4}[AtX]_{H_2O}$	species	$K = [AtXY^-]/([AtX] \cdot [Y^-])$
AtBr	0.040	$AtICl^-$	9
AtI	5.5	$AtIBr^-$	120
		$AtBr_2^-$	320
		AtI_2^-	2000

By Benzene

The extraction of At^0 by benzene was throroughly investigated by Johnson et al. [3] starting with a solution of At in the organic solute and by Appelman [4] starting with a solution of At in the aqueous phase. Both experimental procedures demonstrated that successive equilibrations of the initial At containing phase with fresh portions of the other phase yielded increasing partition coefficients (Table 10/43, p. 275, and Fig. 10-1, p. 184). Johnson et al. expressed the assumption that the true distribution coefficient At^0 may be represented as the limiting partition coefficient of these experiments. These authors have found generally higher partition coefficient than Appleman, the respective limiting values for 0.01 M HNO_3 as the aqueous phase are 227 and 70, respectively.

Back extraction of the benzene layer with aqueous redox systems gave the same results as with carbon tetrachloride (Table 10/44). Neuman [32] has stated that the chlorine oxidized astatine, probably At^I, is not extracted. One may conclude from these results that only At^0 is appreciably extracted by benzene but the other valence states are not.

By Diisopropylether

Diisopropylether is reported to extract astatine from mineral acids both in the At^0 state [6, 34] and after oxidation by chlorine [6, 31, 32] and Ce^{IV} [31]. The distribution coefficient of At^0 for 8 M HCl was reported by Neirinckx [34] to be 166. The astatine can be removed completely from the organic layer by sodium hydroxide [6, 32]. Neuman [32] has worked with a solution of At in 6 M HCl through which gaseous Cl_2 was bubbled for 30 minutes. One may assume that under these conditions the At^I valence state is present and the chemical form is $AtCl_2^-$. The extracted species should be the acid $HAtCl_2$. A mean extraction coefficient 85 ± 1 is calculated from the results in 7 to 9 M HCl. Belyaev et al. [6] executed a similar experiment and found a constant distribution coefficient of the same order of magnitude in the concentration range 2 M to 9 M HCl (**Fig.** 10-**38**).

Wang Yung-yue and Khalkin [31] have shown that extraction of oxidized astatine by diisopropylether depends on the oxidation reagent. They started with a solution in which the astatine was "apparently" in a reduced state, presumably At^0. **Fig.** 10-**39**, p. 278, shows that treatment of the activity with HNO_3 at 100 °C yielded a state where the astatine is well extracted in the range $[HNO_3] = 1$ M to 6 M. This is the behavior to be expected for At^0. After chlorine oxidation, the behavior of the astatine is the same as was reported in [6, 32] for At^I. Oxidation with 0.1 M Ce^{IV} in 3 M HNO_3 at 100 °C yielded an astatine species which is extracted only above 4 M HNO_3. This is a behavior definitely different from At^I. This state is also extracted from H_2SO_4 and $HClO_4$ at concentrations higher than 2 M and 3 M (**Fig.** 10-**40**, p. 279). The authors speculated that pentavalent astatine may have formed

Fig. 10-38

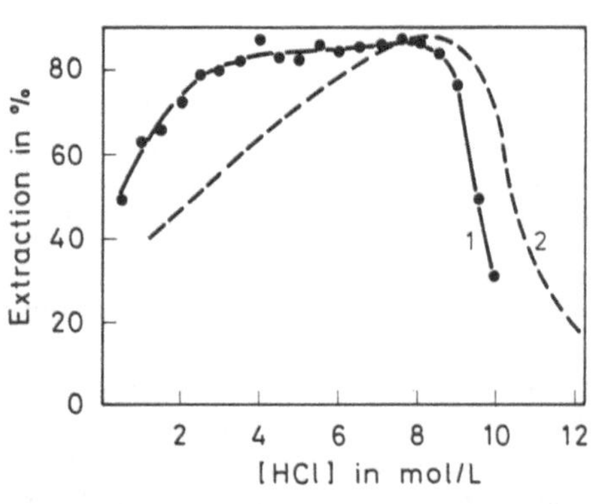

Extraction of chlorine-oxidized astatine by diisopropylether (1) from [9], (2) from [32].

Fig. 10-39

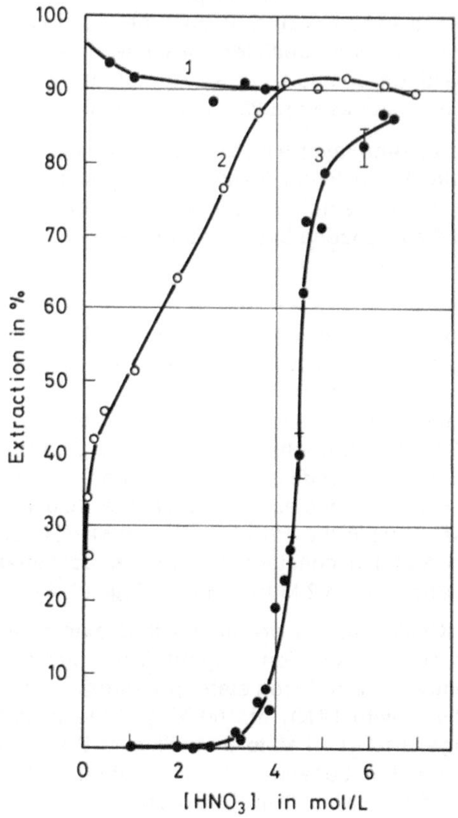

Extraction of astatine by diisopropylether [31]. 1) Treated with 12 M HNO_3 at 100 °C; 2) treated with Cl_2 in 3 M HNO_3; 3) treated with 3 M HNO_3 + Ce^{IV} at 100 °C.

with Ce^{IV}. But they mention that carrier-free ^{131}I is not extracted after Ce^{IV} oxidation. It may be possible that during the experiments trivalent astatine had been formed [11, 35].

Extraction by Miscellaneous Solvents

Visser et al. [1] conducted a series of extraction experiments starting with a solution of astatine 0.05 M in NaOH and 0.2 M in SO_3^{2-}. This solution undoubtedly contained At^- but the researchers claim that upon mixing with acids the astatide is rapidly oxidized by atmospheric oxygen to At^0. Carbon tetrachloride, hexane, and n-butylether were used.

The results are shown in Table 10/46. In every case the extraction yield was appreciably increased upon addition of thiophene to the organic solvent but never exceeded 90%. The species extracted by CCl_4 and hexane without thiophene showed no mobility on SiO_2 thin layer plates and these experiments cannot be regarded to represent regular results. Two extracted species were shown by thin layer chromatography with n-butylether. One has the same R_f value as At^0 and the other being an oxidized form. It is further mentioned that the At activity could be extracted to 65 to 80% by 4-methylpentanone-2 (hexone), 2,4-pentandione (acetylacetone), and tributylphosphate (TBP) from solutions having pH 7.

Fig. 10-40

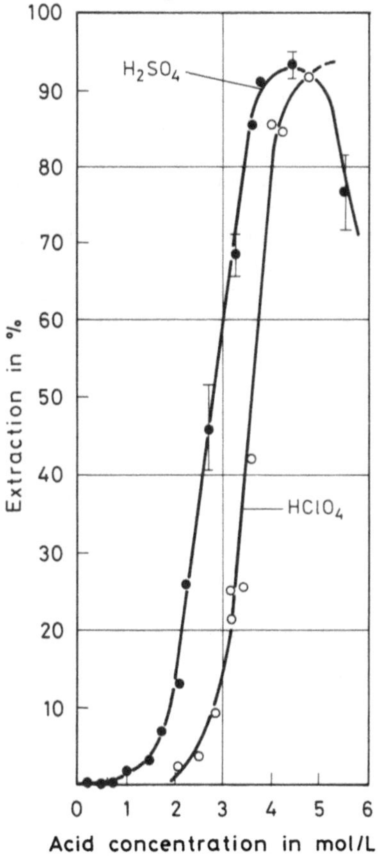

Extraction of astatine with diisopropylether after oxidation with 3 M HNO_3 + 0.1 M Ce^{IV} [31].

Hexone also extracted the astatine from alkaline solutions (85%). Back extraction occurred with cysteine and Na_2S.

Wang Yung-Yue, Khalkin [31] reported that astatine, oxidized with 0.1 M Ce^{IV} in 3 M HNO_3 for one hour at 100 °C, is extractable by a methylisobutylketone and nitromethane mixture as described for diisopropylether.

Table 10/46
Extraction of Astatine Oxidized by Atmospheric Oxygen [1].

aqueous phase	organic solvent					
	CCl_4 yield in %		hexane yield in %		n-butylether yield in %	
	a	b	a	b	a	b
0.2 M H_2SO_4	10	40	20	40	65	80
1.4 M HNO_3	25	90	60	90	90	90
1 M $HClO_4$	25	90	60	90	90	96
6 M HAc	10	—	10	—	80	—

a = pure organic solvent; b = organic solvent and 15 μL thiophene per mL.

References on pp. 288/9

10.3.6 Electrochemistry

Electrolytic Deposition

The electrolytic behavior of astatine was investigated by Johnson et al. [3] in an electrolysis apparatus (**Fig.** 10-**41**) which allowed the direct measurement of the activity deposited onto thin gold windows from the aqueous solution. These windows consisted of a gold foil of 8 to 10 mg/cm² thickness, that covered the two horizontal arms of a t-shaped electrolysis cell made of glass. On the outside of the windows an ionization chamber was connected. The experiments consisted in measuring the deposition rate at different potentials and finding the critical deposition potential by extrapolation to zero deposition rate. Both cathodic and anodic deposition was observed in a small potential range (Table 10/47), the latter not being reversible. At^0 dissolved in sulfuric acid and oxidized with hot persulfate gave no consistent cathodic deposition.

Fig. 10-41

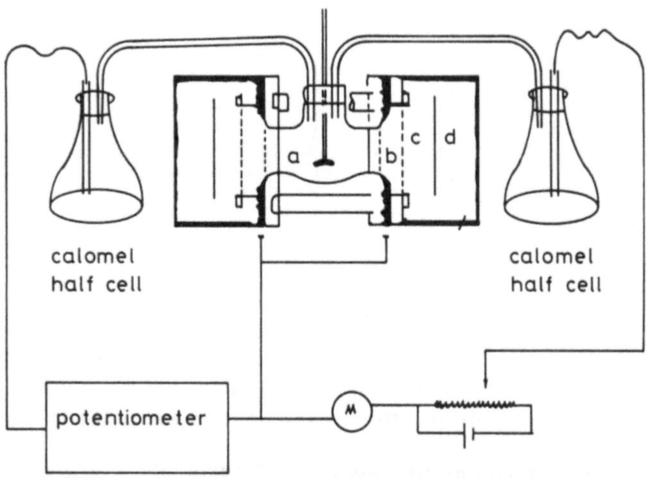

Schematic diagram of the electrolysis apparatus used by Johnson et al. [3]. (a) t-shaped glass cell with 10 mL volume; b) gold windows, 1 cm diameter; c) thin Al leaf; d) brass disk, connected to an electrometer.)

Table 10/47
Critical Deposition Potential of Astatine Measured Versus the Normal Hydrogen Electrode [3].

cathodic deposition solution composition	E in V (±0.025)	anodic deposition solution composition	E in V (±0.02)
0.066 M HNO_3	1.225	0.066 M HNO_3	1.46
1.0 M HNO_3	1.24	0.1 M HNO_3+0.1 M $K_2S_2O_8$	1.445
0.075 M H_2SO_4+0.1 M $Na_2Cr_2O_7$	1.20		
0.066 M HNO_3+3 mg Au	1.22		

Migration in Free Electrolyte

The measurement of migration in an electrical field has proved to be one of the most powerful experimental techniques in aqueous astatine chemistry. It was the main tool to establish without doubt the intermediate valence states, their ionic forms and some of their chemical equilibria.

It was first used by Johnson et al. [3] in 1949 who started with a solution of elemental astatine in nitric acid. They reported that the activity migrates to the anode both in acid and alkaline solution whether if a redox reagent (SO_2, $K_2S_2O_7$, Br_2, NaOCl) is present or not. Therefore every valence state formed behaved as an anion under the experimental conditions used.

Later, Wang Fu-Chiung et al. [29] found that upon treating an astatine solution 0.4 M in HNO_3 and 0.005 M in $H_2Cr_2O_7$ for one to two hours at 20 to 25 °C a species was formed which moved to the cathode in a 1.25 M HNO_3 electrolyte. It was concluded from this and other experiments that a cationic oxidized state of astatine exists, which is now known to be At^{I}. Do Kim Tyung et al. [36] have resolved the difference between both these findings by demonstrating that At^{I} behaves as a cation in noncomplexing acids like $HClO_4$ but as an anion in the presence of complex forming anions as chloride. They report further that At^0 in 0.25 M $HClO_4$ does not migrate in the electric field.

I. Dreyer and R. Dreyer with collaborators have published a whole series of articles wherein electromigration measurements are applied for the chemical investigation of astatine. Their experimental setup is shown in **Fig.** 10-**42**. The main parts are a horizontally oriented thermostated capillary glass pipe with two open vertical stand pipes of the introduction of the sample and two scintillation detectors at a distance of 100 mm. After introducing the astatine sample through one of the stand pipes and connecting to a direct current voltage equal to about 30 V/cm, the activity moves and passes sequentially to the two detectors (**Fig.** 10-**43**, p. 282). The time difference of the counting maxima of the two detectors is the basic result of the experiments from which the electrophoretic velocity and mobility can be calculated.

Fig. 10-42

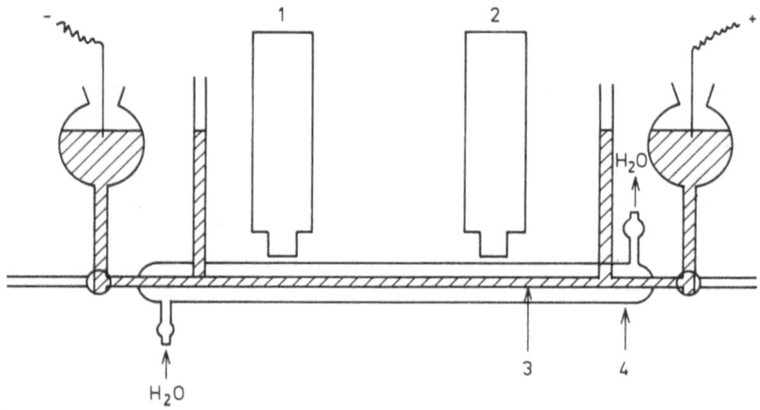

Setup for electrophoretic investigation of astatine [37]. — 1, 2) Scintillation detector heads at 100 mm distance; 3) glass pipe 4 mm internal diameter; 4) thermostatic mantle.

References on pp. 288/9

Fig. 10-43

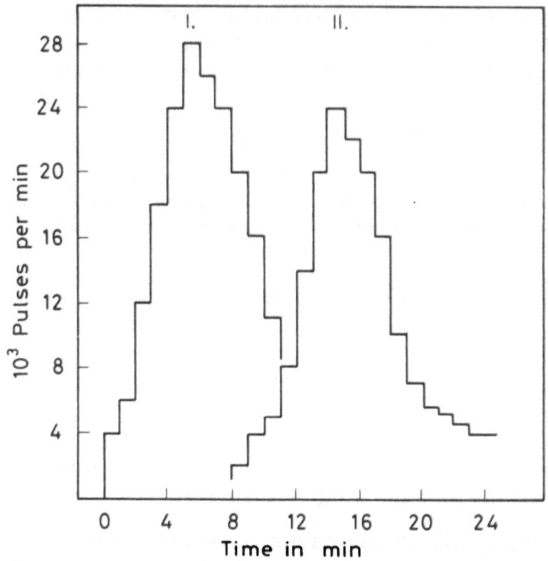

Example of the result of an electrophoresis experiment with the setup of Fig. 10-42, At^- in 0.04 M KNO_3, 25 °C, 30.5 V/cm [37]; counting rate of first (I) and second (II) detector.

Milanov et al. [38] have developed a more sophisticated setup which records the distribution of the activity along the migration tube automatically with high reproducibility.

Fig. 10-**44** shows, as an example of the result obtainable, the time-dependent electromigration curves of an At^I preparation [20]. At the entry point of the activity into the electromigration tube, there is immobile astatine in balance with mobile astatine. The authors interpreted this as an adsorption of At^I onto colloidal particles of silicic acid, which are formed from the glass surfaces by the intense radiation of the solutions. Using this measuring technique, Dreyer et al. [39] were able to detect the existence of instable intermediates in the reaction of At^- with SCN^-, one of which was the complex $At(SCN)_2^-$ formed by oxidation of the At^-.

A summary of electrophoretic mobilities measured under a variety of conditions is given in Table 10/44, p. 276. From these data, the positive discrimination of perastatate was possible [37] and for the first time, the trivalent state of astatine was discovered by Dreyer, Dreyer, Khalkin [11]. It formed upon addition of 1 to 3 mg $Na_2S_2O_8$ to about 0.1 mL 0.5 M $HClO_4$ containing At^0 and heating to 100 °C. It is assumed from the mobility that the chemical form is AtO^+. Rendering the solution alkaline converts the At^{III} cation to a nonmoving species which is further converted to an anion by heating the solution. The authors [11] assume the following equilibria

$$AtO^+ \underset{H^+}{\overset{OH^-}{\rightleftharpoons}} At(OH)_3 \underset{H^+}{\overset{OH^-}{\rightleftharpoons}} AtO_2^-$$

A behavior analogous to this to some extent was observed with At^I. In neutral and alkaline solution a nonmoving species is also formed but is converted to astatide upon heating to 100 °C for one hour [11]. The formation of At^- may be due to a self-disproportionation

Fig. 10-44

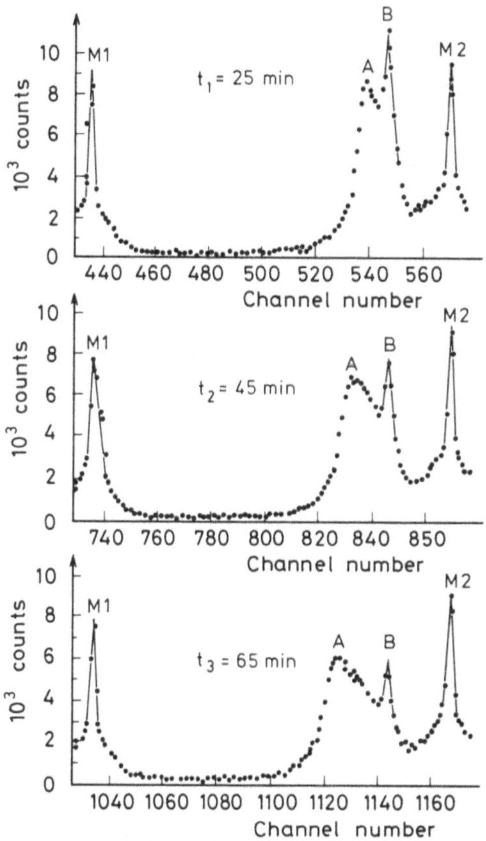

Electromigration curves of At^0 in 0.4 M $(Na,H)ClO_4$ + 0.0001 M $K_2Cr_2O_7$ at 25 °C [20]. M1, M2: Radioactive markers limiting the measuring base 190 mm long; A: mobile At; B: immobile At.

according to

$$AtOH + OH^- \rightarrow At^- + H_2O + {}^1/_2O_2$$

as was proposed in [40]. The intermediate neutral species AtOH (hypoastatic acid) was proved by Milanov et al. [20] who investigated the dependency of the electrophoretic mobility on the concentration of free acid in $NaClO_4/HClO_4$ mixtures. They interpreted their data with the protonation equilibrium

$$AtOH + H^+ \rightleftharpoons AtOH_2^+$$

The dissociation constant of $AtOH_2^+$ measured in the interval $0.63 < pH < 1.68$ was calculated from the results to be $K_D = 0.032 \pm 0.005$. The calculated pH of half-dissociation is 1.5 (25 °C, ionic strength 0.4).

The cationic behavior of At^I in electromigration is only observed in $HClO_4$ and HNO_3. Adding a halide even in small concentrations or taking dilute HCl, HBr as the electrolyte yields anionic species [35] which were identified as AtX_2^- complexes. The same was found

References on pp. 288/9

to occur with At^{III} to give complexes $AtOX_2^-$ ($X = Cl^-$, Br^-, I^-). All these species could be characterized by their electrophoretic mobility (Table 10/48).

Table 10/48
Ionic Mobilities u of Astatine Species in Aqueous Solution.

ion	electrolyte	$u \cdot 10^4$ in $cm^2 \cdot s^{-1} \cdot V^{-1}$	Ref.	ion	electrolyte	$u \cdot 10^4$ in $cm^2 \cdot s^{-1} \cdot V^{-1}$	Ref.
				At^{III}			
astatide ion				AtO^+ (?)	5	3.3 ±0.2	[11]
At^+	1	6.1 ±0.2	[11, 37]	AtO_2^-	2	4.6 ±0.02	[35]
	2	7.2 ±0.2	[35]	$AtOCl_2^-$	0.02 M HCl	3.3 ±0.2	[35]
	0.1 M NaOH	7.2 ±0.02	[40]	$AtOBr_2^-$	0.02 M HBr	4.6 ±0.2	[35]
	0.01 M NaOH	7.15±0.1	[41]	$AtOI_2^-$	0.01 M KI	4.35±0.2	[35]
At^I	4	1.9 ±0.2	[11]	At^V			
	0.02 M HNO_3	2.7	[40]				
	6	3.08	[20]	AtO_3^-	1	2.3 ±0.2	[11, 37]
	0.04 M $HClO_4$	2.67	[20]				
$AtCl_2^-$	0.02 M HCl	3.4 ±0.2	[11]				
$AtCl_2^-$	0.005 M H_2SO_4	3.1 ±0.1	[40]	At^{VII}			
$AtBr_2^-$	0.02 M HBr	4.6 ±0.02	[35]	AtO_4^-	1	3.4 ±0.2	[11, 37]
$AtBr_2^-$	0.005 M H_2SO_4	3.78±0.1	[40]				
AtI_2^-	0.01 M KI	4.35±0.02	[35]				
AtI_2^-	0.006 M H_2SO_4	3.82±0.1	[40]				
$At(SCN)_2^-$	0.03 M SCN^-	3.7 ±0.01	[40]				

electrolytes

1: 0.04 M HNO_3, 25 °C
2: 0.02 M $(NH_4)_2CO_3$ + 0.01 M NH_4OH, 25 °C
3: 0.02 M $HClO_4$, 25 °C
4: 0.02 M $HClO_4$ + 0.001 M $K_2Cr_2O_7$, 25 °C
5: 0.02 M $HClO_4$ + 0.002 M $Na_2S_2O_8$, 25 °C
6: extrapolated to pH = 0

The chloride and bromide complexes of At^I are stable up to pH ≅ 10 (**Fig.** 10-**45**) and convert at higher pH to At^- as indicated by the electrophoretic mobility of the end product [35, 40]. Therefore, the halogen complexes of At^I decompose at higher pH according to the disproportionation scheme discussed above. No such reaction takes place in the presence of iodine presumably as a consequence of the rather high stability of the iodo complexes (Fig. 10-45). The halogen complexes of At^{III} also dissociate in alkaline solution but do give the anion AtO_2^-, which was identified as the product of At^{III} hydrolysis at high pH [35].

Electrophoretic measurements were also used to identify the formation of $At(SCN)_2^-$ and $At(CN)_2^-$ [39] and complexes of At^I with thiourea and some of its derivatives [42]. All the species were characterized by their electrophoretic mobility. It was assumed that the sulfur containing ligands form cationic complexes [42].

Using the electrophoretic method Dreyer et al. [37] have proven qualitatively that the same halogen complexes AtX_2^- are formed from both At^- and At^I if NaX (X = Cl, Br) is added to a solution of the uncomplexed astatine species in H_2SO_4 (or $HClO_3$, HNO_3). The

Fig. 10-45

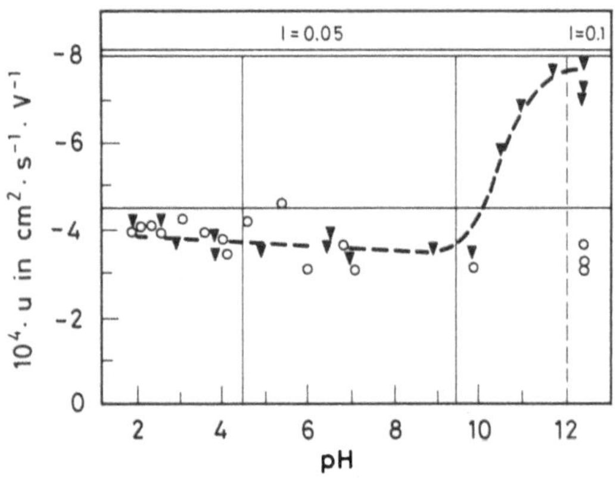

pH dependence of the ionic mobility of AtX^- complexes (0.03 M NaBr (▾) or NaI (o) + (Na, H)ClO_4 + NaOH); I = ionic strength in mol/L.

authors assume that the astatide ion is easily oxidized by atmospheric oxygen to first At^0 and then to At^I.

The equilibrium constants for the reaction $AtX + X^- \rightleftharpoons AtX_2^-$ have been calculated from the mobilities measured at different halogen concentration using the equation

$$K_2 = \frac{u(AtX) - u}{u - u(AtX_2^-)} \cdot [X^-]^{-1}$$

where u(AtX) and $u(AtX_2^-)$ are the mobilities of the pure species indicated in the brackets and u is the measured mobility at concentration $[X^-]$ ([39, 40], Table 10/49).

Table 10/49
Formation Constants of the AtX_2^- Species Calculated from Electrophoretic Measurements.

reaction $AtX + X^- \rightleftharpoons AtX_2^-$		
X	K	Ref.
Br	250	[40]
I	1600	[40]
SCN	420	[39]

ionic strength 0.05
temperature 25 °C

Paper Electrophoresis

Paper electrophoresis was used by several authors to identify the valence state of astatine oxidized or reduced under different conditions. Published data on migration velocities are collected in Table 10/50. It is generally observed that perastatate AtO_4^- is almost immobile on Whatmann No. 1 paper [12, 30] as is IO_4^-. Astatate AtO_3^- and astatide migrate normally with a velocity amounting to about 50 to 70% of the respective iodine species.

Table 10/50
Data on the Electrophoretic Mobility of Astatine and Iodine Species on Whatmann No. 1 Paper.

species	electrolyte (gradient in V/cm)	migration velocity in mm/min X=At	X=I	Ref.	species	electrolyte (gradient in V/cm)	migration velocity in mm/min X=At	X=I	Ref.
X^-	1 (30)	2.9	3.5	[43]	XO_3^-	1 (30)	1.4	2.2	[43]
	2 (30)	6.5	7.2	[10]		2 (20)	≅1.2	—	[30]
	2 (50)	5.0	5.8	[12]		3 (33)	1.5	—	[30]
	2 (20)	2.9	0	[30]	XO_4^-	2 (50)	0	0	[12]
	3 (20)	1.5	2.25	[30]		2 (20)	≦0.2	—	[30]
	3 (33)	2.9	—	[30]		3 (33)	≦0.2	—	[30]

Electrolyte 1) 0.1 M Na_2SO_4, 14 °C; 2) 0.1 M Na_2SO_4, room temperature?; 3) dimethylformamide/3 M NH_4OH = 1:2.

10.4 Miscellaneous Investigation Methods

Evaporation from Solution

Because of the analogy to iodine and the measured boiling point of 230 °C [44], it is expected that elemental astatine is rather volatile. This may cause problems for the handling of astatine, especially during the preparation of samples for radioactivity counting. In accordance with the expectation, Johnson et al. [3] reported a striking volatility of At^0 from glass, but in contrast to that low losses from gold or platinum plates. Appelman [4] has investigated this effect by drying a droplet of several solvents or aqueous solutions to which 5 to 50 μL of an At^0 stock were added under an infrared lamp. The results collected in Table 10/51 show a rather good retention from aqueous solutions by platinum but a low retention by certain other metals. At is retained poorly with organic solutions, even by platinum. Wang Fu-Chiung et al. [25] did similar experiments with similar results. Their data are included in Table 10/51.

Table 10/51
Retention of Astatine upon Evaporation of its Solution on Metal Plates [4].
Table 10/51 [continued]

platinum plates: composition of solution	platinum plates: relative retention in %	other metals: composition of solution and plate material	other metals: relative retention in %
3 M HCl[a)]	1.00 ± 0.01	3 M HNO_3-Ag	0.4
12 M HCl	1.0	3 M HNO_3-Te	0.5
1 M HCl	1.0	3 M HNO_3-steel	0.1
0.5 M HCl	1.00[d)]	3 M HCl-Te	0.5
0.2 M HCl	0.9	0.5 M HCl-Ag	1.00[d)]
16 M HNO_3	0.5	0.5 M HCl-steel	0.85[d)]

Table 10/51 [continued]

platinum plates		other metals	
composition of solution	relative retention in %	composition of solution	relative retention in %
3 M HNO_3	0.7	0.5 M HCl-Cu	0.89 [d]
1.5 M HCl+1.5 M HNO_3	1.0	0.5 M HCl-PVC	0.67 [d]
15 M NH_4OH	0.9	0.5 M HCl-Ta	0.12 ± 4 [d]
0.1 M NaOH	0.9	0.5 M HCl-glass	<0.05 [d]
0.01 M NaOH	0.9		
water	0.8 ± 0.1		
benzene, CCl_4 [a]	0.64 ± 0.07		
CCl_4 [b]	0.14		
CCl_4 [c]	1.0		

[a] Same with silver plates; [b] organic phase washed several times with 0.1 M $HClO_4$+0.1 M Fe^{III}+0.001 M Fe^{II} before evaporation; [c] organic phase washed several times with 0.1 M $HClO_4$+0.001 M Fe^{III}+0.1 M Fe^{II} before evaporation; [d] from [25].

Distillation with Solvent

Johnson et al. [3] have investigated the codistillation of astatine orginally present as At^0 with several solvents (Table 10/52). Distillation yields were low from inert and oxidizing acid solutions but almost quantitative under reducing conditions in sulfuric acid. This was recommended by Belyaev et al. [6] for separating astatine from impurities. They report 90% yield in the distillate when 1 M H_2SO_4+0.5 M $FeSO_4$ is distilled to 85% of its volume. Meyer, Roessler [45] have worked out a rapid and quantitative separation of astatine from ^4He bombarded bismuth by distilling At together with chloroform from concentrated sulfuric acid.

Table 10/52
Distillation of Astatine Solutions [3].

initial solution 10 mL	% loss in first 2 mL distillate	% remaining in still after 9 mL distillate
16 M HNO_3	0.2	99
5 M HNO_3+3 mg I^-	2.0	98
12 M HCl	0	84
$H_2SO_4/H_2O=1:1$	0.6	99
60% $HClO_4$	0.1	99
0.5 M H_2SO_4+0.05 M Fe^{II}	69	7
CCl_4	2	82
benzene	7	62

Mass Spectrometry of Astatine Compounds

Several molecular ions of astatine compounds (Table 10/53) have been identified by time of flight mass spectrometry [46] and in a plasma ion source/mass separator apparatus [7]. The detection of the molecular ion At_2^+ is of special interest. The researchers neglect

References on pp. 288/9

the possibility of its formation from At_2 molecules by ionization but assume the formation according to the reaction

$$At + At^+ \rightarrow At_2^+$$

at the outlet of the plasma ion source. The experiments have been done with an ion stream equal to only 10^7 At atoms per second and the At_2^+ yield was 0.03% of the mass 211 collected in the separator (acceleration voltage 25 kV).

Table 10/53
Astatine Molecular Ions Observed by Mass Spectrometry.

ion	Ref.	ion	Ref.	ion	Ref.
At_2^+	[7]	$AtOH^+$	[7]	$AtBr^+$	[7, 46]
AtH^+	[7, 46]	$AtOH_2^+$	[7]	AtI^+	[7, 46]
AtO^+	[7]	$AtCl^+$	[7, 46]	CH_3At^+	[46]

References

[1] G.W.M. Visser, E.L. Diemer (Radiochim. Acta **33** [1983] 145/51). – [2] D.R. Corson, K.R. MacKenzie, E. Segrè (Phys. Rev. [2] **58** [1940] 672/8). – [3] G.L. Johnson, R.F. Leininger, E. Segrè (J. Chem. Phys. **17** [1949] 1/10). – [4] E.H. Appelman (UCRL-9025 [1960] 1/113; N.S.A. **14** [1960] No. 11501). – [5] A.H.W. Aten Jr., T. Doorgeest, U. Hollstein, H.P. Moeken (Analyst **77** [1952] 774/8).

[6] B.N. Belyaev, V. Yun-Yui, E.N. Sinotova, L. Nemet, V.A. Khalkin (Radiokhimiya **2** [1960] 603/13). – [7] N.A. Golovkov, I.I. Gromova, M. Janicki et al. (Radiochem. Radioanal. Letters **44** [1980] 67/78). – [8] E.H. Appelman (J. Am. Chem. Soc. **83** [1961] 805/7). – [9] A.H.W. Aten, J.G. van Raaphorst, G. Nooteboom, G. Blasse (J. Inorg. Nucl. Chem. **15** [1960] 198/9). – [10] M. Bochvarova, Do Kim Tyung, I. Dudova, Y.V. Norseev et al. (Radiokhimiya **14** [1972] 858/65; Soviet Radiochem. **14** [1972] 889/95).

[11] I. Dreyer, R. Dreyer, V.A. Chalkin [Khalkin] (Radiochem. Radioanal. Letters **36** [1978] 389/98). – [12] V.A. Khalkin, Yu.V. Norseev, V.D. Nefedov et al. (Dokl. Akad. Nauk SSSR **195** [1970] 623/5; Dokl. Chem. Proc. Acad. Sci. USSR **190/195** [1970] 855/7). – [13] G.A. Brinkman, J.T. Veenboer, A.H.W. Aten Jr. (Radiochim. Acta **2** [1963] 48). – [14] M. Lefort, G. Simonov, X. Tarrago (Bull. Soc. Chim. France **1960** 1726/7). – [15] W. Garrison, J. Gile, R. Maxwell, J. Hamilton (Anal. Chem. **23** [1951] 204/5).

[16] M. Lefort, G. Simonoff, X. Tarrago (Compt. Rend. **248** [1959] 216/8). – [17] Yu.V. Norseev, Chao Tac-Nan, V.A. Khalkin (Radiokhimiya **8** [1966] 497/504; Soviet Radiochem. **8** [1966] 461/6). – [18] W.A. Chalkin [Khalkin], E. Herrmann, J.W. Norseev, I. Dreyer (Chemiker-Ztg. **101** [1977] 470/81). – [19] W.A. Chalkin [Khalkin], E. Herrmann (Isotopenpraxis **11** [1975] 333/40). – [20] M. Milanov, V. Doberenz, V.A. Khalkin, A. Marinov (J. Radioanal. Nucl. Chem. II **83** [1984] 291/9).

[21] V.M. Vakhtel, G.V. Vinel, Ts. Vylov, I.I. Gromova et al. (Radiokhimiya **18** [1976] 886/93; Soviet Radiochem. **18** [1976] 792/8). – [22] V. Doberenz, Dang Duc Nang, R. Dreyer, M. Milanov et al. (Radiochem. Radioanal. Letters **52** [1982] 119/28). – [23] J. Merinis, Y. Legoux, G. Boussières (Radiochem. Radioanal. Letters **11** [1972] 59/64). – [24] B. Eichler (Radiochem. Radioanal. Letters **22** [1975] 147/56). – [25] Wang Fu-Chiung, K. Meng-Hua, V.A. Khalkin (Radiokhimiya **4** [1962] 94/8; Soviet Radiochem. **4** [1962] 81/5).

[26] Do Kim Tyung, I.V. Dudova, V.A. Khalkin (Radiokhimiya **15** [1973] 548/53; Soviet Radiochem. **15** [1973] 552/6). — [27] J.K. Roessler, W. Tornau, G. Stoecklin (J. Radioanal. Chem. **21** [1974] 199/209). — [28] Yu.V. Norseev, V.A. Khalkin (J. Inorg. Nucl. Chem. **30** [1968] 3239/43). — [29] Wang Fu-Chiung, Yu.V. Norseev, V.A. Khalkin, Ch'ao Tao-Nan (Radiokhimiya **5** [1963] 351/5; Soviet Radiochem. **5** [1963] 318/22). — [30] I. Dreyer, R.Dreyer, Yu.V. Norseev, V.A. Chalkin [Khalkin] (Radiochem. Radioanal. Letters **33** [1978] 291/300).

[31] Wang Yung-Yue, V.A. Khalkin (Radiokhimiya **3** [1961] 662/6; Radiochemistry [USSR] **3** [1961/64] 258). — [32] H.M. Neumann (J. Inorg. Nucl. Chem. **4** [1957] 349/53). — [33] E.H. Appelman (J. Phys. Chem. **65** [1961] 325/31). — [34] R.D. Neirinckx, J.A. Smit (Anal. Chim. Acta **63** [1973] 201/4). — [35] I. Dreyer, R. Dreyer, V.A. Chalkin [Khalkin], M. Milanov (Radiochem. Radioanal. Letters **40** [1979] 145/54).

[36] Do Kim Tyung, I.V. Dudova, V.A. Khalkin (Radiokhimiya **14** [1972] 766/7; Soviet Radiochem. **14** [1972] 790/2). — [37] I. Dreyer, R. Dreyer, V.A. Chalkin [Khalkin] (Radiochem. Radioanal. Letters **35** [1978] 257/62). — [38] M. Milanov, W. Doberenz, A. Masinow, V.A. Khalkin (J. Radioanal. Nucl. Chem. II **82** [1984] 101/9). — [39] R. Dreyer, I. Dreyer, M. Pfeiffer, F. Roesch (Radiochem. Radioanal. Letters **55** [1982] 207/14). — [40] R. Dreyer, I. Dreyer, F. Roesch, G.-J. Beyer (Radiochem. Radioanal. Letters **54** [1982] 165/76).

[41] R. Dreyer, I. Dreyer, F. Roesch (Z. Chem. [Leipzig] **22** [1982] 54/6). — [42] R. Dreyer, I. Dreyer, F. Roesch, S. Fischer (Z. Chem. [Leipzig] **23** [1983] 346/7]. — [43] G.A. Nagy, V.A. Halkin, J.V. Norszejev [Norseev] (Magy. Kem. Folyoirat **73** [1967] 191/5). — [44] K. Otozai, N. Takahashi (Radiochim. Acta **31** [1982] 201/3). — [45] G.-J. Meyer, K. Roessler (Radiochem. Radioanal. Letters **25** [1976] 377/90).

[46] E.H. Appelmann, E.N. Sloht, M.H. Studier (Inorg. Chem. **5** [1966] 766/9).

Table of Conversion Factors

Following the notation in Landolt-Börnstein [7], values which have been fixed by convention are indicated by a bold-face last digit. The conversion factor between calorie and Joule that is given here is based on the thermochemical calorie, cal_{thch}, and is defined as 4.1840 J/cal. However, for the conversion of the "Internationale Tafelkalorie", cal_{IT}, into Joule, the factor 4.1868 J/cal is to be used [1, p. 147]. For the conversion factor for the British thermal unit, the Steam Table Btu, Btu_{ST}, is used [1, p. 95].

Force	N	dyn	kp
1 N (Newton)	1	10^5	0.1019716
1 dyn	10^{-5}	1	1.019716×10^{-6}
1 kp	9.8066**5**	9.8066**5** $\times 10^5$	1

Pressure	Pa	bar	kp/m^2	at	atm	Torr	lb/in^2
1 Pa (Pascal) = 1 N/m^2	1	10^{-5}	1.019716×10^{-1}	1.019716×10^{-5}	0.986923×10^{-5}	0.750062×10^{-2}	145.0378×10^{-6}
1 bar = 10^6 dyn/cm^2	10^5	1	10.19716×10^3	1.019716	0.986923	750.062	14.50378
1 kp/m^2 = 1 mm H_2O	9.8066**5**	0.98066**5** $\times 10^{-4}$	1	10^{-4}	0.967841×10^{-4}	0.735559×10^{-1}	1.422335×10^{-3}
1 at = 1 kp/cm^2	0.98066**5** $\times 10^5$	0.98066**5**	10^4	1	0.967841	735.559	14.22335
1 atm = 76**0** Torr	1.0132**5** $\times 10^5$	1.0132**5**	1.033227×10^4	1.033227	1	760	14.69595
1 Torr = 1 mm Hg	133.3224	1.333224×10^{-3}	13.59510	1.359510×10^{-3}	1.315789×10^{-3}	1	19.33678×10^{-3}
1 lb/in^2 = 1 psi	6.89476×10^3	68.9476×10^{-3}	703.069	70.3069×10^{-3}	68.0460×10^{-3}	51.7149	1

Work, Energy, Heat	J	kWh	kcal	Btu	MeV
1 J (Joule) = 1 Ws = 1 Nm = 10^7 erg	1	2.778×10^{-7}	2.39006×10^{-4}	9.4781×10^{-4}	6.242×10^{12}
1 kWh	$3.\mathbf{6}\times10^{6}$	1	860.4	3412.14	2.247×10^{19}
1 kcal	4184.**0**	1.1622×10^{-3}	1	3.96566	2.6117×10^{16}
1 Btu (British thermal unit)	1055.06	2.93071×10^{-4}	0.25164	1	6.5858×10^{15}
1 MeV	1.602×10^{-13}	4.450×10^{-20}	3.8289×10^{-17}	1.51840×10^{-16}	1

1 eV/mol = 23.0578 kcal/mol = 96.473 kJ/mol

Power	kW	PS	kp m/s	kcal/s
1 kW = 10^{10} erg/s	1	1.35962	101.972	0.239006
1 PS	0.73550	1	**75**	0.17579
1 kp m/s	$9.80\mathbf{665}\times10^{-3}$	0.01333	1	2.34384×10^{-3}
1 kcal/s	4.18**40**	5.6886	426.650	1

References:

[1] A. Sacklowski, Die neuen SI-Einheiten, Goldmann, München 1979. (Conversion tables in an appendix.)

[2] International Union of Pure and Applied Chemistry, Manual of Symbols and Terminology for Physicochemical Quantities and Units, Pergamon, London 1979; Pure Appl. Chem. **51** [1979] 1/41.

[3] The International System of Units (SI), National Bureau of Standards Spec. Publ. No. 330 [1972].

[4] H. Ebert, Physikalisches Taschenbuch, 5th Ed., Vieweg, Wiesbaden 1976.

[5] Kraftwerk Union Information, Technical and Economic Data on Power Engineering, Mülheim/Ruhr 1978.

[6] E. Padelt, H. Laporte, Einheiten und Größenarten der Naturwissenschaften, 3rd Ed., VEB Fachbuchverlag, Leipzig 1976.

[7] Landolt-Börnstein, 6th Ed., Vol. II, Pt. 1, 1971, pp. 1/14.

Key to the Gmelin System of Elements and Compounds

HCl

$ZnCl_2$

System Number	Symbol	Element
1		Noble Gases
2	H	Hydrogen
3	O	Oxygen
4	N	Nitrogen
5	F	Fluorine
6	**Cl**	**Chlorine**
7	Br	Bromine
8	I	Iodine
	At	Astatine
9	S	Sulfur
10	Se	Selenium
11	Te	Tellurium
12	Po	Polonium
13	B	Boron
14	C	Carbon
15	Si	Silicon
16	P	Phosphorus
17	As	Arsenic
18	Sb	Antimony
19	Bi	Bismuth
20	Li	Lithium
21	Na	Sodium
22	K	Potassium
23	NH_4	Ammonium
24	Rb	Rubidium
25	Cs	Caesium
	Fr	Francium
26	Be	Beryllium
27	Mg	Magnesium
28	Ca	Calcium
29	Sr	Strontium
30	Ba	Barium
31	Ra	Radium
32	**Zn**	**Zinc**
33	Cd	Cadmium
34	Hg	Mercury
35	Al	Aluminium
36	Ga	Gallium

$CrCl_2$

$ZnCrO_4$

System Number	Symbol	Element
37	In	Indium
38	Tl	Thallium
39	Sc, Y La–Lu	Rare Earth Elements
40	Ac	Actinium
41	Ti	Titanium
42	Zr	Zirconium
43	Hf	Hafnium
44	Th	Thorium
45	Ge	Germanium
46	Sn	Tin
47	Pb	Lead
48	V	Vanadium
49	Nb	Niobium
50	Ta	Tantalum
51	Pa	Protactinium
52	**Cr**	**Chromium**
53	Mo	Molybdenum
54	W	Tungsten
55	U	Uranium
56	Mn	Manganese
57	Ni	Nickel
58	Co	Cobalt
59	Fe	Iron
60	Cu	Copper
61	Ag	Silver
62	Au	Gold
63	Ru	Ruthenium
64	Rh	Rhodium
65	Pd	Palladium
66	Os	Osmium
67	Ir	Iridium
68	Pt	Platinum
69	Tc	Technetium[1]
70	Re	Rhenium
71	Np, Pu...	Transuranium Elements

Material presented under each Gmelin System Number includes all information concerning the element(s) listed for that number plus the compounds with elements of lower System Number.

For example, zinc (System Number 32) as well as all zinc compounds with elements numbered from 1 to 31 are classified under number 32.

---

[1] A Gmelin volume titled "Masurium" was published with this System Number in 1941.

A Periodic Table of the Elements with the Gmelin System Numbers is given on the Inside Front Cover